Livestock in a Changing Landscape

Volume 2

About Island Press

Since 1984, the nonprofit Island Press has been stimulating, shaping, and communicating the ideas that are essential for solving environmental problems worldwide. With more than 800 titles in print and some 40 new releases each year, we are the nation's leading publisher on environmental issues. We identify innovative thinkers and emerging trends in the environmental field. We work with world-renowned experts and authors to develop cross-disciplinary solutions to environmental challenges.

Island Press designs and implements coordinated book publication campaigns in order to communicate our critical messages in print, in person, and online using the latest technologies, programs, and the media. Our goal: to reach targeted audiences—scientists, policymakers, environmental advocates, the media, and concerned citizens—who can and will take action to protect the plants and animals that enrich our world, the ecosystems we need to survive, the water we drink, and the air we breathe.

Island Press gratefully acknowledges the support of its work by the Agua Fund, Inc., The Margaret A. Cargill Foundation, The Nathan Cummings Foundation, Betsy and Jesse Fink Foundation, The William and Flora Hewlett Foundation, The Kresge Foundation, The Forrest and Frances Lattner Foundation, The Andrew W. Mellon Foundation, The Curtis and Edith Munson Foundation, The Overbrook Foundation, The David and Lucile Packard Foundation, The Summit Foundation, The Summit Fund of Washington, Trust for Architectural Easements, The Winslow Foundation, and other generous donors.

About the French Agricultural Research Centre for International Develoment

Most of the research conducted by the French Agricultural Research Centre for International Development (CIRAD) is in partnership in developing countries. CIRAD has chosen sustainable development as the cornerstone of its operations worldwide. This means taking account of the long-term ecological, economic, and social consequences of change in developing communities and countries. CIRAD contributes to development through research and trials, training, dissemination of information, innovation, and appraisals. Its expertise spans the life sciences, human sciences, and engineering sciences and their application to agriculture and food, natural resource management, and society.

About the Food and Agriculture Organization of the United Nations

The mandate of the Food and Agriculture Organization of the United Nations (FAO) is to raise levels of nutrition, improve agricultural productivity, better the lives of rural populations, and contribute to the growth of the world economy.

About the Swiss College of Agriculture

The Swiss College of Agriculture (SHL) in Zollikofen is the specialist Swiss institution for agriculture, forestry, and food business within Bern University of Applied Sciences. It offers three BSc programs and one MSc program in agriculture, forestry, and food science and management. In addition SHL conducts applied research and renders services in Switzerland and around the world. At SHL, sustainability is at the foundation of the degree programs and projects. With up to 500 students, the campus close to Bern is a manageable size. Thus the synergies between education, research, and services can be optimally leveraged.

About the International Livestock Research Institute

The Africa-based International Livestock Research Institute (ILRI) works at the crossroads of livestock and poverty, bringing high-quality science and capacity building to bear on poverty reduction and sustainable development. ILRI is one of 15 centers supported by the Consultative Group on International Agricultural Research (CGIAR). It has its headquarters in Nairobi, Kenya, and a principal campus in Addis Ababa, Ethiopia. It has other teams working out of offices in West Africa (Ibadan, Bamako), Southern Africa (Maputo), South Asia (Delhi, Guwahati, and Hyderabad in India), Southeast Asia (Bangkok, Jakarta, Hanoi, Vientiane), and East Asia (Beijing).

About the Livestock, Environment and Development Initiative

The Livestock, Environment and Development (LEAD) Initiative, established in 2000, is an interinstitutional consortium within the FAO. The work of the initiative targets the protection and enhancement of natural resources affected by livestock production while at the same time maintaining a focus on strengthening food security and alleviating poverty.

About the Scientific Committee on Problems of the Environment

The Scientific Committee on Problems of the Environment (SCOPE) was established in 1969 as a nongovernmental organization by the International Council for Science (ICSU). Through its program of scientific assessments, SCOPE brings together natural and social scientists and civil society experts and practitioners from around the world to look beyond the horizons of disciplines and institutions at emerging cross-cutting environmental challenges and opportunities. Its assessments have guided and shaped many of today's global environmental programs and actively foster science-policy dialogue and decision processes.

About the Woods Institute for the Environment at Stanford University

The Woods Institute for the Environment at Stanford University harnesses the expertise and imagination of leading academics and decision makers to create practical solutions for people and the planet. The institute is pioneering innovative approaches to meet the environmental challenges of the twenty-first century by sponsoring interdisciplinary research; infusing science into policies and practices of the business, government, and not-for-profit communities; and developing environmental leaders for today and the future.

Bern University of Applied Sciences
Swiss College of Agriculture SHL

Volume 1
Drivers, Consequences, and Responses

Volume 2
Experiences and Regional Perspectives

Livestock in a Changing Landscape

Volume 2

Experiences and Regional Perspectives

Edited by

Pierre Gerber Harold A. Mooney Jeroen Dijkman Shirley Tarawali Cees de Haan

Washington | Covelo | London

Cataloging-in-Publication Data is available from the Library of Congress.

ISBN (paper) 978-1-59726-673-4
ISBN (cloth) 978-1-59726-672-7

Printed on recycled, acid-free paper

Manufactured in the United States of America

10 9 8 7 6 5 4 3 2 1

Contents

Preface

Livestock in a Changing Landscape is a collaborative undertaking facilitated by the Livestock, Environment and Development Initiative (LEAD), an interinstitutional effort coordinated by the UN Food and Agriculture Organization (FAO), the Scientific Committee on Problems of the Environment (SCOPE), the Swiss College of Agriculture (SHL) of the Bern University of Applied Sciences, the French Agricultural Research Centre for International Development (CIRAD), the International Livestock Research Institute (ILRI-CGIAR), the CGIAR-coordinated System-wide Livestock Programme consortium, and the Woods Institute for the Environment at Stanford University.

This book brings together the mosaic of patterns and draws the variability of "changing landscapes" in which the livestock sector operates. The companion volume resulting from this endeavor—*Livestock in a Changing Landscape: Drivers, Consequences, and Responses*—gives a global perspective on the livestock sector trends. The two volumes together provide a full picture of the impacts of livestock on the environment, social systems, and human health, both globally and locally, and the various approaches that are being or could be undertaken to alleviate negative impacts.

Scientists and professionals from around the world contributed to address such a complex task. Natural and social scientists, agriculturalists, livestock specialists, economists, and industry representatives engaged in the process of sharing knowledge and developing common analyses and visions about some of the key features of the "changing landscape." The integrated chapters in this volume result from their effort to overcome disciplinary divides and draw broad trends from vast but fragmented literature and databases.

The lead authors of these publications met twice—once to agree on the outline of the proposed effort—and a second time, along with many other colleagues, to present preliminary findings and receive critiques and to interact in an attempt to ensure that these issues were being addressed in an integrated manner. The editors attended both meetings and engaged in continuous exchanges with the authors throughout the development of the manuscript. Paul Harrison, Carolyn Opio, and Henning Steinfeld greatly contributed to the finalization of this volume and are cordially thanked by the editors for their support.

Executive Summary

The main technical volume and the case study volume take a detailed, comprehensive, and integrated look at the drivers of the global livestock sector; at the social, environmental, and health consequences of livestock production; and at the variety of responses to opportunities and threats associated with the sector. What emerges is indeed a picture of "Livestock in a Changing Landscape," where economic, regulatory, and environmental contexts are changing rapidly, as are the modes of production. Benefits and costs are distributed rather unevenly, and a variety of diseases and environmental threats are the source of growing concern.

Throughout history, livestock have been kept for a variety of purposes, with the almost exclusive focus on food use of livestock in modern agricultural systems being a relatively recent development. But in many developing countries, livestock are still a critical support to the livelihoods of people who live in or near poverty, and it is here that nonfood uses remain predominant. These include the use of animals for work and as a source of fertilizer (manure), as a means to store wealth, and as a buffer to hedge against the vagaries of nature and other emergencies. Livestock are often the only way to use marginal land or residues and waste material from food and agriculture. Livestock, or symbols of them, also play an important role in religious and cultural lives.

The nonfood uses of livestock, however, are in decline and are being replaced by modern substitutes. Animal draft power is replaced by machinery, and organic farm manure by synthetic fertilizer. Insurance companies and banks replace the risk management and payment functions of livestock. The many purposes for which livestock are kept are vanishing and being replaced by an almost exclusive focus on generating food for humans—meat, eggs, and milk. Hides and fiber still play a role, but these pale in comparison to food uses.

Over recent decades there has been a demand-led rapid expansion of production and consumption of animal products, the so-called livestock revolution. Demand for meat, for example, is projected to double between 2000 and 2050. Population and income growth coupled with urbanization has driven this demand. On the supply side, the livestock revolution has been fed by inexpensive, often subsidized grains, cheap fuel, and rapid technological change. This is particularly evident in poultry, pork, and dairy production. The rapid development of the livestock sector has occurred in a global environment that has been favorable to capital and market liberalization and to rapid technology flows. The recent decades have also been a period of neglect with regard to the environmental and livelihood consequences of livestock production. The response to changing disease patterns and public health threats has been slow and inadequate in many places. Similarly, the need for ever-growing amounts of animal-derived food in the diets of affluent consumers is increasingly questioned on health and environmental grounds.

In industrial countries, however, demand for livestock products is stagnant or even in gradual decline, particularly among the educated and wealthy, where concerns about health aspects, environmental issues, and animal welfare have become more widespread. Some saturation has even started to occur in developing countries, and recent price increases in livestock products may further dampen demand. Yet in most of sub-Saharan Africa and parts of South Asia, the livestock revolution has yet to occur. Overall, however, demand for livestock products may soften, and the recent economic downturn may limit expenditures, particularly for expensive livestock products such as beef. Competition for land is ever more acute, and prices for feed have reversed their decade-long declining trend. Other critical inputs to livestock production, such as water, energy, and labor, are also becoming more expensive. A still more daunting challenge is the fact that land-based livestock production is particularly exposed to the vagaries of climate change.

Increases in livestock production have come from increases in animal numbers and in yields per animal. The latter has been particularly important in dairy, pork, and poultry production and has been facilitated by the rapid spread of advanced production technologies and greater use of grains and oil crops in animal feeds.

Livestock production is practiced in many different forms. Like agriculture as a whole, two rather disparate systems exist side by side: in one case, livestock are kept in traditional production systems in support of livelihoods and household food security; at the same time, commercially intensive livestock production and associated food chains support the global food supply system. The latter provide jobs and income to producers and others in the processing, distribution, and marketing chains and in associated support services.

Social Issues

Livestock have an overwhelming importance for many people in developing countries; close to a billion poor people derive at least some part of their livelihood from livestock in the absence of viable economic alternatives. Livestock production does not require ownership of land or formal education, can be done with little initial investment, and can be transformed into money as and when required. As part of a livelihood strategy, livestock not only provide food, energy, and plant nutrients; they also have an asset function, in that they hedge against risks and play numerous sociocultural roles. The poor are mainly found in small mixed farming systems (less than two hectares) and in pastoral, dryland, or mountain areas. Projections indicate that small farms will continue to be a prominent feature in rural areas in the next decades.

Many poor producers have not benefited from livestock sector growth, however, as shown in China or India, where in the context of rapidly increasing demand for livestock products, large-scale industrial production units have grown rapidly, displacing, at least partially, the smallholder production system. Similarly, there is a risk of overuse of natural resources, leading to long-term environmental damage. On the contrary, many producers have been marginalized and excluded from growing markets. The exclusion of small producers from growing markets is often the result of heightened market barriers in the form of sanitary and other quality standards and of unfavorable economies of scale.

Soaring demand for animal products and competitive pressures in the sector have led to the emergence and rapid growth of commercialized industrial production. Here, large-scale operations with sophisticated technologies are based on internationally sourced feed and cater to the rapidly growing markets for poultry, pork, and milk. The traditional middle-level mixed family farms, while important in many places, are often relegated to the informal market and gradually squeezed out as formal market chains gain hold. In some areas, however, small producers have the potential to contribute in a sustainable way to increased production, provided the constraints of high transaction costs and product quality can be addressed.

At the same time, emerging industrial production benefits from considerable economies of scale, and labor requirements decline dramatically with growing intensification. The trend to reduce human–animal contact for health reasons further reduces employment opportunities. As a consequence, the modern livestock sector provides dramatically fewer people with income and employment than the extensive traditional sector did. This is only in part compensated by employment in the agroindustries associated with the livestock sector, such as feed mills and other input suppliers, slaughterhouses, dairy plants, and processing and retailing facilities.

Another feature of the changing landscape is that the locus of production is shifting. Because of infrastructure weaknesses in many developing countries, growth in the livestock sector is often limited to the outskirts of major consumption centers. Both intensive livestock production and associated agroindustries tend to be located here; as a result, employment and income opportunities move away from rural areas where traditional livestock raising is located. The dramatic growth in livestock production has therefore not led to broad-based rural growth and has not been tapped by the majority of smallholder producers. However, rapid intensification and industrialization have helped urban consumers as prices declined and accessibility increased.

Effective responses to these social issues have been largely absent. Although there have been many efforts to upgrade smallholder practices and, to a certain extent, the institutional framework in which they operate, this has had very little perceptible impact. Rather, many producers, particularly in rapidly developing economies, have abandoned livestock farming as an activity. As long as growing secondary and tertiary sectors can absorb excess labor, this does not necessarily need to be deplored.

Evidence shows that smallholder production can remain competitive where market barriers can be overcome and transaction costs can be reduced. This requires different institutional formats, such as cooperatives, that manage input supplies and marketing as well as the provision of knowledge. Elsewhere, contract farming, particularly in pig and poultry production, has had some success in allowing producers to remain in business, even though these farmers were seldom poor. One example of this approach is Nestlé, a large company that demonstrates how global markets can work for a large number of smallholders. The provision of credit to small producers has also proved effective, at least in part.

Environment and Natural Resources

Livestock affect the global climate, water resources, and biodiversity in major ways. Livestock occupy over one-fourth of the terrestrial surface of the planet, on pasture and grazing lands, of which a significant part is degraded. Expansion of pasture occurs in Latin America at the expense of forests. Concentrate feed demand occupies about one-third of total arable land. Pasture use and the production of feed are associated with pollution, habitat destruction, and greenhouse gas emissions. Livestock are also an important contributor to water pollution, particularly in areas of high animal densities. Both extensive and intensive forms of production contribute to environmental degradation and destruction.

Livestock, through associated land use and land use change, feed production, and digestive processes and waste, affect global biogeochemistry in major ways, particularly the carbon and nitrogen cycles. A large part of this alteration manifests itself in livestock's contribution to global greenhouse gas emissions in the form of carbon dioxide, methane, and nitrous oxides.

Extensive production is practiced by many poor producers who use low-cost or no-cost feed in the form of natural grasslands, crop residues, and other waste materials. However, a large part of the world's pastureland is degraded—releasing carbon dioxide, negatively affecting water cycles, altering vegetation growth and composition, and generally affecting biodiversity. Forest conversion to pasture and crops has important consequences for climate change and biodiversity. Pastoralists, in particular, are threatened as they lose access to traditional grazing areas as the threat of climate change increases, especially for Africa.

The environmental problems of intensive production are also associated with the production of concentrate feed and the disposal of animal waste. Feed production usually requires intensively used arable land and a concomitant use of water, fertilizer, pesticides, fossil fuels, and other inputs, affecting the environment in diverse ways. Even if increased feed production has mostly been achieved through intensification, the expansion of the area dedicated to production of crops such as soybean is now a major driver of deforestation in the Amazon region.

Because only a third of the nutrients fed to animals are absorbed, animal waste is a leading factor in the pollution of land and water resources, as observed in case studies in China, India, the United States, and Denmark. Total phosphorus excretions are estimated to be seven to nine times greater than that of humans, and livestock excreta contain more nutrients than are found in the inorganic fertilizer used annually. Through growing feed crops and managing manure, the livestock sector also emits nitrous oxides (a particularly potent greenhouse gas) and methane.

Policy makers have largely ignored environmental issues related to livestock, often because of the large role that livestock play in sustaining livelihoods and rural life. Some industrial countries have made sustained efforts, especially since the 1980s, to control waste management, with a particular focus on water pollution. Some countries, such as Denmark, have successfully reduced nitrogen leaching from animal operations. However, climate change issues related to livestock remain largely unaddressed.

Human and Animal Health Issues

Human health is affected by the livestock sector through the impact of animal source food products on human nutrition and through diseases and harmful substances that can be passed on by livestock and livestock products.

With regard to nutrition, animal source food products can be of great benefit for people who suffer from undernourishment and for those who need a diet higher in fats and protein, such as children and pregnant women. Consequently, livestock products play a large role in efforts that target improving nutrition in poor and middle-income countries—although other sources of protein and essential micronutrients might also be available at lower environmental cost. At the same time, livestock products can contribute to unbalanced diets, leading to obesity and unhealthy physical conditions. These meat products are often singled out when consumed in excessive quantities as causes or contributing factors to a variety of noninfectious diseases, including cardiovascular disease, diabetes, and certain types of cancer. Increasingly, this is being addressed by public education programs, but until now these have had little measurable impact.

Animal diseases have the potential to adversely affect people by reducing the quantity and quality of food and other livestock products. Transboundary animal diseases tend to have the most serious consequences. Many diseases are zoonotic, with the ability to be transmitted from animals to humans, threatening both human and animal well-being. Zoonotic disease outbreaks often take the form of grand-scale emergencies, requiring rapid action to prevent food supply systems and markets from collapsing.

Changing ecology, the increased mobility of people and movement of goods, and the shifting and often reduced attention by veterinary services have led to the emergence of new diseases, such as highly pathogenic avian influenza (HPAI), and to the reemergence of traditional ones such as tuberculosis. Emerging diseases are closely linked to changes in the production environment and livestock sector structure.

The focus of health concerns differs, depending on whether diseases are considered from a poverty/livelihood point of view or from a global food supply perspective. Certain diseases, such as foot-and-mouth disease, may have minor implications for smallholder production, but their presence excludes entire countries from international trade in livestock products. The control of diseases therefore tends to have far-reaching and often contrasting impacts on constituents in the livestock sector; for example, movement restrictions and sanitary controls may effectively exclude smallholders from markets and may deprive them of livestock as a livelihood option.

The coexistence of functional modern operations alongside traditional operations, in addition to environmental factors, contributes to the emergence and spread of diseases. HPAI has flourished in particular where backyard systems interact with wildlife and are connected via market channels to production systems of medium and high intensity. This explosive mix has its fuse in the structure of livestock sectors, and sector heterogeneity often results from sector protection.

The Challenges

The challenges posed by the livestock sector cannot possibly be solved by a single string of actions, and any actions require an integrated effort by a wide variety of stakeholders. These need to address the root causes in areas where the impact of livestock is negative. They also need to be realistic and take into account the livelihood and socioeconomic dimensions. For example, although in some quarters reduced intake is touted as the most effective way to address negative impacts, most people are shifting, or would likely shift if their incomes allowed, toward consuming more livestock products. Accepting individual food choices, and considering social and economic realities, several principal considerations emerge.

First, given the planet's finite land and other resources, there is a continuing and growing need for further efficiency gains in resource use of livestock production through price corrections for inputs and the replacement of current suboptimal production with advanced production methods. There appears to be little alternative to intensification in meeting the bulk of growing demand for animal products. Although niche products from extensive systems may be of importance in some markets, the bulk of animal protein supply will need to come from intensified forms of production in order to reduce substantially the requirements for natural resource use, such as nutrients, water, and land. The trends toward larger scales of production are determined by economies of scale and scope and are to a large extent inevitable. The current trend toward monogastric production and crop-based animal agriculture will likely continue and possibly accelerate if competition for land accentuates. Ruminants and roughage-based production will continue to decrease in importance; and though not mutually exclusive, this will need to balance the production of animal products with the provisioning of environmental services, including carbon sequestration, water resources, and biodiversity. Particularly in marginal areas, this balance is critical because the value of environmental services will predictably grow sharply, limiting the use of these areas for livestock production.

Second, livestock are a suitable tool for poverty reduction and economic growth in poor countries and areas that are not fully exposed to globalized food markets. Smallholder dairy production and certain types of cooperatives and contract farming provide opportunities for smallholder livestock production. In rapidly growing and developed areas, economies of scale and market barriers will continue to push smallholders out of production, and alternative livelihoods need to be sought in other sectors. Intensification and efficiency gains from economies of scale mean that fewer and fewer people will depend on livestock production for their livelihood. Policies need to support the transition process so as to avoid social hardship and prevent rapid loss of livelihoods in cases where alternatives cannot be provided. There is more promise for growth linkages in associated agrofood industries than in primary production. Notable exceptions are dairy production in favored environments and, to some extent, poultry. In these cases, there is scope for smallholder development and a reasonable chance to compete, at least for some time to come. For the most part, however, livestock production is overwhelmingly important in sustaining livelihoods but less in providing a pathway out of poverty. Social safety nets and exit options are required for those left behind.

Third, food chains are getting longer and more complex in response to challenges and opportunities from globalization. As a consequence, food safety and quality requirements need to be applied to food chains, and sanitary requirements will be linked to them. The specific location of production and its disease status will likely become a secondary issue. Similarly, private standards are likely to be predominant over public ones. As a consequence, animal health and food safety policies will need to be applied to segments of food chains rather than territories. Public policies must foster international collaboration, upgraded monitoring of disease threats, and early reaction.

The changing landscape for the livestock sector as determined by social, economic, and biophysical aspects requires the urgent attention of policy makers, producers, and consumers alike. The institutional void and the systemic failures apparent in widespread environmental damage, social exclusion, and threats to human health need to be addressed with a sense of urgency. Only then can the livestock sector, with its vast diversity of people, critical issues, and constraints, move forward on a responsible development path.

1

Introduction

Pierre Gerber, Harold A. Mooney, Jeroen Dijkman, Shirley Tarawali, and Cees de Haan

The global livestock sector is undergoing major structural changes. Over the last century, livestock keeping evolved from a means to harness marginal or secondary resources to produce goods and services for local consumption, to a component of global food chains, which are driven primarily by consumer demands.

Today, animal food chains are reorganizing on a global scale to supply markets using increasingly standardized production technologies. Underlying this technical standardization is the globalization of the livestock sector and its institutions. Animal production companies and consortiums manage the whole food chain on a transcontinental scale. Trade liberalization, private and public standards, as well as transport infrastructure have enabled an increasing share of animal feed and products to enter global trade. Increasingly, producers located in various continents compete for the same markets. The term *livestock revolution* is generally used to embrace these trends.

These global trends are described in *Livestock in a Changing Landscape: Drivers, Consequences, and Responses,* which introduces the drivers and consequences of the livestock sector on a global scale, as well as the general principles in terms of institutional development and policy instruments dealing with these issues.

This companion volume provides regional overviews and draws experience from specific contexts and more detailed case studies. The chapters describe how drivers and consequences of livestock sector change play out in specific geographical areas and for different players, and how public and private responses are shaped and implemented. Some chapters in the volume also explore in greater detail some of the specific environmental and social issues analyzed in the *Drivers, Consequences, and Responses* volume.

Focus is on the responses to change and, in particular, on the environmental and social consequences of the sector's transformation. The volume investigates how the livestock sector is reorganizing on a regional scale and what consequences and responses this process generates. It addresses questions such as the following:

- What shape do global trends take at the regional scale and what remains from traditional forms of production and marketing?
- How do global food systems influence the development of local livestock sectors?
- What are the consequences on a local scale? Which are the communities involved and how are the issues understood and tackled?
- What is the effectiveness of current responses and what are the lessons learned?

It does so in an integrated way, analyzing how changes along the food chain connect to changes in the environmental, health, and social contexts. It provides both historical and analytical information for readers from the academic and research communities as well as for policy advisers and development professionals.

Seven different regions are included: Latin America, East Africa, West Africa, China, India, the EU (Denmark), and the United States. In four of the selected regions (East and West Africa, India, and China), drivers, consequences, and responses to livestock sector changes are analyzed, addressing environmental, public health, and social issues. The other three case studies focus on a specific issue—soil and water pollution in Denmark, deforestation in Brazil and Costa Rica, and nutrient issues in the US dairy industry—and describe how these issues are being addressed by the public and private sector as well as by nongovernmental organizations (NGOs) and other civil society organizations.

This volume thus covers a wide range of livestock sector development scenarios, from pre–livestock revolution regions with slowly evolving traditional livestock systems in East and West Africa to one that is currently starting to experience the exploding demand, characteristic of the onset of the livestock revolution (e.g., India), to those with the livestock revolution in full swing (e.g., Latin America, China), and those that are in a postrevolution era (e.g., European Union, United States) where environmental and food safety issues are increasingly in focus. Regional level changes are, however, much more complex than such a broad classification of linear development stages would indicate. A wide diversity of local cultures, economic conditions, and environments shape the sector's changes. Global trade plays an even greater role in blurring the picture, bringing local and global products onto the same markets. This complex blend of long-term practices and recent trends makes each region, and thus each chapter in this volume, specific.

We conclude with an insider's look into one of the major private agrifood companies, both shaping and responding to the global market.

Evolving Traditional Livestock Systems: East and West Africa

A case study from the Horn of Africa opens the series of regional analyses and introduces the reader to the complexity of livestock sector change. Several patterns of change coexist in this region, determined by economic, demographic, and social conditions on the one hand and by agroecological contexts on the other. Although the pastoral production systems are well adapted to highly fluctuating environmental conditions and potentially compatible with wildlife conservation, their productivity is seriously constrained by decreasing resource access and increasing levels of civil strife and weather-related emergencies. Over the medium term, the fate of these pastoral populations is therefore uncertain. In some of the densely populated areas of the East African highlands, livestock are progressively disappearing from mixed crop–livestock systems because of population increase and limited land resources. In parallel, the expansion of urban markets fosters the development of relatively intensive systems, especially dairy, at the periphery of urban areas. The authors review and comment on past and present policies and development programs that have attempted to address these issues. They identify cross-cutting challenges for policy makers, including disease control, access to land, development of safe regional markets, and improvement of productivity.

West Africa is undergoing similar trends to those observed in the Horn of Africa. Growing and urbanizing populations, rampant poverty, and institutional weaknesses, including deterioration of animal health services, are among the common trends. There is increasing land degradation and expansion of cropping, which exacerbate conflicts over access to land and put increased pressure on pastoral systems. Relatively intensive systems, mostly in poultry, tend to develop at the periphery of major urban areas. These systems compete with growing imports, especially for urban markets in coastal areas. The authors provide a detailed analysis of the economic and social factors at play in shaping the livestock sector. They suggest that the future of the livestock subsector in West Africa lies in properly managing a threefold approach based on sustainable management of rangeland, shared pastoral resources, and transhumance; enhancement of livestock productivity; and improvement of market access for local producers.

Livestock Systems at the Onset of the Livestock Revolution: India

India shows all the signs of being at the onset of its livestock sector boom: incomes are growing, and consumption levels of animal products in rural areas have been rapidly catching up with urban settings. Featuring some elements of the livestock revolution, the livestock sector is responding to increasing demand by growing in size and changing structure. Throughout the country, specialized poultry and dairy operations are burgeoning close to urban centers, although dairy continues to be dominated by smallholders. Environmental problems have emerged, and environmental regulations have been established, although the authors report problems of implementation. But the main questions in India relate to the future of smallholders in the sector's development, the economic spinoffs of the livestock sector growth, and the implications for issues of equity and poverty. The authors propose a set of policy interventions to address these issues, including the environmental regulation of large production units, the improvement of production technology among smallholders, and the empowerment of producer organizations.

Systems in Rapid Growth: Brazil and China

The Brazilian livestock sector is one of the most dynamic in the world. National consumption plus booming exports of animal products and feedstuffs are driving the sector's development. As a consequence, and also for other reasons such as land tenure policies, forestland is rapidly being converted to pastures and feed crops. What can be done is shown by the experiences of Costa Rica, where deforestation trends have been halted and reversed. The chapter compares the drivers and rates of deforestation in these two countries and analyzes the relative effectiveness of policies to curb land conversion. It highlights the strength of the Costa Rica policy setup, initiated two decades ago and relying in particular on the discontinuation of direct subsidies to production, on forest conservation measures, and on the implementation of a national payment for environmental services scheme. Yet the authors recognize the recent progress made in Brazil in its effort to put an end to "open frontiers" and see hope in the possible payment for the reduction of emissions from deforestation and degradation (REDD) currently being discussed in the framework of the United Nations Framework Convention on Climate Change (UNFCCC).

China is the country that probably best illustrates the livestock revolution at its postclimax stage. The author explains how China has been leading the global livestock revolution over the past two decades. Population growth, urbanization, and increasing incomes have been the main factors fueling past livestock sector growth. The sector's growth rate in China has been one of the fastest in the world, and the most impressive in absolute terms. Substantial structural changes have occurred, with a move toward larger, intensified production units and greater reliance on commercial—and often imported—feed. This has caused a concentration of livestock production and processing near the seaports of the Eastern seaboard and has also given rise to a real East–West dichotomy. The author highlights the factors that have driven this geographical gradient. Today, the Chinese livestock sector is at a turning point: continuing growth in population and per capita consumption, and the need to create employment in the rural areas mean that it still needs to grow, and at the same time it must increasingly respond to new societal requirements such as for a cleaner environment, improved food safety, sustainable use of natural resources, and better labor conditions.

Systems in the Post–Livestock Revolution: United States and Europe

The structural changes undergone by the dairy sector in the United States of America are examined next. Here the post–livestock revolution sector consolidation is associated with dramatic changes in production and management practices, and the sizes of milk production units. This results in substantial environmental issues but also in social and health concerns. The drivers and consequences of these transformations and current and potential responses, including policy instruments, institutional mechanisms, and technology options are discussed. The authors highlight the need to consider the comparative advantages of federal, state, and local governments when implementing policies to guide the sector. They recommend moving beyond the sole regulation of point-source pollution in large units to a broader set of instruments allowing better control of nutrient flows across the various sizes of production units. Improving the balance between livestock densities and the adsorptive capacity of the land is identified as the core strategy to address pollution problems.

The case study from Europe, with experiences of Denmark as the example, provides insights about how policies can be put in place to control nutrient-based pollution from animal production. In Denmark and in the European Union more generally demand has stabilized and the sector's growth is limited. Changes in the sector are increasingly guided by environmental and public health concerns rather than by production output and prices. In Denmark, nitrogen-based pollution has been effectively reduced, at an estimated annual cost of 2 Euros per kg reduction of nitrogen leaching into aquifers. Environmental policy development has followed an iterative process, adjusted and improved by trial and error, supported by comprehensive policy analysis. The process is continuous because policies need to evolve within the sector and the broader economic and policy context. The authors also draw attention to the delay between action and the actual measurement of results, related to the pace of natural biophysical processes involved, and the particular attention to communication that this delay requires.

A Private Sector Perspective: Changes in the Dairy Food Chain

Chapter 9 introduces Nestlé's response to changes in the private sector–driven dairy food chain. Two authors from the company identify four main drivers influencing Nestlé's activities in processing raw agricultural materials and marketing food products. First is consumer demand, which is growing and changing as dietary habits and expectations about the functional and emotional aspects of food evolve. Next is the increasing complexity of food chains, requiring increased traceability and risk management. Emerging diseases and an increased focus on the sustainability of agriculture production are other major factors. The authors provide examples of how the company has developed specific business models to respond to key objectives related to milk sourcing, such as food safety, quality improvement, animal disease control, environmental sustainability, and productivity improvement.

The volume concludes with an analysis that draws comparisons across regions and consolidates major experiences and lessons. A unique Executive Summary was also prepared for this volume and for the *Drivers, Consequences, and Responses* volume. It is included in both volumes to allow for a stand-alone consultation.

2

Horn of Africa

Responding to Changing Markets in a Context of Increased Competition for Resources

Joseph M. Maitima, Manitra A. Rakotoarisoa, and Erastus K. Kang'ethe

Abstract

The livestock sector in the Horn of Africa is evolving rapidly as a result of internal and external influences affecting production, marketing, and utilization of livestock and livestock products within the region. These changes are related to increases in human population; reductions in the sizes of crop production units; changes in accessibility to land; climate variability; changes in livestock disease challenges; and changes in market opportunities.

These changes have affected different production systems in different ways. In pastoral production systems they have resulted in reduced productivity, shifts in herd composition and size. In mixed crop–livestock production systems population increase, reduced accessibility to land, and agricultural intensification have brought about more integration of livestock with agriculture, leading to well-developed dairy production in certain periurban areas. Demand for livestock products has been high, and in some places this local demand is coupled with an export market of live animals and milk.

Livestock production in the Horn of Africa has suffered a great deal from high levels of insecurity in many parts of the region, which have caused loss of stocks, markets, and livelihoods, especially among pastoralists. In the past, the political economy in several countries has put pastoral livestock production areas at a big disadvantage compared to urban and crop-farming areas. However, pastoral communities have adapted to the new challenges.

This chapter reviews these trends and discusses policy and societal responses to overall changes in the livestock sector in the Horn of Africa. We draw implications about identifying effective policy interventions that have been or could be put in place in different regions that could eventually contribute to improving people's livelihoods.

Introduction

Background

For the purpose of this chapter, the region of the Horn of Africa comprises the following countries; Djibouti, Eritrea, Ethiopia, Kenya, Somalia, Sudan, Tanzania, and Uganda. Most definitions of the Horn of Africa exclude Tanzania, but because of its ties with the other East African countries in the region it must be included here.

These countries are characterized by diverse but interlinked cultures and agroecological and economic conditions. They belong to different and sometimes overlapping political and socioeconomic blocks. The East African Community (EAC) includes Kenya, Uganda, Tanzania, and now Burundi and Rwanda; the Intergovernmental Authority on Development (IGAD) comprises Ethiopia, Eritrea, Kenya, Uganda, Somalia, Djibouti, and Sudan. Tanzania is the only country included in this chapter that belongs to the Southern Africa Development Cooperation (SADC).

Role of the Livestock Sector

Livestock production is an important component of the farming system in the Horn of Africa, contributing to the livelihoods of an estimated 40 million poor people (Upton et al., 2005). The livestock sector contributes significantly to providing income, employment opportunities, and food security; provides agricultural inputs through manure and draught power; serves as a risk-hedging asset; and plays important social and cultural roles. The use of animal draught power and biogas as clean energy alternatives to fossil fuels shows the contribution livestock can make to improving the environment.

The livestock sector is also a major component of country agricultural activity, ranging from 20% of agricultural gross domestic product (GDP) in Uganda up to

52% in Kenya and 88% in Somalia. Its contribution to the overall economy is also high, exceeding 50% of total GDP in Somalia and 8 to 17% in the other countries (see Table 2.1).

At the local level, livestock is a primary income source for many smallholders. For instance, in Kenya more than 30% of household income among smallholders comes from milk (SDP 2004a, SDP 2004b, Bebe et al., 2003). As in most of the developing world, the role of livestock in food security often occurs on the farm. Livestock contribute to sustained food security beyond meeting basic nutritional needs (Simpkin 2005). The ownership of livestock allows producers to maintain a diversity of assets that decreases nutritional vulnerability during times of drought or other shock. In emergencies, livestock can be easily moved or hidden and can give families access to protein and energy in a pattern that has become known as "conflict foods" production (Knips 2004, Simpkin 2005). In semiarid regions the prices of sheep and goats per head are comparatively higher than those of a bag of sorghum or millet, with important seasonal variations (Figures 2.1 and 2.2). Through income generation, the livestock sector also provides food security to non–livestock owners (traders, transporters, butchers) working in the sector.

In unstable currency markets and in the context of weak banking systems, livestock also provide a saving option. Livestock are indeed somewhat less perishable than other living assets, thereby providing resources (e.g., money, eggs, milk) during times of hardship. Livestock also play a social role in rural communities in many countries in the Horn of Africa as a symbol and store of wealth and a source of gifts or sacrifice during traditional ceremonies such as burial proceedings and tribal weddings.

Table 2.1. Livestock contribution to agriculture and to the economy, 2000

Country	Share of Ag. GDP in Total GDP (%)	Share of Livestock GDP in Ag. GDP (%)	Share of Livestock GDP in Total GDP (%)
Ethiopia	52.3	32.5	17.0
Kenya	19.9	52.4	10.4
Somalia	65.5*	88.2*	58.0
Tanzania	45.0	27.9	12.6
Uganda	42.5	19.8	8.1

* 1990 figure, last available.

Source: FAO 2008, Knips 2004.

Composition of Livestock

A wide variety of livestock types are kept in different parts of the Horn of Africa. The charts in Figures 2.3 (in the color well) and 2.4 show that cattle are the most common and have the widest distribution across the region, followed by goats and sheep. Chicken are also widely produced, by traditional subsistence methods in rural areas and by intensive production in periurban areas. Turkeys are kept to a much lesser extent, mainly by small-scale producers, especially in parts of Uganda and near urban centers in other regions. Pigs are reared both by small-scale producers in villages and at commercial levels in periurban centers. Donkeys are kept for traction in many areas, especially in semiarid and arid lands. In the highlands of Ethiopia, crossbreeds of horse and donkey are kept for use in transportation. Camels are kept by most pastoralists for meat and milk and for

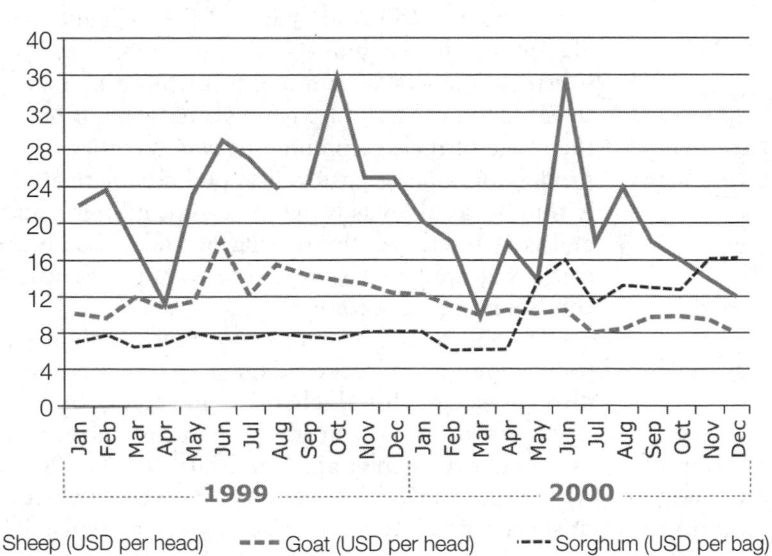

Figure 2.1. Sorghum prices vs. sheep and goat prices in Sudan (1999–2000). *Source:* FAO 2000.

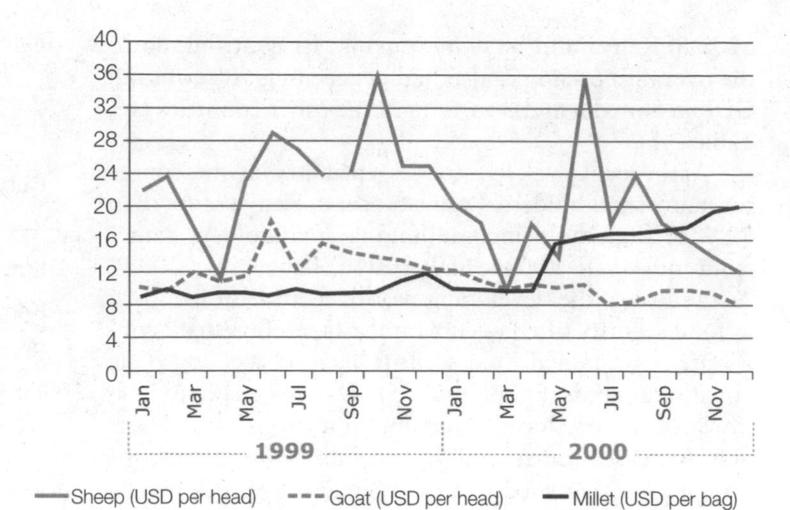

Figure 2.2. Millet prices vs. sheep and goat prices in Sudan (1999–2000).
Source: FAO 2000.

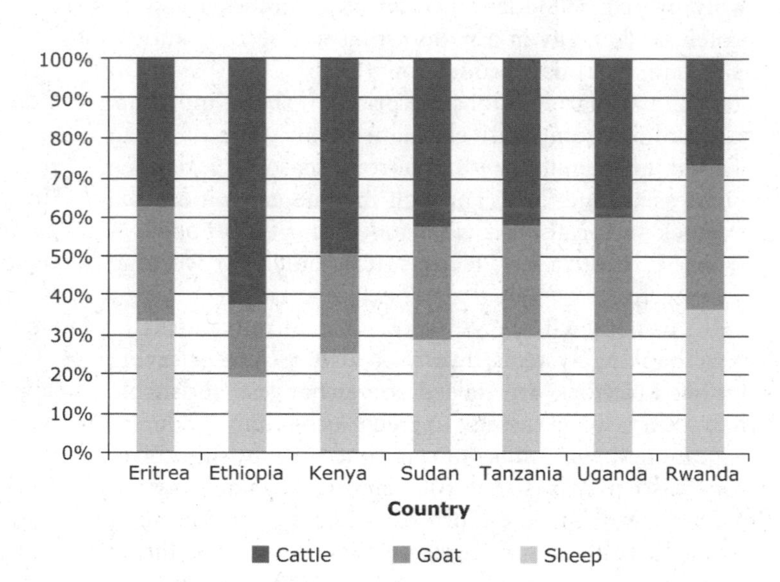

Figure 2.4. Distribution of cattle, sheep, and goats in the Horn of Africa.
Source: ILRI GIS (www.ilri.org/gis).

transporting household belongings during migrations. Rabbits are reared on a relatively small scale for food and cash within the mixed crop livestock production systems.

Main Characteristics of Livestock Production Systems

Livestock production systems in the Horn of Africa vary mainly with the ecological potential of the area and can be characterized as follows:

1. Pastoralism (livestock only)
2. Mixed crop–livestock production systems (mixed irrigated and mixed rainfed)
3. Landless livestock production systems in rural and periurban areas.

Figure 2.5 (in the color well) shows the distribution of livestock production systems according to agroecological potential and categorized into livestock-only, mixed irrigated, and mixed rainfed.

Pastoralism dominates the vast dry savanna rangelands. Geographically, pastoral systems are found in arid and semiarid areas, except in Ethiopia, where pastoralism is practiced in some parts of the highlands (Seré et al., 1996). The pastoral system is dominated by nomadic livestock keepers who move from place to place with their stock in search of pastures and water, often within the range of their communal or clan territory. Livestock production among pastoralists and commercial ranches within pastoral areas is mainly for meat, hides, and skin and is characterized by rearing of indigenous breeds of cattle that are better suited to tolerate the prevalent dry conditions and diseases.

Despite their lower technological capabilities, pastoral communities have adapted to climate variability through sociocultural changes, including in livestock management, usually in response to climatic events such as drought (Galvin et al., 2001, Little et al., 2001). East African pastoralists have adopted a diversity of strategies to sustain production in the face of climatic challenges. These include moving livestock according to vegetation and water availability, changing the species composition of herds to take advantage of the heterogeneous nature of the environment, and diversifying economic strategies to include agriculture and wage labor, These strategies are crucial for pastoralists' own livelihoods and important for national economies because pastoralists are responsible for providing a large share of livestock to markets in the region. In some cases some members of a pastoral community choose to migrate out of the pastoral system on a short-term or permanent basis, and this eases seasonal and drought-induced stresses among community members (Galvin 1992, Galvin et al., 1994).

Many pastoralists have become more sedentary, practicing crop cultivation in areas close to rivers and swamps so as to diversify their means of production, improve food security, and increase household income. Migration to work in towns is now common, especially among younger generations, whereas looking after livestock is left to the older generations. Scarcity of water resources has led to interventions by governmental agencies and donor communities to dig bore holes, dams, and water pans to provide water for animals and people.

Mixed crop–livestock production is practiced in the wetter subhumid and semiarid areas where rainfall is high enough to support cultivation of cereals and root crops. The geographical distribution of mixed crop–livestock systems in the Horn of Africa follows closely the elevation patterns. It is found mainly in areas with rainfall between 800 and 1500 mm per year (Maitima et al., 2004) but varies from place to place depending on other factors like soils and vegetation types and altitude.

In most humid areas of East Africa, the typical livestock production system is dairy production, and the main breeds are exotic cattle (Friesian, Guernsey, and Ayrshire) that produce higher milk yields than the indigenous breeds. This type of production system is practiced by sedentary farmers on individually owned land. Cattle are usually fed in small enclosures, supplemented by cut and carry of feed from other areas (Bourn et al., 2005).

The mixed crop–livestock system has been expanding in the Horn of Africa owing to the conversion of rangelands to cultivation (Olson et al., 2004). In some areas where rangelands have been converted to croplands there are mixtures of indigenous breeds and crossbreeds of indigenous and exotic species. Livestock products of most value in the mixed crop–livestock systems are meat and milk. Farmers in the mixed system also benefit from animal manure and draught power.

Landless livestock production is becoming more and more important with younger generations who have not yet been allocated land by their parents, and with people living in urban centers who produce livestock on small plots and undeveloped land owned by others. Although a few landless farmers have free-grazing animals, most animals are fed by tethering in communal areas like roadsides and under fruit trees in home compounds (Bourn et al., 2005, Canagasaby et al., 2005). Pigs and traditional chickens are kept within home compounds.

Urban and periurban industrial livestock keeping is a growing form of production within (intraurban) or on the fringe (periurban) of towns, cities, or metropolises, raising a variety of livestock at different scales. The animals raised are mainly dairy cattle, chicken, and pigs (Canagasaby et al., 2005). It is estimated that 200 million urban Africans will be partly dependent on urban agriculture for their food by 2020 (Urban Harvest 2004). Surveys have indicated for instance that one fifth of households in Nairobi, Kenya, and a half in Kampala, Uganda, were engaged in urban agriculture in the 1980s. Recent figures show higher proportions, especially in small and medium-sized towns. Horticulture and grazing are widely practiced along roadsides, stream banks, and in public and private vacant or abandoned land areas.

The impacts of farming system changes on the welfare of the region's livestock keepers, most of whom are poor, are difficult to assess without a prior investigation on the links and interactions among the drivers and their consequences. Such an investigation is important to help policy makers and researchers to identify appropriate responses to benefit the poorest farmers in the Horn. The information reported in this chapter is intended to fill this need.

Drivers of Change

Livestock production in the Horn of Africa has undergone significant changes in the recent past due to a number of factors. These include changes in land use, growth in human population, technology adoption, socioeconomic forces (including market growth and liberalization, gender roles, and land policies), and the impact of livestock diseases and the interventions to control them.

Population Growth and Urbanization

Countries in the Horn of Africa have some of the highest rates of population growth in the world (Table 2.2). Kenya and Uganda, for instance, have witnessed rapid and steady growth in population for the last few decades, well above the average population growth of sub-Saharan Africa. This growth and the heavy dependence on land-based livelihoods have created an enormous amount of pressure on land leading to continuous subdivision of land into smaller production units, rapid land use changes, loss of vegetation cover due to conversion of natural vegetation to farms and grazing lands, loss of communal grazing and other communal lands, and increasing demand for food products.

Although human population has been increasing in almost the whole of the Horn of Africa, livestock population has either remained at the same level or, in some areas, has declined over time, especially on a per capita basis. For example, Jabbar et al. (2003) show that Ethiopia's per capita production of livestock and livestock products and export earnings from livestock have

Table 2.2. Some statistics for the Horn of Africa*

Countries	GNI per Capita, PPP (USD)	GNI per Capita, (USD)	GDP Annual Growth (%)	Population Annual Growth (%)	Malnutrition Prevalence, weight for age[1]	Poverty Indicator[2]
Ethiopia	870	280	11.3	2.6	42 (in 2000)	44.2 (in 2000)
Kenya	1,580	770	3.6	2.6	16 (in 2003)	n.a.
Tanzania	1,230	440	7.5	2.9	17 (2005)	35.7 (in 2001)
Uganda	1,140	420	9.5	3.3	19 (in 2001)	37.7 (in 2003)
Sub-Saharan Africa	1,991	1,082	5.0	2.5	27	n.a.

* The figures are from the year 2008 data, unless otherwise indicated.

[1] = Percentage of malnourished children under 5 years old

[2] = Poverty headcount ratio at national poverty line in percentage of population

n.a. = not available

PPP = purchasing power parity

GNI = gross national income

Source: World Development Indicators database, World Bank, 2009.

declined since 1974. The authors further report a general decline in the number of households owning different types of livestock and a significant drop in average livestock holdings per household in 1999 compared to 1991. High population growth in many countries in the Horn of Africa has created a scarcity of land within the high potential arable areas, resulting in migration into the pastoral areas. Land and resource disputes and conflicts in the region have sometimes led to full-blown conflicts over access to wetter, more fertile areas for farming and livestock keeping. However, in recent decades many subhumid areas that had remained unoccupied for various reasons (mountain slopes, sacred forests, forest reserves, etc.) have been allocated or invaded by farmers from arid lands in search for fertile lands for cultivation. This has led to massive deforestation around all the major mountains in East Africa such as Mounts Kenya, Kilimanjaro, and Elgon, the Mau escarpment, and others—leading to serious conflicts between communities, politicians, and conservations (Olson et al., 2004).

The growing population has spurred the already high demand for livestock products and put additional pressure on the inelastic domestic supply of livestock in each country. Because the per capita consumption of livestock products in some countries in the Horn of Africa was already among the highest in sub-Saharan Africa (as an example milk consumption per capita in Kenya reached 140 kg in 2004—four times the average in sub-Saharan Africa) high rates of population growth will increase consumption significantly. This trend of increasing livestock consumption is expected to continue as the countries' urban areas and economies expand fast.

Increased urbanization in many countries in the Horn has grown out of the development of markets for goods and services (e.g., tourism, manufacturing) with improved facilities and infrastructure that attract a growing number of migrants. Urbanization has also created livestock activities in periurban areas to satisfy urban demand in product quantity, quality, and form (e.g., ready-to-consume, or packaging for individual serving sizes). Urbanization has therefore increased the need for more efficient distribution and delivery systems in the livestock sector of the Horn, and this can be seen in the shelves of bustling supermarkets and other food outlets in the major cities.

Economic Growth and Opportunities

Although the countries in the Horn of Africa include some of the poorest populations in the world, their economies have grown at an impressive pace in the past few years (see Table 2.2). This growth has occurred despite social and political crises, including wars over the last three decades. The growth has been highest in Ethiopia and Uganda (Table 2.2). Such a high increase in income drives a surge in per capita consumption of livestock products long considered a luxury for the poorest. The high population growth of the region has increased livestock product demand even further.

Income growth, especially in the growing urban areas, is likely to shift consumer preferences toward safe and high-quality products. Development of the tourism industry, which contributes significantly to local and national economies, especially in countries such as Kenya and Tanzania, has enhanced this shift toward safer and higher-quality livestock products.

Economic growth in the Horn of Africa has also created resources to increase access to public services and development activities and has improved urban and rural infrastructure, communication, and education. Although the full effects remain difficult to assess immediately, these improvements have undoubtedly had some positive effects on the livelihoods of farmers and workers in the livestock sector.

Gender and age groups in the Horn of Africa, as in many other parts of Africa, have different roles in livestock production. But recent economic development has brought about some societal changes that have affected gender and age group roles in livestock production. For instance, an increase in school enrollment or emigration to the cities has reduced the availability of young boys, who traditionally herd livestock. The task is then increasingly shared among other gender and age groups. Similarly, in many pastoral communities small ruminants are traditionally kept mostly by women, whereas large ruminants are mainly kept by men. But as men become more involved in cash cropping and in urban employment, women are increasingly taking over more responsibilities such as keeping and feeding large animals, in addition to their more traditional duties of milking and caring for young and sick cattle (Wangui 2003).

Environmental and Climatic Changes

Environmental changes in the Horn of Africa have been characterized by increasingly frequent and sometimes prolonged droughts (Nicholas 2002, Verschuren et al., 2000). These droughts have adversely affected livestock production, especially in the arid and semiarid regions where livestock production predominates. They have resulted in high livestock mortality, leading to high levels of poverty and hunger among livestock keepers—especially the pastoralists whose livelihoods depend almost entirely on livestock products. They have also affected the availability and distribution of livestock feed resources, resulting in reduced livestock numbers and changes in the types of animals kept by the pastoralists.

Climate change, as observed in the increasing aridity and spreading of deserts, has reduced primary productivity and water availability in many parts of the Horn of Africa leading to reduced carrying capacities of grazing systems. Figure 2.6a shows the dramatic shrinkage in surface areas of glaciers on Mounts Kilimanjaro, Ruwenzori, and Kenya due to global warming (Oludhe 2005)—over time this will reduce the availability of dry-season water for agriculture and livestock. Figure 2.6b shows the declining trends of June, July, and August rainfall patterns around Khartoum from 1961 to 2000 (Oludhe 2005).

Climate change poses many and diverse challenges to pastoralists. Although many parts of the Horn of Africa are experiencing drier conditions, northeastern Kenya is predicted to be slightly wetter than present in the coming decades and will have a much higher vegetation cover index than present (Andresen et al., 2008). This would promote replacement of the vast open grasslands by bushy shrubs, which may be less valuable to grazers such as cows, the livestock type preferred by pastoralists. Indeed bush encroachment is already reported in some parts of northern Kenya, where invasive shrub species like *Prosopis juliflora* are becoming common in grazing lands, reducing the quality of pastures.

Even in years of normal rainfall, there are often anomalies in rain distribution: too much rain may occur within short periods of time, resulting in floods, whereas other periods are characterized by long spells of drought. These variations affect both the distribution and the quality of pastures available for livestock in pastoral areas and other affected regions where farmers practice open grazing systems.

Market Liberalization and Global Policies

Livestock product markets in the Horn of Africa have become more and more open to global and regional markets. For example, in Kenya, liberalization of the milk and dairy market began in the mid-1980s and continued until the early 1990s. A number of measures were aimed at reducing or eliminating government control and regulations over breeding services, milk marketing, and dairy processing. Because Kenya's dairy farming had been overtaxed for decades, elimination of government intervention in 1992 had a significant impact in improving income at the farm and especially postfarm levels (Owango et al., 1998, Ngigi 2005). Ethiopia, Uganda, and Tanzania have conducted similar domestic and border policies that have led to a more open livestock market. The World Bank's Structural Adjustment Programmes have pushed for reduction in Kenyan government subsidies to agriculture and for privatization of many agricultural services. The privatization of veterinary services, however, has had serious negative impacts on the livestock sector in rural areas, often making such services unavailable except within the urban and periurban areas, where such negative effects were somewhat attenuated (FAO 1999).

In Ethiopia, an important recent change due to more open markets has been the development of capacity for exporting chilled and frozen meat to the Middle East. This involves the establishment of new, privately owned and operated abattoirs. This situation provides an opportunity for producers to directly supply animals for export. However, lack of education and experience may limit participation of pastoralists or other rural poor livestock producers in such opportunities.

In Somalia, livestock production and livestock exports have been the backbone of the economy, providing broad financial benefits to pastoral households as well as supporting livelihoods at the ports. Compared to other nomadic livestock systems the Somali system is relatively more market oriented. Despite political and economic crises, wars, and the absence of clear trade policies, Somalia has remained a major player in cross-border livestock trade in the Horn of Africa. Rough estimates from unofficial sources indicate that over 90% of Somalian livestock exports per year go to the Arab gulf countries, most of them to Saudi Arabia.

The opening up of markets in livestock products (meat and dairy) has spurred competition between local

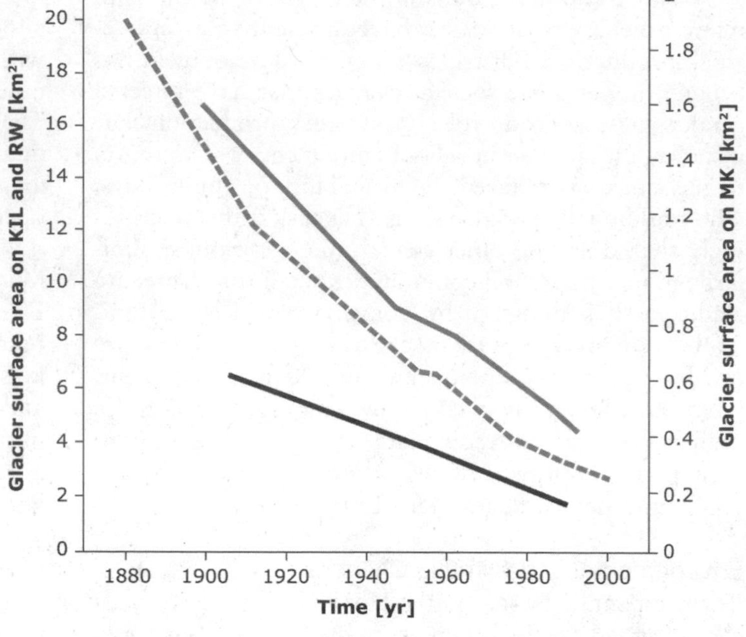

Figure 2.6a. Gradual disappearance of mountain glaciers.
Source: Oludhe 2005.

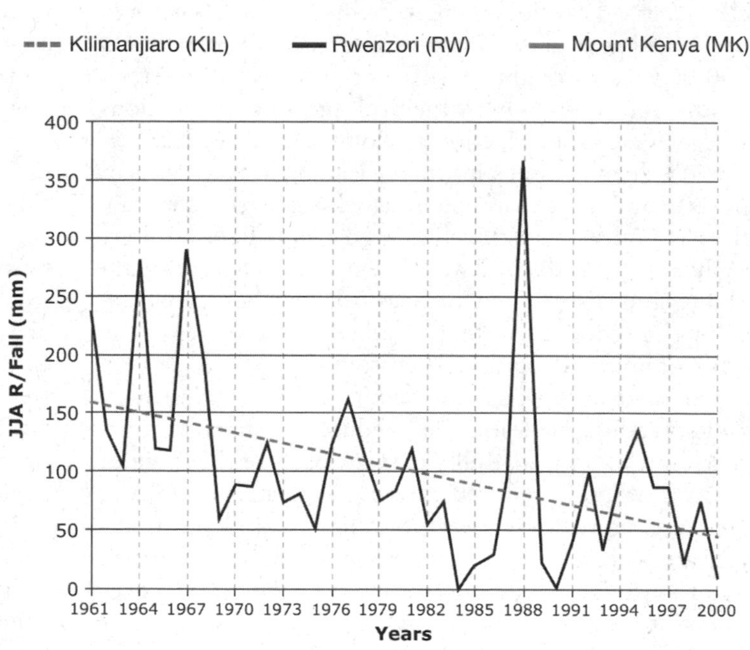

Figure 2.6b. Rainfall trends in Khartoum.
Source: Oludhe 2005.

and outside firms. It has also increased product variety, thus enhancing consumers' choice and welfare. The rises in consumers' incomes in both regional and international markets, combined with the increasing availability of competing products, have induced a strong shift in consumer demand and preference toward higher-quality products with a high level of accessibility. The needs to increase productivity and quality and to address shifting consumer preferences have promoted new market arrangements such as vertical integration and contract farming.

Increased trade openness (especially cross-border trade of live animals) has also had some drawbacks because it has increased exposure to animal diseases and accelerated the spread of these diseases in importing countries. The increased risk of animal diseases hampers the export of processed livestock products, especially under current tight sanitary requirements in major foreign markets. More importantly, although the livestock sector in the Horn of Africa region has been opened to imported products, it suffers from limited access to regional and international export markets because of livestock policy distortions in developed countries. These distortions have depressed livestock producer prices in the Horn of Africa as producers try to compete in both domestic and international markets against developed countries' heavily subsidized livestock production and export.

Openness in the Horn of Africa has gone beyond trade in goods. It also includes financial openness that has brought inflows of capital investment, especially into Ethiopia, Kenya, Tanzania, and Sudan. For the livestock sector, these investments have both direct effects (e.g., improvement in livestock infrastructure in Sudan) and indirect effects (e.g., credit access and improved communication networks).

Land Reform Policies

Issues of land tenure have dominated the Horn of Africa's political arena in livestock production. Whereas land tenure in almost all the crop–livestock mixed production systems is largely private, pastoral land is usually held under an access system of communal control. Land among the pastoralists is held by the community: landownership is a relationship between individuals and groups or tribes consisting of a series of rights and duties with respect to the use of land.

Where land reform policies have been introduced, they have become a major driver of change in livestock systems. For instance, transfer of land from large-scale to smallholder farmers has occurred during the postindependence period in countries such as Kenya and Uganda (Ngigi 2005). This transfer has increased the role of small-scale livestock keepers in livestock production. For instance, in a survey conducted in Kenya, 80% of smallholder respondents reported that they own their land.

High population growth in many countries in the Horn of Africa has created a scarcity of land within the high potential arable areas, resulting in migrations into the pastoral areas. Migrants move to wetter areas where they can practice crop farming. Introduction of croplands within the wetlands of the pastoral areas has created conflicts over access by livestock and wildlife to key resources such as water and pastures. These resources are critical for the survival of animals and humans, especially during periods of drought (Campbell et al., 2003).

Donors have also pushed for land tenure liberalization—in accordance with the World Bank's "security of property" paradigm. So far Ethiopia, Tanzania, and Uganda have resisted pressure for wholesale liberalization. However, all three found it necessary to move in that direction in order to foster an "enabling environment." Tanzania and Uganda enacted new land laws for this purpose, whereas Ethiopia modified its land policy (Fortin 2005, Harrison 2004).

Land reform policies have affected pastoralists' access to land in several ways:

1. Establishment of wildlife protection areas and private game and livestock ranches has denied pastoralists access to grazing lands, resulting in human–wildlife conflicts in many regions.
2. Land reform policies have tended to favor cropping and have ignored rangeland areas on the account that they have low productivity, especially of crops (Knips 2004).
3. Many governments promote sedentary livestock systems and fail to recognize pastoral livestock production as the best way to exploit rangeland resources.
4. Communal landownership, still practiced by pastoral communities, has denied individual legal rights of ownership, leaving land open for common use and overuse.
5. The traditional institutions and experiences of local communities in the management of grazing lands are rarely considered by central decision makers when they are creating policies and laws concerning resources that support livelihoods of local communities.

Animal Health

Livestock diseases have significant impacts on animal health and affect productivity, herd structure, and human health. Via product safety they also affect access to national and export markets. This section will review a few of these diseases in the Horn of Africa.

Rift Valley Fever

Rift Valley fever (RVF) is an acute febrile viral disease of cattle, buffalo, sheep, goats, camels, and humans caused by *Phlebovirus* (family Bunyaviridae). The disease causes livestock and human mortalities. The spread of the disease greatly affects pastoral livelihoods—in areas with RVF outbreaks, consumption of livestock products and animal movements are restricted to prevent transmission to humans. It is associated with climatic conditions of high rainfall following long periods of drought (Davies et al., 1985). These climatic conditions lead to high populations of *Aedes* and *Culex* mosquitoes, which spread the disease rapidly among livestock and humans (Linthicum et al., 1985).

Brucellosis

Brucellosis is a bacterial zoonosis caused by bacteria of the genus *Brucella*. The main hosts of *B. mellitensis*, *B. abortus*, and *B. suis* are goats, cattle, and pigs, respectively. Prevalence rates among cattle in Africa have been reported to range from 3 to 41% (Nakoune et al., 2004, Domingo 2000).

Drier environmental conditions found in most of the pastoral areas of the Horn of Africa tend to reduce infection and disease prevalence, whereas wetter conditions tend to protect the organism from the natural decontaminating effects of heat and sunlight. Movement of infected animals across borders, common in the Horn of Africa due to cattle rustling, interborder trade, and migration due to wars and insecurity increases the infection rate in susceptible herds. Animal management systems such as extensive pastoral grazing systems in Kenya (Omer et al., 2000); mixed breeds of cattle in Djibouti (Omer et al., 2000); and the practice of keeping sheep alongside goats in Uganda (Kabagambe et al., 2001), have been shown to contribute to the prevalence of disease. Once established in a herd, the disease leads to frequent abortions that affect herd structure, cause infertility, and affect humans via consumption of animal products. By raising questions about product safety, it reduces market access and affects livelihoods.

Trypanosomiasis

Trypanosomiasis is a disease of cattle and humans caused by protozoan parasites of the genus *Trypanasoma. T. vivax* and *T. congolense* are important as cattle pathogens, whereas *T. brucei rhodesiense* and *T. brucei gambiense* cause acute and chronic human trypanosomiasis. The vectors for these trypanosomes are tsetse flies of the genus *Glossina. G. morsitans, G. austeni, G. pallidipes, G. swynnertoni*, and *G. longipennis* are classified as savanna species inhabiting grasslands where cattle are traditionally reared. However, they are capable of adapting to other ecological niches. The riverine *Glossina* species (*G. palpalis, G. tachnoides*, and *G. fuscipes*) are important vectors for bovine and porcine trypanosomiasis and chronic human disease due to *T. gambiense*. Asymptomatic infected wild ungulates serve as blood hosts, helping to maintain the trypanosomes and tsetse populations that transmit the disease to other animals and humans.

The disease in cattle causes economic losses due to low productivity (meat, milk, and traction power), treatment costs, and mortalities.

Bovine Tuberculosis

Bovine tuberculosis (TB) is caused by *Mycobacterium bovis*. Although cattle are considered to be the primary hosts of *M. bovis*, the disease has an exceptionally wide mammalian host range, which includes humans (O'Reilly and Daborn 1995).

M. bovis is a robust pathogen and may survive in the external environment in buildings, on transport vehicles, on pastures, and in slurry under certain climatic conditions for months and years (Wray 1975). Manure fertilization of arable land is a common practice in developing countries. *M. bovis* surviving in soil and slurry serves as a source of pasture and vegetation contamination and therefore a potential source of infection to animals and humans.

The losses caused by the disease derive from condemnation of infected carcasses, potential of transmission of disease to humans via consumption of meat and milk, and loss of external market access for export of live animals.

Contagious Bovine Pleuropneumonia

Contagious bovine pleuropneumonia (CBPP) is a highly infectious septicemic disease of cattle caused by *Mycoplasma mycoides* var. *mycoides* (small colony type). The agent is very susceptible to environmental temperatures. Transmission of the disease in the arid and subhumid tropics is from infected to susceptible animals by aerosol. Frequent droughts in the Horn of Africa lead to pastoralists' movements in search of pasture and water. The stress of the movements may activate disease conditions in the carriers and convert them into active cases, causing spread of the disease in susceptible herds. International trade is another contributor to the spread of CBPP. Kouba (2003) reported that, of 117 Office International des Epizooties (OIE) List A disease cases communicated through international trade, CBPP accounted for 7.7%. This fraction would be higher if all countries fully reported their cases. Even where reporting is advanced incidence is probably much higher than reported.

The disease leads to loss of livelihood due to high mortalities and economic losses due to external trade embargo.

Rinderpest and Peste des Petits Ruminants

Rinderpest is an acute, highly contagious viral disease of ruminants and swine caused by *Morbillivirus* (family Paramyxoviridae). Now probably eradicated, it used to occur in many strains with considerable variation in virulence between them, but all are immunologically identical (Blood et al., 1994). In small ruminants, a disease similar to rinderpest, peste des petits ruminants (PPR), occurs. The disease is caused by a virus closely related to the rinderpest virus. The disease is more severe in goats than sheep, with case fatality rates of 55 to 85% in goats and 10% in sheep (Blood et al., 1994).

Herd migrations in search of pastures and water bring uninfected animals into contact with infected herds, and explosive outbreaks occured. Introduction of new animals into herds may serve as a source of infection. In cattle and small livestock, the disease leads to high mortalities causing losses to pastoral livelihoods. Exportation of animals and meat to external markets was prohibited from areas where the disease was thought to be endemic.

Tick-Borne Diseases (Theilerioses)

Theilerioses are tick-borne diseases caused by *Theileria* species occurring in cattle, sheep, and goats as well as in wild and captive ungulates. *Theileriae* are found throughout the world. Important pathogens of cattle are restricted to certain geographical areas of the world. *T. parva* is the most important member of the group occurring in East and Central Africa causing East Coast fever (ECF).

ECF is an acute disease of cattle caused by *T. parva parva* transmitted by the brown ear tick. ECF occurs in a less virulent form as corridor disease or January disease transmitted by *T. parva lawrencei* from buffaloes to cattle by *Rhipicephalus appendiculatus* and *R. zambeziensis*. Corridor disease occurs in eastern and southern Africa where there is contact between buffalo ticks and cattle.

T. velifera is associated with mild theileriosis and is transmitted by *Amblyomma* ticks. The disease causes economic and livelihood losses due to loss of production (milk), high treatment costs, and mortalities in susceptible herds.

Livestock Sector Changes and Related Impacts

Main Shifts in Production Systems

The livestock sector in the Horn of Africa has been influenced by economic, social, demographic, and environmental changes at regional and global levels. These changes bring opportunities for livestock keepers, but they have also had adverse impacts on production and marketing systems in the sector.

Population and income growth have increased demand for livestock products (Delgado et al., 1999, Simpkin 2005, Knips 2004) leading to higher production and profits for farmers, but they have also led to shrinking land parcels that can lead to land degradation, unless land management is improved. Moreover, urbanization and climate changes further reduce grazing areas and freshwater availability.

Countries in the Horn of Africa now experience a greater openness in livestock product markets along with innovations in production and distribution systems. These changes have led to new modes of livestock production and marketing characterized by contract farming, vertical integration, and the supermarket revolution (Simpkin 2005, Knips 2004). Harsh competition has also pushed toward higher productivity and quality. But open markets and increased standards raise the challenges for disease control, sanitary barriers, and intensification. They have also limited the access of resource-poor livestock keepers to these new production systems.

Another trend is that urbanization induces the development of periurban livestock production and creates new opportunities for farmers living at the edge of the cities. Moreover, the livestock manure from periurban centers is often used as organic fertilizer for greening some urban households and cities and for improving soil structure in rural areas. However, periurban livestock also increase water, noise, and air pollution.

Pastoral Systems

Approximately two thirds of the area of eastern Africa is inhabited by pastoral groups whose livelihoods depend on rearing livestock in the pastoral production system. Many factors, including the following, affect the production capacity of pastoral systems and lead to insecurity of livelihoods:

- The changing agroecological conditions and physical characteristics of range resources
- Encroaching economic interests that seriously challenge the sustainable management of fragile rangelands
- Rights to access specific land resources at different times of need
- Political marginalization of pastoral interests in framing national policies
- Wars and civil unrests that create mass movements of livestock and people
- Transboundary diseases that threaten livestock populations.

Insecurity is thus a permanent factor of life for pastoralists, and changes in their security situation can serve as a driver for changes in ranges, herd sizes and composition, and so on.

Since the 1900s the frequency and distances of herd movements have declined, and various forms and degrees of settlement have occurred among pastoralists. Spontaneous settlement is caused by a variety of factors: long droughts; encroachment of other land uses (Mkutu 2004, Leloup 1994); lack of infrastructure and social services; disease control policies (Morton 2001); shifting ownership; breakdown of customary pastoral social hierarchies; and insecurity. Governments sometimes promote settlement to intensify and commercialize animal production, to provide cheaper meat to urban areas, or to facilitate social control, administration, and delivery of social and livestock services (Pratt et al., 1997). Involuntary settlement of pastoralists by governments has also been reported in cases of dam construction, famine, and civil war (Larsen and Hassan 2003). In Kenya's Kajiado district and other Masai pastoral areas, there was a shift in the early 1970s from free grazing areas to group ranches with access rights limited to members of the group, but later most of the group ranches were subdivided into individual land holdings (Campbell et al., 2003).

The increasing role played by absentee investors/owners is another major trend. They contract pastoralists to herd their livestock, while often putting restrictions on livestock movements to facilitate control. For example, absentee livestock owners in the Sahel are increasing in numbers and are estimated to own 50% of livestock (Fafchamps and Gavian 1996).

Periurban Systems

A number of surveys have shown that livestock keeping is increasing in eastern African cities (Mosha 1991, Lee-Smith and Lamba 1991, Egziabher 1994). The major reasons cited for livestock keeping in urban and periurban areas are provision of income, employment (direct and indirect), improved nutrition, food security, improved social capital, and financial security (access to credit and "bank on the hoof"). Land sizes where livestock is kept are diminishing: they average 0.75 hectares in Nairobi (Kang'ethe et al., 2005) and 0.42 ha in Addis Ababa (Tegegne et al., 2002). This has led to wide adoption of the stall-feeding system. The numbers of livestock kept vary, but the general trend in urban and periurban areas is that there are more chickens than cattle, sheep, and goats (Tegegne et al., 2002, Foeken 2005, Mlozi 2005, Owour 2003, Onim 2002, Kang'ethe et al., 2005). A species analysis of the market value of all the livestock in Kisumu and the Kibera district of Nairobi (Kang'ethe et al., 2007) showed that, although poultry are numerically the most common species kept, they constitute a much lower share of the market value of the livestock. In Kisumu, poultry accounted for only 9% of the total livestock market value compared to cattle, which constituted about 70%. In Kibera, where there were no cattle and pigs, poultry contributed 25% of total value, less than half of the contribution by goats, whereas the rest was contributed by sheep, ducks, geese, and rabbits.

Changes in Farming Practices

Nutrient Cycling, Land Use Practice, and Irrigation

For centuries, when population pressures were low in many parts of the region and availability of good arable land and grazing areas was high, farmers and herders in the Horn of Africa practiced nutrient cycling to sustain crop–livestock systems. During this time, land was used for cropping or grazing for an average of 2 to 4 years, followed by a period of rest (fallow) of 7 to 20 years, allowing soils to rebuild their fertility (Gachimbi et al., 2003). This land use practice was very efficient and required low inputs to sustain agricultural production.

Increasing population pressure and land subdivision have resulted in land use intensification and shortened fallow periods. Among the pastoral communities, grazing orbits have reduced, leading to keeping of smaller herds with more sheep and goats, whereas among the mixed crop–livestock systems more farmers are abandoning open grazing and adopting tethering or the more commercial dairy-based zero-grazing.

The increase in irrigation has led to increased cultivation in the drylands, reducing the amount of land available for grazing (Githaiga 2004, Olson et al., 2004, UNEP 2002). Large-scale irrigation has also resulted in environmental problems such as waterlogging and salinization, water pollution, eutrophication, and unsustainable exploitation of groundwater aquifers that degrade the drylands' provision of environmental services. In such irrigation systems, rivers are often disconnected from their floodplains and other inland water habitats, and groundwater recharge has been reduced. These human-induced changes have in turn had an impact on the traditional migratory patterns of wildlife and the species composition of riparian habitats. They have opened up paths for exotic species, changed coastal ecosystems, and contributed to an overall loss of freshwater biodiversity and inland fishery resources (UNEP 2002). Within the livestock sector, reduced availability of water has increased livestock–wildlife and human conflicts because humans, livestock, and wild animals now have to compete for water from fewer places. On the whole, there is a decline in biodiversity and services provided by inland water systems in drylands.

Fragmentation and Intensification

Changing land use policies have led to fragmentation of land and agricultural activity in the Horn of Africa as ownership shifted from groups to individuals. To make up for the lost revenue out of the shrinking land size, individual owners intensify their activities. Intensification of agriculture in these fragmented lands has promoted the type of livestock that do not need large spaces to grow, such as pigs and dairy cows. These policies have therefore encouraged the smallholders' contribution to livestock production in the Horn of Africa. For instance, the share of smallholder dairies in Kenya's milk sales rose from 35% in 1975 to 80% in 1995 (Knips 2004). Although most of the production from these smallholders still feeds into growing informal markets—as exemplified by the Kenya milk production where only 14% of milk production is processed (SDP 2004a, 2004b)—it has created revenues and employment within and outside the sector. These benefits from land reform policies have often not been balanced with better management practices for resources such as soil, pasture, and water.

Technology Adoption and Productivity

On average, livestock productivity in the Horn of Africa remains low, even by developing country standards (see Table 2.3), and this situation has not changed much in the last few years. The only exception is milk production in Kenya, where productivity has increased, in part because of the success of local projects such as the Smallholder Dairy Programme. The causes of slow growth in livestock productivity in the Horn include animal diseases, reduced grazing areas and lack of access to technology. Tribal and regional conflicts and lack of clear strategies on livestock production in some cases (Somalia) have also weakened livestock productivity.

Animal diseases reduce productivity through reduced fertility and degraded quality of livestock products, and especially animal mortality. A disease such as RVF, for example, can increase mortality by 10% in calves, causing a huge loss in productivity. CBPP increases mortality by 50 to 90% and morbidity by 75 to 90%. Saaed et al. (2004) reported mortality rates of 21% in sheep caused by PPR in the northern state of Sudan. During the 1997 rinderpest outbreak among wildlife, Kock et al. (1999) estimated the mortality in buffaloes at 80%, whereas Abu Elzein (2004) reported 100% mortality rates in captive gazelles in Saudi Arabia, which had been imported from Sudan. Trypanosomiasis can decrease lambing and kidding rates by 37%, whereas brucellosis can reduce fertility due to large calving intervals and abortions (McDermott and Coleman 2001).

Table 2.3. Productivity for selected livestock products and countries, 2005

Country	Beef (carcass) Kg/animal	Milk Kg/animal
Ethiopia	108.4	200
Kenya	164	511
Somalia	110	348
Tanzania	107	174
Uganda	150	350
Canada	326	7596

Source: FAO 2008.

Moreover, the shrinking size of grazing areas has hampered the pastoral livestock access to required nutrition, and this has reduced their growth rates. Limited access to key resources such as feed grass and freshwater has further slowed animal growth. In addition, adoption of improved technologies and access to animal health services remain low in most communities. In the Amhara region of Ethiopia only 19 to 25% of households used purchased feed, whereas 33% used animal health services in 1991, rising to 55% in 1999. However, in 1999, returns to livestock were negative due to high mortality and loss of stock (Jabbar et al., 2003). Technology adoption in livestock production is lowest by far among pastoralists, especially in the fields of breed improvement, feeding, and processing of livestock products for markets. However, in the pastoral areas, there has been some significant improvement in technology adoption such as vaccination and control of epidemics for dealing with livestock diseases (UNEP 2002).

The quest to improve productivity in the Horn has induced the adoption of new production technologies such as artificial insemination to create more productive crossbreeds (Bebe et al., 2003). However, the levels of adoption remain low and vary widely between countries. For instance, 23% of Kenya's cattle herd are improved breeds, whereas the proportion in Uganda is only 4% (Knips 2004). Furthermore, changes in local and national government policies can greatly improve productivity levels in the Horn of Africa. The elimination of local and national taxation of livestock products and export has generated more revenues (Owango et al., 1998, Bebe et al., 2003) allowing livestock keepers to access indispensable inputs to increase productivity and production.

Impacts on Natural Resources and Environment

Increased Competition for Land and Livestock Feed Resources

Scarcity of livestock feed resources has been the major consequence of the changes in climate, social, and political scenarios described above. Reducing grazing orbits for most pastoralists has reduced availability of livestock grazing areas (Maitima et al., 2004, Olson et al., 2004, Reid et al., 2004). Climate variability and change have also resulted in changes in grass species composition. Within crop–livestock production systems, increase in land use intensification has also resulted in the shrinkage of grazing areas, forcing farmers to adopt new livestock feeding strategies like cut and carry of feed materials, tethering of animals around fruit trees, and setting aside certain areas (paddocks) for grazing (Bourn et al., 2005). Many farmers have been constrained to feed animals with weeds harvested from cropped areas and crop residues. Relatively rich farmers depend on industrially produced animal feeds. However, this requires high capital and is only economic when rearing exotic breeds (Canagasaby et al., 2005). As a result farmers in crop–livestock systems are changing their breeds from indigenous to exotic types, whose milk productivity fetches higher returns and can justify industrial feed products.

Urbanization and especially expansion of agriculture systems have reduced land availability. Moreover, government land reforms (e.g., in Kenya) have often changed land use from pastoral to agropastoral or in some cases to cropland (Olson et al., 2004). Since the 1920s, vast areas of natural rangelands in arid and semiarid regions have been taken over by cropping systems, private livestock and game ranches, nature reserves, and infrastructure. The rangelands most often encroached for these purposes include the best dry season grazing areas with the easiest access to water. As a result of encroachment, the total area and overall diversity and condition of the remaining rangelands have declined, while fragmentation (resulting from encroachment and land reform) has limited their accessibility.

Historically, communities in the pastoral areas have avoided cultivation of crops. However, over the last 25 years, they have rapidly converted semiarid grazing areas to agricultural croplands. The Masai in Kenya provide a useful example. Part of their motivation to turn to cultivation has been to protect the land from encroachment by other ethnic groups because farmers have more secure land tenure than livestock keepers (Campbell et al., 2003). The adoption of crop growing has also allowed them to capitalize on the cash market for grain, diversifying their income by growing maize and beans, while at the same time expanding their livestock herds.

Water Availability

Climate variability due to global warming, increased competition between crops and livestock, and increased irrigation have all contributed to reducing water availability. Reduced water availability leads to crop failures, lack of water and pasture for animals, and lack of food for humans and animals, leading to increased morbidity and mortality rates. Because these consequences are more severe in the arid and semiarid areas where pastoralists operate, climate variability and lack of water are responsible for loss of pastoral livelihoods and household assets, as well as increases in malnutrition and diseases for humans and animals.

Environmental Degradation and Pollution

Overgrazing has long been considered the primary cause of desertification in Africa. Overgrazing and land degradation can occur when livestock are forced to stay in a restricted area. However, under traditional mobile pastoralism, land degradation from overgrazing is often temporary. Because herds can visit different areas from one year to the next, stressed pastures get a chance to recover.

Recent work has shown that changing human activities are a major cause of land degradation (UNEP 2002). Increased population growth, privatization of rangeland, and encroachment of cropland have meant more frequent visits to the same pastures and less complete recovery. Because of the lack of natural feed resources and competition for land, farmers have cleared and burned forests and other dry vegetation on an increasing scale to extend pastoral lands and to stimulate fresh growth of grasses. Forest cutting for energy or for construction and craftsman use has also helped increase grazing land. These practices have not only aggravated land degradation but have also caused loss of biodiversity (UNEP 2002). The reduction of forest areas has also disturbed microclimates in livestock areas and increased runoff, helping to cause severe drought or flooding (Reid et al., 2000).

Expansion and intensification of livestock farming and processing have increased air pollution (odors, gas emissions) and noise pollution, especially in urban areas. Livestock activities can also cause water pollution (pathogenic microorganisms, eutrophication). Livestock depend on water from a variety of naturally occurring sources such as streams, springs, rivers and lakes, or human-made sources. Pollution occurs when animals concentrate around the water sources and contaminate with dung, urine, and soil from increased surface erosion, especially where those sources are shared by other users. Runoff from heavily manured fields and discharges from intensive production units, abattoirs, and processing plants into streams and rivers can have big impacts on aquatic systems, in particular eutrophication of water bodies and consequential algal blooms, composition of fish populations, ecological balance, and water quality (Bourn et al., 2005).

Demographic pressures including migration contribute to unsustainable development and food insecurity. The massive displacement and migration of millions of people from conflicts and/or environmental disruptions and their resettlement in concentrated densities in marginal areas are major contributors to land degradation and famine. These pressures are particularly pronounced in resource-poor areas.

Impacts on Food Security

The prevalence of malnutrition in the Horn of Africa remains high (Table 2.2). Livestock products play an important role in meeting food security, especially for the vulnerable (children and elderly) and for the low-income food insecure population—nearly 80% of Africa's population rely on agriculture for their income, and 75% of these farmers own livestock. But the rising demand for livestock products due to rapid population and income growth, coupled with the slow supply response to such high demand, has slowed improvements in food security in most countries. Per capita consumption of livestock products in the Horn of Africa has changed very little over the last 10 years, with the exception of milk consumption in Kenya, which had substantially declined during the 1990s and then rose to the original levels by the year 2000 (Figures 2.7a, 2.7b, and 2.7c). The consumption of livestock products in the Horn of Africa is in general far lower than consumption in developed countries. Moreover, these levels of consumption vary considerably by income level. The poor living in landlocked areas consume less livestock products.

Moreover, in areas where animals are important for draught power, reduced animal health leads to reduction in crop yields due to reduction in manure production and tillage. This tends to increase food insecurity and reliance on relief food (Mullins et al., 2000).

In times of severe drought, the typical emergency management response has been food and humanitarian aid to the affected areas. Unfortunately, resources allocated to caring for the symptoms have undoubtedly been larger than those directed toward addressing the root causes.

Impacts on Livelihoods and Poverty

Drivers of changes in and outside the livestock sector have enhanced the sector's direct role as provider of income and food among communities in the Horn of Africa. Moreover, livestock continue to form the basis for farmers' social and economic perspectives and can help make them less vulnerable to shocks. The increase in market opportunities arising from upward shifts in demand for live animals and processed livestock products has benefited some smallholder and landless livestock keepers, especially under local development projects.

At national and regional levels, poverty reduction is also affected by spillover effects of livestock sector development on other sectors such as transportation, feed production, and packaging industries. Indeed, development in the livestock sector has generated new employment and income for formerly unemployed youths in both rural and urban areas.

Concerns have remained over the continuing overexploitation of land and natural resources accompanying urbanization, population growth, and economic growth. Deforestation, overgrazing, and abusive intensification have affected the availability and quality of land and

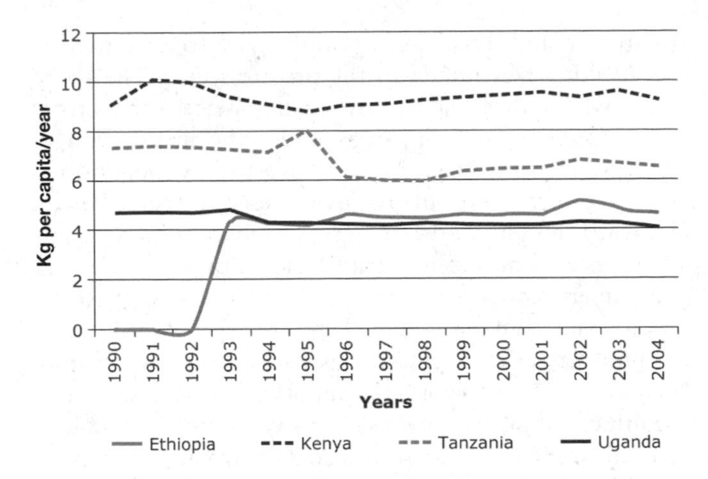

Figure 2.7a. Trends in per capita consumption of meat.
Source: FAO 2008.

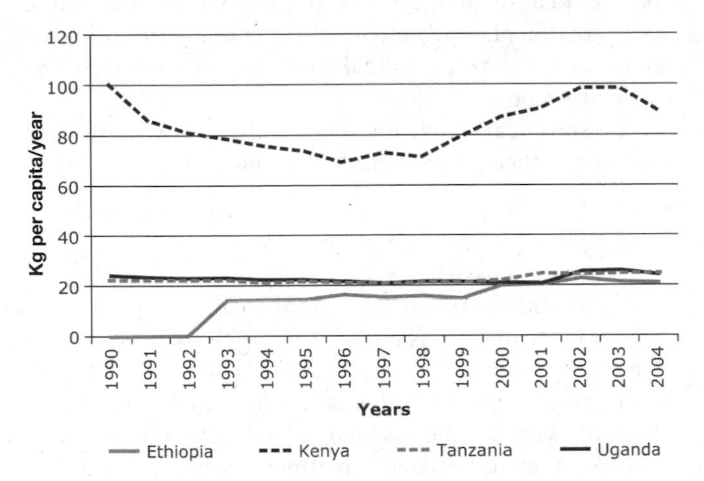

Figure 2.7b. Trends in per capita consumption of milk.
Source: FAO 2008.

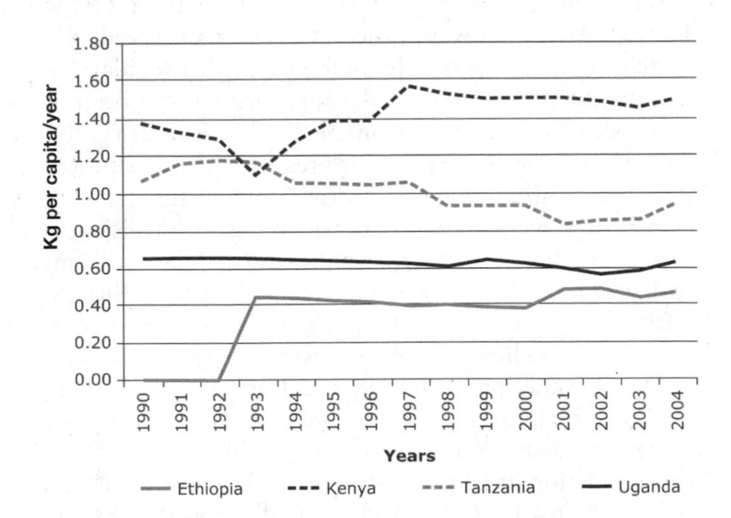

Figure 2.7c. Trends in per capita consumption of eggs.
Source: FAO 2008.

freshwater. Land use policies and measures to mitigate the impacts of climate change (e.g., biofuel production) put even more pressure on these scarce resources. As a result, access to inputs has become more difficult, and input prices have risen, thus reducing profits for livestock keepers. Moreover, the resulting degradation of environment and resources affects the health and general well-being of inhabitants, including livestock keepers.

The burden of treatment costs for infected animals and people is a major constraint to livestock productivity and livelihoods. For instance, the evolution of multi-drug-resistant strains of *M. bovis* puts the resource poor, who are less able to afford vaccines or treatment, at a disadvantage. Where cost-recovery measures are introduced, charging for vaccines may damage rapport with communities and erode trust in government services. Farmers may see the loss of cattle and the costs of vaccine as severe burdens, which threaten their livelihoods and social well-being and even, in extreme cases, their survival.

Impacts on Input and Output Markets

The increased demand for quantity and quality of livestock products, due to growth in population, urbanization, income, and trade, has led to changes in livestock input and output markets. At the production level, contract farming and vertical integration have been more widely adopted because these structures facilitate both input and output deliveries and allow controls over productivity and product quality. This has been the case for dairy production in Kenya and Uganda. Urban and periurban livestock production has also grown to efficiently meet increasing urban demand. However, these changes, especially vertical integration, have left behind the more remotely located farmers who continue to produce and sell their products through informal channels. Changes in product delivery systems have also adapted to these needs, as witnessed by the rapid growth of supermarkets in the Horn of Africa. Urban consumers increasingly buy meat and especially dairy products from supermarket chains or independent supermarkets (Ayieko et al., 2006).

Livestock prices have trended upward internationally because demand for livestock products has grown faster than supply. In addition, the scarcity and high prices of input resources (especially feed grains) contribute highly to rising prices of livestock products. This trend can be seen in most of the major cities in the Horn of Africa for meat and dairy products. Precisely how much of the rise in consumer prices trickles back to farm prices remains unclear. Knips (2004) noted that high taxation of livestock products before reaching market outlets in Ethiopia, Kenya, and Sudan has considerably reduced the producer share of livestock prices and margins.

Although there has been no specific study on wages in the livestock sector of the Horn, they seem to be highly affected by the availability of labor, especially in rural areas. In pastoral areas, despite their high population growth, the migration of labor to growing urban centers within rural areas reduces labor availability and increases the opportunity costs of household labor. Ironically, the abundant labor in urban areas as a result of population growth and migration remains relatively unskilled to handle the increasingly technology-driven urban and periurban livestock sector (from production to processing). These constraints on the level and quality of labor may constrain supply, leading to increased farm prices of livestock products.

At the national level, the lack of infrastructure and inadequate processing facilities in the Horn of Africa still limit the commercialization of livestock products, especially meat and dairy from rural areas. Urbanization and changes in urban consumer preferences benefit mostly large livestock farmers and producers located near urban centers.

At the international level, most countries in the Horn of Africa are net exporters of live cattle. Intraregional (or cross-border) trade is frequent—for example, between Ethiopia and Somalia or between Sudan and Uganda. For livestock *products*, however, subsidies on livestock production and export in rich countries have reduced income opportunities for pastoralists. Moreover, strict regulations and lack of technical support to control livestock product quality have impeded exports to international markets. The only exception is Sudan, which has increased its meat exports to Gulf countries.

Also, except for dairy products from Kenya and hides and skins from Ethiopia, the increase in local demand and the lack of infrastructure and quality and disease control has further limited the expansion of livestock product exports. The Horn of Africa, for instance, has remained a net importer of poultry meat and dry milk (Knips 2004, FAOSTAT 2007). The growth in productivity and production has not kept up with high domestic demand for live animals and processed livestock products and has constrained export growth. Moreover, market opportunities for livestock export are narrowed by occurrences of animal diseases (including rinderpest, RVF) that affect the product quality and safety and cast a negative reputation for some of the Horn of Africa's livestock export products.

Responses

This section deals with case studies of efforts made in response to the changes, as well as the consequences of those efforts for the benefit of the livestock sector. The case studies address responses to issues of animal diseases, land use and natural resource constraints, poverty reduction, and competitiveness of the livestock sector in the Horn of Africa.

Animal Diseases

Animal health interventions have evolved from government and nongovernmental organization (NGO) treatment and vaccination campaigns, to community-based initiatives linked to the private sector. The speed with which these shifts are taking place varies from country to country. In areas under the jurisdiction of the governments (e.g., Eritrea) livestock vaccination is still largely carried out by government veterinarians. In Kenya, Ethiopia, and Djibouti, vaccination increasingly involves community-based animal health workers under the supervision of government veterinarians. In south Sudan and Somalia where there are few or no governmental services, vaccination is carried out by community animal health workers supported by NGOs. Animal health community workers play a very important role in remote areas and in areas affected by armed conflicts.

Most interventions to manage emergencies in animal health focus on internal and external parasite control. Anthelmintic drenches and acaricides are used to control ticks and manage tick-borne diseases in all livestock species because they are cheap and easy to administer. Some treatments have included injectable antibiotics for all species and trypanocidal injections, particularly for cattle and camels.

Lessons learned from the trends and evolution of animal health services include the following:

- Free provision of drugs or drug subsidies should be avoided because they create dependency on government assistance.
- Combining destocking programs with emergency veterinary programs enables livestock owners to pay for services.
- The private sector or existing community animal health workers should be involved in all veterinary services, including mass treatment during disease outbreaks.

It has been realized that linking community animal health workers to the private sector is the key to achieving sustainable animal health service delivery in most of the Horn of Africa. However, success has been observed only in the higher potential areas (e.g., Kenya's highlands). Within low potential areas the approach has not been so successful for several reasons, including insecurity and lack of sufficient private sector presence. Therefore most past interventions have aimed at building the capacity of private veterinarians to operate in pastoral regions.

A number of Pan-African livestock disease control programs have been implemented in the Horn of Africa over the last few decades, with variable success rates. One of these initiatives attempted to control and eradicate rinderpest through the Inter-African Bureau of Epizootic Diseases, founded in 1950. The aim of this initiative was to vaccinate all cattle of all ages in phases every year for three successive years. Twenty-two countries were involved, of which 17 had rinderpest. By the end of 1979, only one country, Sudan, admitted to having the disease.

However, this initiative failed to completely eradicate rinderpest, opening the way to a subsequent pandemic of the disease. The Organization of African Unity (now the African Union) launched the Pan-African Rinderpest Campaign (PARC) in 1987 covering all the countries in the Horn of Africa. This initiative also faced difficulties because countries were unable to sustain high enough levels of immunity. It was evident to PARC's leadership that to achieve a continentwide success, a series of technology-based procedures had to be developed, standardized, and systematized for the village level.

After PARC ended, the Pan African Programme for the Control of Epizootics (PACE), overseen by the African Union with Global Rinderpest Eradication Programme (GREP) Secretariat hosted at the United Nations Food and Agriculture Organization (FAO), and the International Atomic Energy Agency (IAEA). Technical Cooperation (TC) Programme helped to transfer technologies for reducing risk from transboundary livestock diseases and those of veterinary public health, with the Joint FAO/IAEA division of the IAEA providing technical expertise and assistance in the battle against rinderpest. There is now increasing confidence that the disease is no longer present in its last reserves (Pakistan, Sudan, and Yemen) and has thus been eradicated.

Vaccine development to control theileriosis has been tried, using stabilized live sporozoites, followed by treatment. A number of sporozoite strains have been tried, including Marekebuni stock (Morzaria et al., 1985) and Mundali stock (Berkvens et al., 1992). The limitations of this approach are failure to produce protective immunity, problems in using live parasites, ensuring product safety for use in animals, and severe ECF reactions following immunization. Newer forms of recombinant vaccines are now being tried that could help to overcome some of the limitations of infection and treatment vaccines.

Land Use and Natural Resources Constraints: Sample Successes and Failures of Past Policies

Although the pressure on food availability for the region as a whole is growing, individual country analyses indicate that there is still considerable scope within the region for increases in productivity, and that regional trade can be a contributing factor in stimulating this production. In the short term, the Greater Horn region will require considerable assistance to raise production levels, but policies that promote sustained growth in agriculture through intensification and emphasize comparative advantages can contribute to increasing regional and national food security.

Responses to Problems of Drought and Regional Conflicts over Land Resources

The formation of IGAD was one response adopted by governments in the Horn of Africa to the problem of drought and famine. IGAD was established in 1986 by heads of the member states (Djibouti, Eritrea, Ethiopia, Kenya, Somalia, Sudan, and Uganda) with a mandate to address the severe drought and other natural disasters that cause widespread famine in the region. Initially, as a result of its limited role and focused program area, IGAD did not address conflict and related issues. In addition, some organizational and structural problems made the organization ineffective.

The many conflicts in the region made the efforts to address the problems of drought and famine more difficult. Internal conflicts in Sudan, the secession of Eritrea from Ethiopia, the civil war that led to the collapse of Somalia and other conflicts around border areas among neighboring countries all contributed to suffering and famine. Establishing an organization that could address the conflicts of the region was vital. Following this realization, IGAD was reconstituted to give it a mandate to address issues facing the Horn of Africa in a broader perspective (IGAD 2008).

The Conflict Early Warning and Response Mechanism (CEWARN) was born out of the reconstituted IGAD in 2002. Its objectives are to support member states in the following:

- Preventing cross-border pastoral conflicts
- Enabling local communities to play an important role in preventing violent conflicts
- Enabling the IGAD secretariat to pursue conflict prevention initiatives
- Providing members with technical and financial support.

So far, through CEWARN, IGAD is working on capacity building and awareness about early warning signs of conflict (IGAD 2008).

The Famine Early Warning System Network (FEWS) (http://www.fews.net), of the United States Agency for International Development (USAID) is presently working in all the countries in the Horn of Africa. FEWS relies on secondary data produced by host governments for its analyses, with the exception of satellite imagery (Normalized Difference Vegetation Index [NDVI] and Meteosat/Rainfall Estimation), which it receives directly from the National Aeronautics and Space Administration (NASA) and the National Oceanic and Atmospheric Administration (NOAA) every 10 days. Geographic information systems (GISs) are used to do spatial analyses of available data that are included in the system's regular reporting. FEWS works closely with national Early Warning Systems and Market Information Systems (where these exist), and with Ministries of Rural Development or Agriculture, the World Food Programme (WFP), and certain NGOs. The institution also undertakes frequent field trips to assess availability and access conditions affecting food security, often in tandem with WFP, host government partners, and NGOs. FEWS produces a widely disseminated monthly food security report.

FEWSNET has enabled governments and development agents to gain access to spatial information on climate and general environmental situations within the region and on how the region compares with other regions. FEWSNET, however, remains at regional IGAD level, whereas changes and responses to these changes are localized to individual productive systems. There is a need for FEWSNET to focus on individual productive systems at small localized scales.

The Livestock Early Warning System (LEWS) (http://www.brc.tamus.edu/lews), also operating in East Africa, is intended to provide an additional 6 to 8 weeks advanced notice ahead of the current early warning systems. The project combines predictive and spatial characterization technologies with the formation of a network of collection and measurement sites in East Africa. The system is based on near-infrared spectroscopy (NIRS) and fecal profiling technology supported by advanced grazing land and crop models. The foundational technology consists of the African GIS dataset used by the Spatial Characterization Tool, providing spatial analysis of weather, soils, terrain conditions, and human and livestock populations. LEWS involves the linkage of several new technologies capable of predicting the current nutritional status of free-ranging animals and the impact of weather on forage supply and crop production among carefully selected sets of households reflecting a variety of effective environments across diverse landscapes of East Africa.

LEWS has also been quite effective in informing pastoral communities on changes in spatial distribution of pastures for livestock within the pastoral areas. This needs to be institutionalized in the policy making process to regulate movement of livestock together with movement of wildlife to avoid seasonal conflicts of different users.

Policies to Regulate Landholding Sizes and Streamline Ownership to Maintain Productivity

Land tenure in the Horn of Africa is a sensitive and complex issue. At independence from colonial powers, the countries in the region established quite different tenure reforms, all aimed at improving productivity. For example, in Ethiopia, all land became public land, with leasing or sale of land being forbidden, whereas in Kenya, the government promoted private ownership (Bruce et al., 1996). In both Kenya and Ethiopia, fragmentation of land parcels through subdivision has reduced the average farm size to less than 1 ha in many areas. As a result, fallow periods have been reduced or are omitted altogether, in order to produce sufficient quantities to meet the needs of the family. In spite of these policies, the countries of the subregion have all suffered impediments to large-scale agricultural development, and the majority of the populations are small-scale farmers (Bruce et al., 1996).

Subdivisions of land have made household productive units very small and hardly able to support livelihoods. Although this applies to all categories of land uses, it is more severe within the high-potential agricultural areas where land is individually owned. Due to scarcity of land for different uses (cultivation, herding, wildlife conservation, as well as land for urban growth) many policies have been developed to help in defining land use and user rights. In Kenya, for example, policies to govern land use started prior to independence. In 1940 the African Land Development policy (ALDEV) was introduced, confining Africans to the homelands and leaving vast tracts of land for white farmers. In 1980 the Arid and Semi-Arid Lands (ASAL) program focused on improvement of the arid and semiarid lands and may have contributed to the formation in 1990 of group ranches among Masai herders. However, in 2000 the group ranches were converted to individual holdings (Campbell et al., 2003).

The Constitution of the Federal Republic of Ethiopia of 1994 does recognize pastoralists' rights to free land for grazing and cultivation. In Eritrea, following the long war with Ethiopia, pastoralists have gradually adopted agropastoralism. In the highlands they practice rain-fed agriculture and in the plains irrigated agriculture. In Sudan, after independence the government began to establish large agricultural schemes on land traditionally used by nomads. Even though commercial ranchers were granted rights of passage through mechanized farms, pastoralists were denied this. Traditional patterns of movement were altered and traditional production systems changed. The system as it stands is bound to give rise to conflict between nomadic pastoralists and mechanized farmers but also between transhumant groups and tenants and between herders and other groups. Because the system rests on the expansion of mechanized farming, government policy favors large-scale producers. Legislation on land tenure reflects misconceptions about the pastoral system. New schemes are favoring large-scale irrigation usually taking up dry season grazing areas.

Governments are recognizing that central control of land and agricultural resources is limited by capacities and resources, and that land policy reform needs to encourage the formation of farms of viable size, for sustainability and growth of agricultural output (FAO 2001). In addition, just as state ownership has not yielded the anticipated growth in agricultural production; private ownership has also shown little benefit to increasing production, largely as a result of market failures. Therefore, market reform must go hand in hand with tenure reform (Bruce et al., 1996). Policy makers are also reforming attitudes toward communal land tenure and access and realizing that, under certain conditions, communal systems can provide security of tenure, environmental and production sustainability, and conflict avoidance (Bruce et al., 1996). However, this transformation in attitudes has been slow and is still experiencing opposition in some countries. In Kenya, for example, individual titling is still regarded as the political and social ideal, and, therefore, claims to communally owned land are often thrown out of court. This has led to land grabbing or illegal occupancy in some areas, notably in urban areas and state forests (Warner et al., 1999). Means for strengthening the voice of community groups include the decentralization of political power and the formation of natural resource use councils consisting of community members (Warner et al., 1999). In Uganda, the new Land Act of 1998 combines the objectives of agricultural productivity and equity by promoting democratization and good governance with some redistribution of land rights. Implementation of the Land Act (1998) has been hindered by lack of an overall land policy and by insufficient strategic planning, limited resources and capacity, and widespread corruption (Warner et al., 1999).

Fencing of Wildlife Conservation Areas in Kenya

One intervention to reduce conflicts between herders and wildlife conservationists over land was the introduction of programs to fence areas used as national parks so that wildlife could be contained within the park. This solved the problem to some extent in some areas where fences were introduced. However, in other areas, especially where there were common resources shared by livestock and wildlife, fencing was perceived as a way of preventing herders from accessing these resources. On the other hand, it was realized that among the small parks most wildlife spend most of their time in human settlement areas outside the park boundaries. In some locations, wildlife utilization fees have been introduced to compensate agropastoralists for *not* cultivating their land, so that livestock and wildlife can share the land without crops. Classic examples of this are the Amboseli National Park and Masai Game Reserve in Kenya (Western et al., 1998).

Uganda's Decentralization Policy and Ethiopia's Land Privatization

Uganda has successfully implemented decentralization of governance, including formulation of policies, to districts and lower administrative units. These include policies to govern and regulate utilization of natural resources. In districts that are predominantly pastoralist, like those in or bordering the Karamojong area, decisions about using grazing lands and associated key resources are made by the local people through their local administrative bodies. If properly used, this move can reduce intertribal conflicts over use of natural resources. On the other hand it can limit herd movements within tribal territorial boundaries, leading in some cases to overstocking and serous land degradation (Kisamba–Mugerwa 2001). In Ethiopia, following a change in governance to the present parliamentary system, landownership changed to allow individuals to own land. This change has attracted investors to make arrangements with local communities to utilize the land for economic purposes (Muduuli 2001).

Reducing Rural Poverty through the Livestock Sector

Policy Intervention and Aims

As in many countries in the developing world, changes in consumption and production patterns have mainly benefited large-scale livestock keepers and manufacturers but have so far eluded many small livestock keepers in rural areas. The widespread poverty in many livestock areas of the Horn is a reminder that the employment and revenues generated by the large-scale livestock keepers and growth in the sector have yet to reach these smallholders. A wide range of policy interventions have been put in place but these need to be reinforced for livestock production to become one of the key pathways to reduce poverty. These interventions include the following:

- Ensuring that the poorest and most vulnerable, especially children in rural areas, eat enough to survive and become active (e.g., school meal programs)
- Enabling the poorest to own and manage basic assets and to reduce their vulnerability, for instance by distribution of live animals (Heifer International) and land (land settlement schemes)
- Facilitating livestock keepers' access to essential inputs (including land) and financial credit and providing them with means to deal with production and marketing risks
- Encouraging research and extension to increase technology adoption (e.g., artificial insemination)
- Securing both domestic and foreign markets (including niche markets for small animals such as for rabbits or poultry) and eliminating price distortions that harmed small keepers.

Regional Level Intervention

Numerous livestock-related projects define and apply regional strategies to help poor livestock keepers in the Horn of Africa to get out of poverty. Because of the regional dimension of animal health issues in the Horn, regional interventions are mainly centered on combating animal diseases. For instance, the Pro-Poor Livestock Policy Initiative (PPLPI) run by FAO and funded by the United Kingdom's Department for International Development covers the world's most vulnerable regions, including East Africa. In Uganda, for instance, the project focuses on the dairy sector and aims to build a stakeholder network that engages the poor as partners sharing rights and responsibilities as well as benefits from coordinated actions such as combating animal diseases. The African Union's Inter-African Bureau of Animal Resources (AU/IBAR), among its numerous projects coordinated with various donors, intervenes in eradicating rinderpest, reducing the risk of avian influenza, and providing support for sanitary standards for several countries in the Horn of Africa.

International or regional research institutions operating in the region, such as the International Livestock Research Institute or the Association for Strengthening Agricultural Research in Eastern and Central Africa (ASARECA), have also provided technical and policy support on many aspects of livestock development for poverty reduction. Supports include designing risk management tools, defining best soil management practices, and sharing genetic materials for some high-quality feed crops. Regional trading and marketing arrangements—such as the Common Market for Eastern and Southern Africa (COMESA), which includes countries in the Horn—have also been used as platforms to improve livestock for the benefit of the livestock sector.

Regional projects have made only small inroads in mitigating the negative impacts of climate changes in the Horn of Africa. This situation needs to be improved because the region's livestock production, which still relies greatly on pastoralism, remains vulnerable to the slightest changes in the ecosystem resources. There is also a need for regional responses to the flows of the informal or undocumented livestock traded across borders within the region, at least to contain the spread of animal diseases to small-scale keepers. All current and future regional projects will benefit from better coordination with local governments and especially communities, so that efforts yield significant gains for the Horn's poor livestock keepers.

National-Level Interventions

National planning for the development of the livestock sector to benefit the poor in the Horn of Africa tends to be specific to key subsectors. In Kenya, Uganda, and Ethiopia, for instance, policy responses have been mostly focused on dairy projects. In Kenya, the Smallholder Dairy Project (SDP), jointly initiated by the Government of Kenya and the Kenya Agricultural Research Institute, has enabled small farmers to diversify their income sources and to connect with the market (Staal et al., 2000, Ngigi 2005, Knips 2004). It has been achieved through a mixed system of dairy farming and cropping in the highlands of Kenya. The direct impacts of such a project on reducing poverty have been seen in the increase in income for small-scale producers involved (Ngigi 2005). Moreover, the project has also increased employment opportunities in farming and manufacturing (SDP 2004b). The Dairy Development Authority of Ethiopia and Uganda and the Livestock Development Authority of Tanzania have conducted similar efforts and have significantly boosted small keepers' income.

Although the pro-poor policy responses focus on other livestock products such as bovine meat production, greater emphasis has lately been put on promoting the production of small animals (including poultry) in both rural and periurban areas. The main interventions consist of developing a network of assistance in animal health to encourage livestock keepers to meet the level and quality of products demanded in urban cities. These interventions were initially aimed to benefit the landless but have been expanded to poor small landowners to provide an additional source of income. But these efforts have so far lacked the risk management tools to deal with the resurgence of small animal diseases such as avian flu and Newcastle disease. This has significantly reduced the benefits of the development of small animal production among the poor keepers.

Increasing the Competitiveness of Livestock Products from the Horn of Africa

Improving Productivity and Product Quality

In the Horn of Africa, policy responses to the shortfall of domestic supply relative to demand consist mostly of finding ways to increase productivity and production levels. Actions to increase the level of productivity have focused on improving animal health care and access to technology. Liberalization of the veterinary services in some countries in the Horn has not solved the livestock production and animal health issues; the governments with help from private companies and various donors have now planned for accessible vaccination services to livestock keepers. There is a consensus, however, that research and extension on vaccines to protect animals from various diseases will be reinforced, and the governments will play a significant role in such a move.

The use of advanced technology, including biotechnology, is still limited but is starting to receive attention among donors and farmers. The success of the artificial insemination in Kenya in improving animal meat and milk productivity has yet to be reproduced (Bebe et al., 2003). With the spotlight that the donors, NGOs, and private companies put on the Horn of Africa, the problem is more about wider access than just the availability of technology.

One of the important policy responses to the awareness of the importance of competitiveness for the Horn of Africa in both domestic and foreign markets has been the promotion of quality control. Nearly all quality control policies are tied to animal health care. Various measures supported by government and donor projects have been put in place. The Kenya Meat Commission since 1950, Tanzania's Livestock Development Authority since 1974, Ethiopia's Livestock Marketing Authority since 1998, and the Sudan's Live Animals and Meat Export Council since the mid-1980s are examples of the long-term efforts to foster high-quality standards for livestock products. In nearly all the countries in the Horn, however, these quality control services are still too centralized and do not yet reach out to provinces and communities where the bulk of the quality problems are found. Similarly, animal nutrition, especially feed quality in rural areas, needs to receive more attention in policy making because animal health and product quality and quantity depend on it.

Targeting Key Subsectors
The livestock sector in the Horn of Africa faces a dilemma between maintaining the usual, relatively stable export markets of live cattle and bovine meat on the one hand, and investing more in relatively risky but promising markets for small animals and advanced processed products (e.g., cheese, delicatessen) on the other. Most of the policy responses in the past have been to play safe and stick with the usual export products and destinations. This is justifiable for countries like Sudan, which has had long-lasting success in exporting of live animals (Knips 2004, Simpkin 2005). But growing efforts have been made recently to identify and try other important opportunities. For example, Sudan has become the leader in the Horn of Africa in bovine and especially ovine meat exports to the Gulf countries. Likewise, as Ugandan, Kenyan, and Ethiopian bovine meat exports decline these countries have started to promote pork and small animals and poultry meat production (Canagasaby et al., 2005). Several NGOs and international donors have backed these efforts, although their efficacy depends on other important conditions such as good infrastructure, strong market institutions, and undistorted market prices.

Building and Maintaining Basic Infrastructure (Roads, Ports, and Other Facilities)
The livestock sector in the Horn of Africa has a desperate need for improved infrastructure to make any effort a success. The prevalent situation at present is summed up in the image of tired animals traveling on long, rough roads under a hot sun before reaching the markets.

The competitiveness of Horn of Africa countries in domestic and international markets is hampered by the deterioration of basic infrastructure, from watering facilities, feeding areas, and shelters, to weighing scales for animals in market outlets. Moreover, exports of livestock and livestock products have been hampered by lack of adequate quarantine and storage facilities. Efforts such as the rehabilitation of holding ground facilities in Kenya were not enough to ensure high export capacity (Knips 2004). However, some recent initiatives, such as the Kadero and Port Sudan quarantine stations in Sudan, have been successful in promoting livestock production and trade.

Investing in infrastructure from farm to harbor is key to an improved marketing system and would benefit all agents in the livestock sector including poor livestock keepers (Simpkin 2005). Better infrastructure to improve collection, storage, and delivery systems of livestock products needs to be built in remote areas so that poor livestock keepers living there can benefit from emerging markets.

Building and Strengthening of Institutions
Projects such as the Pro-Poor Livestock Policy Initiative have offered an institutional framework to empower poor livestock keepers to move toward a market-oriented and competitive livestock sector. There is therefore a need to establish advocacy strategy groups and regional fora to pursue the interests of pastoralists in general. Regional bodies like the African Union (AU) and IGAD are best placed to push forward the agenda of disadvantaged groups. In Kenya, the Pro-pastoralist Parliamentary Group, an initiative of the Kenya Pastoralists' Forum, is an ad hoc committee with advocacy on pastoral land rights as its main agenda. This example needs to be emulated across the region. PARC, Pan African Tsetse and Trypanosomiasis Eradication Campaign, PACE, Farming In Tsetse Controlled Areas, and GREP are other examples of international programs to strengthen local capacity and institutions to tackle transboundary diseases that need replication. All the success of these efforts to strengthen institutions are, however, bound to general economic and political reforms in the countries—such as combating corruption, enforcing the rule of law, empowerment of minority groups, and especially resolution of regional and internal conflicts.

Lessening Distortions in the World Market
Since international markets for livestock products (especially dairy products) remain highly distorted mainly because of developed countries' policies and sanitary barriers, some countries in the Horn have been active in various trade negotiations. But there is little the Horn countries can achieve, individually or as a region, to reverse these distortions in international markets. Renegotiating preferential access terms, especially to European markets under the Economic Partnership Agreement, would be a way forward, but such access may not last forever. Further liberalization of livestock product markets through regional trading arrangements (Common Market for East and Southern Africa, Cotonou Conventions, East African Community, SADC) could also benefit livestock sectors in the Horn of Africa. Trade within the region has been mainly informal, and this has to be addressed to minimize the risks of jeopardizing the quality and safety efforts within each country that damage its export sectors. Nontariff barriers to market access and especially compliance with sanitary regulations among the trading countries need to be addressed.

Conclusions

Despite all the challenges described in this chapter, many significant achievements have been made in various sectors of livestock development ranging from animal health, environment, and livestock production interfaces, to marketing and political aspects. Community–private sector partnerships, for example, have proven to be a better option compared to governments being the sole agents in delivering livestock disease management. However, government participation still remains crucial in controlling disease outbreaks when they occur.

International markets for livestock products demand high sanitary standards, which are proving to be a big challenge, especially to small-scale producers. If the Horn of Africa is to gain and maintain access to foreign markets, especially in Europe, a greater effort needs to be put into containing livestock diseases. Creation of disease-free zones where livestock can be reared for export purposes is a good step toward addressing the problem. For the Horn of Africa to benefit from the increasing demand for livestock products, the challenges of declining land productivity, scarcity of land, and issues of climate variability and change, especially among pastoralists, must be addressed. Adoption of new technologies to improve animal breeds has for a long time been used as a viable way to increase productivity, but this has so far been applicable mainly in crop–livestock production systems.

The future of livestock production in the Horn of Africa will depend on how production systems will be able to cope or adapt to the challenges of climate change. Fortunately, many organizations are now making investments in developing strategies for climate change adaptations in Africa after the realization that the continent will be affected the most and the earliest compared to other regions. The Horn of Africa has been affected by droughts for a long time, and local communities have developed ways to cope with climate variability. If these indigenous knowledge systems are analyzed, to identify, strengthen, and upgrade some of the best practices, solutions or options may be found to reduce risks of livestock production to climate change.

Individuals, communities, governments, and international bodies need to be aware of both the positive and the negative impacts of animal husbandry and livestock production, so that appropriate measures can be taken to maximize benefits and minimize or mitigate adverse consequences. Because the livestock sector largely contributes to people's livelihood in the poorest areas of the Horn of Africa, its development, if well managed, remains a solid pathway to reduce poverty in a sustainable manner.

References

Abu Elzein, E. M. E., F. M. T. Housawi, Y. Basharek, A. A. Gameel, A. I. Al-Afaleq, and E. Anderson. 2004. Severe PPR infection in gazelles kept under semi-free range conditions. *Journal of Veterinary Medicine Series B* 51(2): 68–71.

Andresen, J., J. Olson, E. Massawa, and J. Maitima. 2008. *The Effect of Climate Change and Land Use Change on Climate and Agricultural Systems in Kenya.* Climate Land Interactions Project (CLIP) Policy Workshop, June 2008. Nairobi: ILRI.

Ayieko, M.. D. Tschirley, and M. Mathenge. 2006. Fresh Fruit and Vegetable Consumption Patterns and Supply Chain Systems. In: *Urban Kenya: Implications for Policy and Investment Priorities.* Working paper 16. Egerton, Kenya: Tegemeo Institute of Agricultural Policy and Development, Egerton University.

Bebe, B. O., H. M. J. Udo, G. J. Rowlands, and W. Thorpe. 2003. Small dairy systems in the Kenya highlands: breed preferences and breeding practices. *Livestock Production Science* 82: 117–127.

Berkvens, D. L., D. M. Geysen, and G. M. Leylen. 1989. East Coast Fever Immunization in Eastern Province of Zambia. In: *Theileriosis in East and Southern Africa*, ed. T. T. Dolan, 83–86. Proceedings of a workshop on ECF immunization held in Lilongwe, Malawi. Nairobi: ILRAD.

Blood, D. C., O. M. Radostis, and C. C. Gay. 1994. Disease caused by *Mycobacterium* IV. In: *Veterinary Medicine,* 8th edition, ed. M. Radostis. London: Bailliere Tindall.

Bourn, D., J. Maitima, and B. Motsamai. 2005. Livestock and the environment, 2005. In: *Livestock and Wealth Creation: Improving the Husbandry of Livestock Kept by the Poor in Developing Countries,* ed. A. J. Kitalyi, N. Jayasuriya, E. Owen, and T. Smith, pp. 145–165. London: Department for International Development.

Bruce, J., J. Subramanian, A. Knox, K. Bohrer, and S. Leisz. 1996. *Land and Natural Resource Tenure on the Horn of Africa: Synthesis of Trends and Issues Raised by Land Tenure Country Profiles.* Paris: Sahara and Sahel Observatory.

Campbell, D. J., D. P. Lusch, T. Smucker, and E. E. Wangui. 2003. *Root Causes of Land Use Change in the Loitokitok Area, Kajiado District, Kenya.* LUCID Working Paper Series Number 19. Nairobi: ILRI.

Canagasaby, D., J. Morton, B. Rischkoowsky, D. Thomas. 2005. Livestock Systems. In: *Livestock and Wealth Creation: Improving the Husbandry of Livestock Kept by Resource-Poor People in Developing Countries,* ed. A. J. Kitalyi, N. Jayasuriya, E. Owen, and T. Smith, 29–52. Nottingham: Nottingham University Press.

Davies, F. G., K. J. Linthicum, and A. D. James. 1985. Rainfall and epizootic Rift Valley fever. *Bull. Wld. Hlth Org.* 63:941–943.

De Leeuw, P. N. and J. C. Tothill. 1990. *The Concept of Rangeland Carrying Capacity in Sub-Saharan Africa: Myth or Reality.* Working Paper 29b. London: ODI Pastoral Development Network.

Delgado, C., M. Rosegrant, H. Steinfield, S. Ehui and C. Courbois. 1999. *Livestock to 2020: the next food revolution.* Food, Agriculture, and the Environment Discussion Paper 28. Washington DC: IFPRI/FAO/ILRI.

Domingo, A. M. 2000. Current status of some zoonoses in Togo. *Acta Trop.* 76: 65–9

Egziabher, A. G. 1994. Urban Farming, Cooperatives and the urban poor in Addis Ababa. In: *Cities Feeding People: An Examination of Urban Agriculture in East Africa.* Ottawa: International Development Research Centre.

Fafchamps, M. and S. Gavian. 1996. The Spatial Integration of Livestock Markets in Niger. *Journal of African Economies* 5 (3): 366–405.

FAO. 1999. *The effects of structural adjustment programmes in Africa.* Animal Production and Health Division. Rome: FAO.

FAO. 2000. Special report: FAO/WFP crop and food supply assessment mission to Sudan. Rome: FAO. Available at http://www.fao.org/docrep/004/x9218e00.htm

FAO. 2001. *Non-wood forest products in Africa: a regional and national overview*. FAO Forestry Department Working Paper FOPW/01/1. Rome: FAO.

FAO. 2008. *Pro-Poor Livestock Initiative*. Rome: FAO. Cited 16 July 2008. Available at http://www.fao.org/ag/againfo/programmes/en/pplpi/home.html

FAOSTAT. 2007. Available at: http://faostat.fao.org/

Foeken, D, 2005: *Urban agriculture in East Africa as a tool for poverty reduction: A legal and policy dilemma?*, ASC Working Paper 65. Leiden: African Studies Centre Available at http://www.ascleiden.nl/pdf/workingpaper65.pdf

Fortin, E, 2005: Reforming Land Rights: The World Bank and the Globalization of Agriculture. *Social & Legal Studies* 14 (2): 147–177.

Gachimbi, L. N., J. M. Maitima, and P. Kathuli. 2003. The relationship between land use change, soil fertility and erosion across different agro-ecological zones in South Eastern Slopes of Mt. Kenya. In *Proceedings of the 21st Annual conference of the Soil Science Society of East Africa, December 2003, 137–146.* Nairobi: Kenya Agricultural Institute.

Galvin, K. A. 1992. Nutritional ecology of pastoralists in dry tropical Africa. *Am. J. Hum. Biol.* 4(2): 209–221.

Galvin K. A., D. L. Coppock and P.W. Leslie. 1994. Diet, nutrition and the pastoral strategy. In: *African Pastoralist Systems: An Integrated Approach*, ed. E. Fratkin, K. A. Galvin, and E. A. Roth, 113–132. Boulder, Colorado: Lynne Rienner.

Galvin, K. A., R. B. Boone, N. M. Smith and S. J. Lynn. 2001. Impacts of climate variability on East African pastoralists: Linking social science and remote sensing. *Climate Research* 19: 161–172.

Githaiga, J. M. 2004. *Survey of Water Quality Changes with Land Use Type in the Loitokitok Area, Kajiado District, Kenya*. LUCID Working Paper Series Number 35. Nairobi: ILRI.

Harrison, G. 2004. *The World Bank and Africa: The Construction of Governance States*. London: Routledge.

IGAD. 2008. Intergovernmental Authority on Development. Djibouti: IGAD. Cited 16 July 2008. Available at http://www.africa-union.org/Recs/IGAD_Profile.pdf.

Jabbar, M. A., M. Ahmed, S. Benin, B. Gebremedhin, and S. Ehui. 2003. Livestock, livelihood and land management issues in the highlands of Ethiopia. In: *Policies for Sustainable Land Management in the East African Highlands*, ed. S. Benin, J. Pender, and S. Ehui. Socioeconomics and policy research working paper 50. Nairobi: ILRI.

Kabagambe, E. K., P. H. Elzer, J. P. Geaghan, J. Opunda–Asibo, D. T. Scholl, and J. E. Miller. 2001. Risk factors for *Brucella* seropositivity in goat herds in eastern and western Uganda. *Prev. Vet. Med* 52:91–108.

Kang'ethe, E. K., T. F. Randolph, B. Mcdermott, A. K. Lang'at, V. Kimani et al. 2005. *Characterization of Benefits and Health Risks Associated with Urban Smallholder Dairy Production in Dagoretti Division, Nairobi, Kenya*. Project report to IDRC. Ottawa: IDRC. Grant no. 102019-004.

Kang'ethe, E. K., T. M. Kimani, and G. Kuria. 2007. *Determining Users, Research Priorities to Translation of Research Outcomes into Tangible Benefits*. Scoping Study of Urban Livestock Keepers in Kibera, Nairobi and Kisumu. Report submitted to Aylesford, UK: Natural Resources International Ltd.

Kisamba-Mugerwa, W. 2001. Rangeland management in Uganda. Presentation at International Conference on Policy and Institutional Options for the Management of Rangelands in Dry Areas, May 5–11, 2001, Hammamet, Tunisia. n.p.

Knips, V. 2004. *Review of the Livestock Sector in the Horn of Africa*. Livestock Information, Sector Analysis and Policy Branch (AGAL). Rome: FAO.

Kock, R. A., J. M. Wambua, J. Mwanzia, H. Wamwayi, E. K. Ndung'u, T. Barret, N. D. Koch, and P. B. Rossiter. 1999. Rinderpest epidemic in wild ruminants in Kenya, 1993–97. *Vet. Rec.* 145(10): 275–83.

Kouba, V. 2003. Globalization of Communicable Animal Diseases – A Crisis of Veterinary Medicine. *Acta Vet. Brno* 72: 453–460

Larsen, K. and M. Hassan. 2003. *Sedentarisation of nomadic people: the case of the Hawawir in Um Jawasir, northern Sudan.* DCG report No. 24. Oslo: Drylands Coordination Group.

Lee-Smith, D. and D. Lamba. 1991. The potential of urban farming in Africa. *Ecodecision*, December 1991, pp 37–40.

Leloup, S. 1994. *Multiple Use of Rangelands within Agropastoral Systems in Southern Mali.* University of Wageningen, Netherlands: Dissertation.

Linthicum, K. J., F. G. Davies, A. Kairo and J. Bailey. 1985. Rift Valley fever virus (family Bunyaviridae, genus Phlebovirus). Isolations from Diptera collected during an inter-epizootic period in Kenya. *J Hyg (Lond)* 95(1): 197–209.

Little, P., K. Smith, B. A. Cellarius, D. L. Coppock, and C. B. Barrett. 2001. Avoiding disaster: diversification and risk management among East African herders. *Development and Change* 32:387–419.

Maitima, J., R. S. Reid, L. N. Gachimbi, A. Majule, H. Lyaruu, D. Pomery, S. Mugatha, S. Mathai, and S. Mugisha. 2004. *The Linkages between Land Use Change, Land Degradation and Biodiversity Across East Africa.* LUCID Working Paper Series Number 42. Nairobi: ILRI.

McDermott, J. J. and P. G. Coleman. 2001. Comparing apples and oranges: model-based assessment of different tsetse-transmitted trypanosomosis control strategies. *Int. J. Parasitol* 5–6: 603–609

Mkutu, K. 2004. *Pastoralism and Conflict in the Horn of Africa.* Bradford, UK: Peace Forum/Saferworld/University of Bradford.

Mlozi, M. R. S. 2005. *Urban Animal Agriculture: Its Palliativity and Reasons for Persistence in Tanzanian Towns*. London: RICS Foundation.

Morton, J., ed. 2001. *Pastoralism, Drought and Planning: Lessons from Northern Kenya and Elsewhere.* Chatham, U.K.: Natural Resources Institute.

Morzaria, S. P., A. D. Irvin, E. Taracha, and P. R. Spooner. 1985. East Coast fever immunization in Coast Province of Kenya. In: *Immunization against Theileriosis in Africa*, ed. A. D. Irvin, 76–78. Nairobi: International Laboratory for Research on Animal Diseases.

Mosha, A. C. 1991. Urban farming practices in Tanzania. *Review of Rural and Urban Planning in South and East Africa.* 1, 83–92. Proceedings of a workshop held in Kadoma Ranch Hotel, Zimbabwe, July 20–27, 1992. S. Mubi (ed.). ILCA. Addis Ababa, Ethiopia, pp. 157–164.

Muduuli, M. C. 2001. Uganda's poverty eradication action plan: National sustainable development strategy principles tested. Presentation at International Forum on National Sustainable Development Strategies, 7–9 November 2001, Accra, Ghana.

Mullins, G. R., B. Fidzani, and M. Kolanyc. 2000. At the end of the day: The socioeconomic impacts of eradicating contagious bovine pleuropneumonia from Botswana. *Annals of the New York Academy of Sciences* 916:333–344

Nakoune, E., O. Debaere, F. Kaumanda-Kotogne, B. Selekon, F.

Samory, and A. Talamin. 2004. Serological surveillance of brucellosis and Q fever in cattle in the Central African Republic. *Acta Trop,* 92:147–151

Ngigi, M. 2005. *The Case of Smallholder Dairying in Eastern Africa.* Environment and Production Technology Division Discussion Paper No. 131. Washington DC: International Food Policy Research Institute.

Nicholas, R. A. J. 2002. Contagious Caprine Pleuropneumonia. In: *Recent Advances in Goat Diseases,* ed. M. Tempesta. Ithaca, New York: International Veterinary Information Service.

Olson, J., S. Misana, D. Campbell, M. Mbonile, and S. Mugisha. 2004. *The Spatial Patterns and Root Causes of Land Use Change in East Africa.* LUCID Project Working Paper 47. Nairobi: ILRI.

Oludhe, C. 2005. Coping with Climate Variability and Change in the Greater Horn of Africa: ICPAC's Experience. Presentation at the UNFCCC COP11 Meeting, 28 November to 9 December 2005, Montreal Canada. Nairobi: IGAD Climate Predictions and Applications Centre.

Omer, M. K. E., Skjerve, G. Holstand, Z. Woldehiwet, and A. P. MacMillan. 2000. Prevalence of antibodies to *Brucella* spp. in cattle, sheep, goats, horses and camels in the State of Eritrea: influence of husbandry systems. *Epidemiol Infect* 125:447–453.

Onim, M. 2002. *Scoping Study of Urban and Peri-urban Poor Livestock Keepers in Kisumu,* ed. W. Richards and S. Godfrey. Proceedings of a workshop March 2003, Nairobi.

O'Reilly, L. M. and C. J. Daborn. 1995. The epidemiology of *Mycobacterium bovis* infections in animals and man—a review. *Tuber Lung Dis* 76: 1–46.

Owango, M. O., J. S. Staal, M. Kenyanjui, B. Lukuyu, D. Njubi, D., and W. Thorpe. 1998. Dairy co-operatives and policy reform in Kenya: effects of livestock service and milk market liberalisation. *Food Policy* 23:173–185.

Owour, S. O. 2003. *Rural Livelihood Sources for Urban Households.* A study of Nakuru town, Kenya. ASC Working Paper 51. Leiden: African Studies Centre.

Pomeroy, D. et al. 2003. *Linkages between Changes in Land Use, Land Degradation and Biodiversity in S.W. Uganda.* LUCID Working Paper Series Number 12. Nairobi: ILRI.

Pratt, D. J., F. Le Gall, and C. de Haan. 1997. *Investing in Pastoralism: Sustainable Natural Resource use in Arid Africa and the Middle East.* World Bank Technical Paper No. 365. Washington DC: World Bank.

Reid, R. S., R. L. Kruska, N. Muthui, A. Taye, S. Wotton, C. J. Wilson, and W. Mulatu. 2000. Land-use and land-cover dynamics in response to changes in climatic, biological and socio-political forces: the case of southwestern Ethiopia. *Landscape Ecology* 15(4): 339–355.

Reid, R. S., P. K. Thornton, and R. L. Kruska. 2004. Loss and fragmentation of habitat for pastoral people and wildlife in East Africa: concepts and issues. *African Journal of Range and Forage Sciences* 21(3): 103–113

Saeed, I. K., A. I. Khalafalla, S. M. El-Hassan, and M. A. El-Amin. 2004. Peste des petits ruminants (PPR) in the Sudan: Investigation of recent outbreaks, virus isolation and cell culture spectrum. *Journal of Animal and Veterinary Advances* 3(6): 361–365

SDP. 2004a. *The Demand for Dairy Products in Kenya.* SDP Policy Brief No. 1. Nairobi: Smallholder Dairy Project.

SDP. 2004b. *Employment Generation in the Kenya Dairy Industry.* SDP Policy Brief No. 2. Nairobi: Smallholder Dairy Project.

Seré, C., H. Steinfeld, and J. Groenewold. 1996. *World Livestock Production Systems: Current Status, Issues and Trends.* FAO Animal Production and Health Paper 127. Rome: FAO.

Simpkin, P. S. 2005. *Regional Livestock Study in the Greater Horn of Africa.* Nairobi: International Committee of the Red Cross.

Staal, S. J., C. Delgado, I. Baltenweck, and R. Kruska. 2000. *Spatial Aspects of Producer Milk Price Formation in Kenya: A Joint Household GIS Approach.* Paper presented at the International Association of Agricultural Economics Meeting, Berlin, August 2000.

Tegegne, A., M. Tades, M. Alemayehu, D. Woltedji, and Z. Sileshi. 2002. *Scoping Study on Interactions between Gender Relations and Livestock Keeping Practices in Addis Ababa, Ethiopia.* Aylesford, UK: Natural Resources International Ltd.

UNEP. 2002. *Africa Environment Outlook: Past Present and Future.* London: EarthPrint.

Upton, M., S. Mbogoh, J. Rushton, and S. Islam. 2005. Marketing to Promote Trade and Development. In: *Livestock and Wealth Creation: Improving the Husbandry of Animals Kept by Resource-Poor People in Developing Countries,* ed. E. Owen, A. Kitalyi, and T. Smith. Nottingham: Nottingham University Press.

Urban Harvest. 2004. *Policy Prospects for Urban and Peri-urban Agriculture in Kenya.* Policy Dialogue Series No. 2. Nairobi: ILRI/KARI/CIP.

Verschuren, D., K. R. Laird, and B. F. Cumming. 2000. Rainfall and drought in equatorial East Africa during the past 1,100 years. *Nature* 403:410–414

Wangui, E. E. 2003. *Links between Gendered Division of Labour and Land Use in Kajiado District, Kenya.* LUCID Working Paper Series No. 23. Nairobi, ILRI.

Warner, R., R. Walker, and R. Scharf. 1999. *Strategic Conflict Analysis and Conflict Impact Assessment: A DFID/CHAD Discussion Paper.* London: Department for International Development.

Western, D. et al. 1998. Wildlife Conservation in Kenya. *Science* 280(5369): 1507.

World Bank. 2006. World Development Indicators. Washington DC: World Bank. Available at http://web.worldbank.org/WBSITE/EXTERNAL/DATASTATISTICS/0,,contentMDK:20899413~pagePK:64133150~piPK:64133175~theSitePK:239419,00.html

Wray, C. 1975. Survival and spread of pathogenic bacteria of veterinary importance within the environment. *Vet. Bull.* 45:543–550.

3

West Africa

The Livestock Sector in Need of Regional Strategies

Cheikh Ly, Abdou Fall, and Iheanacho Okike

Abstract

Most livestock in West Africa are kept in traditional smallholdings, in a large diversity of livestock and crop–livestock systems, in humid, subhumid, semiarid, and arid agroecological zones. There is a degree of interaction between livestock keepers and farmers right across these zones, which allows West Africa to be considered as an interdependent whole.

These diverse livestock production systems are evolving. The path and rate at which they evolve are largely determined by the interaction of agroecological, social, economic, demographic, technological, and institutional factors. Changes are occurring in pastoral systems, crop–livestock systems, and stall-fed/intensive urban and periurban livestock production systems.

Following a brief description of the livestock and crop–livestock farming systems of West Africa, this chapter provides a more detailed examination of the key drivers of change, followed by a discussion of the environmental, health, and social and economic consequences of the changes. The final section presents potential responses to the changes in terms of changing economic strategies, livestock policies, institutions, capacity building, and research.

Introduction

West Africa is a solid geographical bloc from Nigeria in the east to Mauritania in the west. Countries are spread from the southern stretch of the Sahara desert and splayed out along the coast line. For the purposes of this chapter, West Africa is limited to the 15 countries that constitute the Economic Community of West African States (ECOWAS).

Most livestock in West Africa are kept in traditional smallholdings, in a large diversity of livestock and crop–livestock systems. The region's agroecological zones range from the humid zone (along the coastline) to the dry Sahelian zone (in the north), and roughly define an increasing gradient of dependence on ruminant livestock for livelihoods. At the tip of this gradient, in the Sahel rangelands of Niger, Burkina Faso, Mali, Chad, Mauritania, Gambia, and Senegal, keeping livestock in extensive pastoral systems (about five cattle/km^2) is the main land use form and the primary means of livelihood for millions of people who depend on them for meat, milk, transport, and manure and as a store of wealth and a source of societal prestige. In climatically favorable years, pastoralists produce and market excess young bulls, but during drought, poorly performing cattle of all sexes and ages are sold. However, some of these young bulls do not head directly to terminal markets and instead provide replacement stock for traction and fattening operations in the adjoining crop–livestock systems of the savanna zones. Availability of crop residues in this system enables farmers to maintain the bulls in very good to excellent body condition through supplementary stall feeding within homesteads. Though this practice is primarily aimed at keeping the bulls fit to provide traction, it ultimately not only yields heavier animals for the beef market (when the bulls are retired from traction) but also pools manure for return to farmlands at the beginning of each planting season. This high degree of interaction between livestock keepers and crop–livestock farmers right across the region's zones allows West Africa to be considered as an interdependent whole (Figure 3.1 in the color well).

These diverse livestock production systems are evolving. The path and rate at which they evolve are largely determined by the interaction of agroecological, social, economic, demographic, technological, and institutional factors. Changes are occurring in pastoral systems, crop–livestock systems, and stall-fed/intensive urban and periurban livestock production systems.

Following a brief description of the livestock and crop–livestock farming systems of West Africa, this chapter provides a more detailed examination of the key drivers of change, followed by a discussion of the environmental, health, and social and economic consequences of the changes and responses made by major stakeholders.

Livestock and Crop–Livestock Production Systems in West Africa: An Overview

West Africa has 42.5 million cattle, 127 million sheep and goats, just under 10 million pigs, about 380 million poultry, and only 1 million horses (Table 3.1).

The Sahelian region is a key region for ruminant livestock production in West Africa. The Sahel (from Arabic for shore or border) with 75 to 150 days as the length of growing period (LGP) is the dry boundary zone between the Sahara to the north and the more fertile region to the south. It encompasses very different areas, with rainfall ranging from more than 1000 mm in the south to 100 mm in the north, and rainfed agriculture progressively giving way to livestock-only production. The former activity extends roughly up to the 300 mm isohyet, while during the rainy season livestock production can use areas up to the 100 mm isohyet. This often leads to a subdivision of the area into the agricultural Sahel and the pastoral, arid, or northern Sahel. This terminology, however, does not mean that there is no livestock production in the "agricultural Sahel"—transhumant herders go south during the dry season into agricultural areas, and more sedentary forms of livestock production have developed on a wide scale. Nor does it mean that there is no agriculture in the northern Sahel—there is agriculture, although it is essentially irrigated or flood-recession agriculture. Agriculture–livestock production relationships actually form the basis of all Sahelian agrarian systems.

Classification

Many typologies have been proposed to classify the different existing systems (Sére et al., 1996, Wint et al., 1999, Dixon et al., 2001, Manyong 2002, Thornton et al., 2002, Kruska et al., 2003). They range widely in the numbers of categories. Jahnke (1982) proposed two broad systems: range–livestock and crop–livestock systems. At the other extreme, Fernández-Rivera et al. (2004) proposed 15 systems for West Africa. In addition to livestock and crop–livestock production systems, emerging intensive, stall-fed, urban, and periurban livestock systems need to be considered. For the purposes of this study, the classification by Fernández-Rivera et al. (2004) is used (Figure 3.2 in the color well). The proposed typology has two major classes of systems—sole livestock and crop–livestock. The sole livestock class has two systems (rangeland-based pastoral and landless), and the crop–livestock class has three subclasses (rain-fed crop–livestock, tree crop–livestock, and irrigated/flooded crop–livestock). Within the three crop–livestock subclasses 13 systems are identified, defined by their dominant crops. The pastoral system is also known as the extensive system, the 13 crop–livestock systems

Table 3.1. Livestock populations (2004)—values expressed in 1000

Country	Cattle	Cattle Milked	Sheep	Goats	Poultry	Pigs	Horses*	Asses**	Camels**	TLU
Benin	1745	210	700	1350	13,000	309	1	0	0	1269
Burkina Faso	5200	1040	7000	8800	24,000	674	27	159	11	4555
Cape Verde	23	9	10	113	450	205	0	4	0	69
Côte d'Ivoire	1111	197	1523	1192	33,000	343	0	0	0	1226
Gambia	328	44	147	265	620	18	17	11	0	215
Ghana	1365	273	3112	3596	29,500	305	3	4	0	1709
Guinea	3400	442	1070	1278	14,000	68	3	1	0	2088
Guinea-Bissau	520	84	290	330	1550	360	2	1	0	410
Liberia	36	6	210	220	6200	130	0	0	0	149
Mali	7500	750	8364	12,036	30,000	68	170	210	329	6104
Niger	2260	460	4500	6900	25,000	40	106	174	294	2528
Nigeria	15,200	1800	23,000	28,000	140,000	6611	205	300	13	15,422
Senegal	3100	310	4700	4000	46,000	315	500	120	3	2943
Sierra Leone	400	85	375	220	7570	52	0	0	0	346
Togo	279	41	1850	1480	9000	320	2	1	0	627
ECOWAS	42,466	5749	56,850	69,779	379,890	9816	1036	985	649	39,658

* Values for 2003.

** In tropical livestock units (TLU).

Source: FAO 2007.

represent the mixed system, whereas the stall-feeding systems in urban and periurban settings are representative of the intensive system. We shall describe the systems (including the specific roles of crops and livestock), assess their feed production potential, and discuss the factors driving their evolution.

Pastoral systems evolve in response to uncertainty of rainfall and the demand for live animals in the highly populated and more urbanized wetter zones. Crop–livestock systems are more labor intensive. In some cases they include animal traction and are characterized by higher use of agricultural inputs such as fertilizers, pesticides, and feed supplements. Landless, stall-feeding systems (which exist alongside or within other systems) evolve primarily in response to demand for meat in urban areas and are frequently associated with religious events such as weddings and funerals. In all these systems the large majority of producers are poor, and their land and animal holdings are small. All the livestock systems are evolving, though the changes and stages vary between systems (Williams et al., 2000, Tarawali et al., 1998, Tiffen 2004, Kristjanson et al., 2001).

Pastoral Systems

The area under pastoral systems constitutes 25.1% of West Africa and contains 4.6 million cattle and 24.8 million sheep and goats, equivalent to 8 million tropical livestock units (TLUs) or 18.1% of all ruminant TLUs in West Africa. Traditional, low-input, transhumant pastoral systems are prevalent, although livestock are also raised in villages by semisettled agropastoralists. Depending on the extent of their mobility, pastoral systems fall into three categories, namely (1) nomadic, (2) semisettled agropastoralists (transhumant), and (3) sedentarized agropastoralists. *Pastoral systems* in this chapter refers to nomadic and transhumant pastoralists.

In pastoral systems, the primary role of livestock is to serve as an asset for risk management (with mobility allowing reduction of feed and water shortages) in addition to the traditional role of providing savings from the wealth stored as animals. As some pastoralists begin crop production and new technologies for water and soil nutrient management are introduced in cropped areas, the role of livestock can change so as to include provision of manure for crop production. As cropping activities increase, herd mobility may be reduced, although the loss of rangeland feed may be compensated by the increased quantities of crop residues that become available within the system.

The pastoral vocation is the result of interaction of climatic, geographic, cultural, technological, political, and socioeconomic factors. Interactions between climate, soils, vegetation, animals, and humans have developed over time through different modes of land appropriation and exploitation, often controlled by traditional pastoral organizations. Pastoral areas have thus been subject to livestock mobility to ensure the livelihood of human and animal populations. This nomadic or transhumant management of livestock and natural resources (pastures and water) has involved minimal inputs of capital and work, and the productive potentials of animals and soils have not been fully exploited. Thus traditional management systems have preserved the natural equilibrium because the movements of animal and human populations have depended on the natural availability of water and grazing resources.

Crop–Livestock Systems

Crop–livestock systems in West Africa cover 2.7 km^2 and account for 83% of the cattle and 75% of the sheep and goats in the region. They are raised on farms planted to cereals (maize, sorghum, millet, rice), legumes (cowpea, groundnuts, soybeans), tubers (cassava, yams). Indigenous breeds of pigs and poultry are also reared as free roaming within the rural community space. The gradient of integration of livestock into farming systems increases in a southerly direction from the Sahel and peaks at the southern fringes of the savanna zones. Toward the South, disease challenges, especially from trypanosomiasis, pose a significant challenge to ruminant production, resulting in a steady decline toward the coastline in the density of ruminant livestock that are susceptible to the disease. Within the savanna zones, agricultural intensification based on crop–livestock interactions is more advanced than elsewhere in the region. This is based on crop residue and manure exchanges between traditional crop farmers and transhumant pastoralists for whom livestock feed scarcity and declining soil fertility, respectively, are major constraints. Crop–livestock integration for crop-based farmers in the savanna regions of West Africa, therefore, involve acquiring more animals and leasing or selling off less fertile parts of their farmlands. On the other hand, livestock-based farmers sell some animals and acquire these plots knowing they have a resource—manure—that can help to restore and sustain their fertility. Thus, in an operational sense, crop–livestock integration involves land-for-livestock and livestock-for-land exchanges that are constantly taking place, tending to the convergence of previously distinct pastoral and crop farming systems under the management of single households.

Livestock production is therefore a natural activity in the region, perfectly adapted to the resource potential and ensuring complementarity and key exchanges with crop production systems. At the same time it provides a livelihood to a large segment of the population. For smallholder livestock keepers, livestock are a source of food and income and an integral part of their livelihoods. They principally contribute draught power and manure to agriculture and serve as both physical and social capital. For these farmers, the management of the natural resource base on which their livelihoods depend is becoming increasingly crucial and challenging as the farming systems evolve under pressure to produce more food, feed, and incomes for the growing human and livestock populations (Tarawali et al., 1998).

Stall-Feeding Systems in Urban and Periurban Areas

In West Africa, stall-feeding is a common and longstanding tradition, probably rooted in fattening of sheep, specifically for the *Tobaski* Muslim festival. It is not restricted to "landless" systems in urban and periurban locations; it is also practiced in most mixed systems. In major livestock markets across the region, it is also common to find some livestock traders who purchase lean animals and then fatten and resell them in the same markets. The boundaries of the stall-fed (landless) urban and periurban livestock system are not well defined because it is the most fragmented system and is also found nested within other systems. However, it has been calculated by Fernández-Rivera et al. (2004) to cover 52,642 km^2 and account for 1.2% of cattle and 2% of sheep and goats in West Africa. The most distinctive characteristic of the system is that it embodies the final stages in the evolution of crop–livestock production systems, when specialized and sophisticated crop and livestock systems begin to emerge.

Stall-feeding systems also importantly include commercial poultry production and to a lesser extent dairy production systems. The emergence of a number of industrial-scale private integrated poultry farms is a main feature of urban and periurban intensive farming systems supplying eggs and broiler chickens to meet the increasing demand for meat in growing African cities.

Key Drivers in West Africa Livestock Sector

Table 3.2, based on integrating data from Fernández-Rivera et al. (2004) and Randolph et al. (2006), provides a framework for analyzing the different livestock and crop–livestock production systems, connecting the role of livestock, crops, principal driving forces, main sources and types of risk, and key opportunities for development. It shows that in pastoral systems, found mostly in the arid and drier semiarid zones of the region where a high risk of crop failure exists, there is a comparative advantage for livestock production, and there is also a ready and growing market for livestock in other adjoining zones. Hence agroecology and market forces are key drivers of pastoral production. Because of high herd mobility this system is well adapted to the unpredictable weather and the risk associated with the constant threat of drought.

In the wetter zones, where crop–livestock systems thrive, the principal driving forces are population pressure, urbanization, rising incomes, and changing consumption patterns—all of which improve market opportunities for agricultural produce. In the cereal–legume-dominated systems, livestock supply manure, milk, traction, and cash, whereas crops are grown for subsistence and for cash in addition to supplying residue for feeding livestock. These systems, found mostly in the wetter semiarid and drier subhumid zones, illustrate the important influences of factors other than agroecological potential in the evolution of crop–livestock systems.

McIntire et al. (1992) proposed that, first of all, agroecology and, secondly, population density are the principal factors that create and drive the diversity of agricultural systems, after which market access, land forms, and incomes, among other factors, begin to exert significant influences. Boserup (1965, 1981) argued that as population pressure leads to smaller land holdings or shortened fallows, it also induces agricultural intensification. In addition, the introduction of appropriate technologies, institutional support, policy reforms, and other socioeconomic factors play key roles (Gabre-Madhin and Haggblade 2004). Randolph et al. (2006) perceive these factors in terms of background drivers external to the livestock sector, and other drivers that are livestock-sector-specific (Figure 3.3). Background drivers include climate change, population growth, urbanization, economic growth, privatization, decentralization, and globalization, whereas livestock sector-specific drivers include land use intensification, increasing risk of zoonotic diseases, diseases of intensification, segmented markets, and increasing demand for livestock products. We shall now examine these various drivers in more detail.

Background Drivers

Climate

In West Africa harsh agroecology and climate are widely accepted as background drivers of livestock production systems. Climate variability has always been present, but climate change is a major new threat (Hillel and Rosenzweig 2002) even if significant uncertainties surround the detailed and regional impacts (Niasse et al., 2004). Most modeling results predict that dry areas in Africa could become hotter and drier, with a decline in precipitation in the range of 0.5 to 40% with an average of 10 to 20% by 2050, whereas its tropical zone is expected to become more humid (Dennis et al., 1995, Faye et al., 2001, Parry 2002). According to Niasse et al. (2004), many of the scenarios for West Africa indicate a generally more pronounced downtrend in river flow regimes and replenishment of groundwater. The suggested impact of these changes implies a reduction in the yield of major crops (maize, early and late millet, sorghum, rice, cowpea) and a reduction in cereal production in particular.

Considerable variation is expected within regions and countries according to the prediction of Obasi and Toepfer on climate change in Africa (2001, cited in DFID 2004). Modeling of the expected impact of climate change on agriculture in sub-Saharan Africa is revealing likely changes in the length of growing period (LGP),

Table 3.2. Main features of the 15 livestock and crop–livestock production systems in West Africa

Production Systems	Characteristics, Determinants, Constraints, and Opportunities for Development				
	Role of Livestock	Role of Crops	Principal Driving Forces	Main Sources or Types of Risk	Opportunities for System Development
Pastoral	Supply of milk, meat, and manure (if crops grown), risk management, savings	Subsistence, crop residues for feed (if crops grown)	Comparative livestock production advantage, ready market for animals in other zones	Climate variability and change—especially risk of drought and contracting grazing land. Conflicts, marginalization. Poor service provision (e.g., for animal health)	Technology (nutrient, water management, crops), policies to promote adoption. Early warning systems. Community-based animal health schemes. Improved regional trade in livestock.
Millet/cowpea/livestock Millet/groundnut/livestock Sorghum/maize/cowpea/livestock Maize/sorghum/livestock Groundnut/rice/livestock Rice/livestock High-value vegetable/rice/livestock Cotton/maize/sorghum/livestock	Supply of manure, milk, and animal traction. Savings, risk management. Cash	Subsistence, cash, residues for livestock feed	Population pressure, urbanization, rising incomes, good market opportunities for crop and livestock products, comparative crop production advantage. Use of improved technologies for cash-crop production	Increasing competitive pressure on smallholder production. Soil degradation. Price, financial. Globalization and trade arrangements especially in the case of systems incorporating cotton	Improving handling, processing, and food safety, especially for meat and milk products. Better access to markets along with technologies to enhance productivity. Increased access to inorganic fertilizers. Improving support services. Developing regional market information systems.
Rice/cassava/maize/livestock Yam/cassava/maize/livestock Cassava/yam/livestock	Supply of manure, milk, and animal traction. Savings, risk management	Subsistence, cash, residues serve mainly as compost		Price, financial, animal health risks	Reducing animal health risk to enhance crop-livestock integration Developing cassava residue value chain for livestock feed. Technologies for feeding and preserving emerging feed sources (e.g., pineapple pulp)
Cocoa/plantain/cassava/livestock Coconut/oil-palm/fruits/livestock	Value-added maintenance of undergrowth, cash	Cash, subsistence	Urbanization, shortage of plantation labor.	Price, financial. Globalization and trade arrangements	
Stall feeding/intensive urban and periurban livestock production	Cash	Cash, subsistence, residue for feed (if grown)	Segmented markets, higher incomes and higher demand for livestock products.	Price, diseases of intensification and crowding, foodborne diseases, zoonoses, feed safety	Facilitating access to markets, providing microfinancing support, improving global competitiveness (e.g., more favorable WTO arrangements), antidumping

Sources: An integration of Fernandez-Rivera et al., 2004 and Randolph et al., 2006.

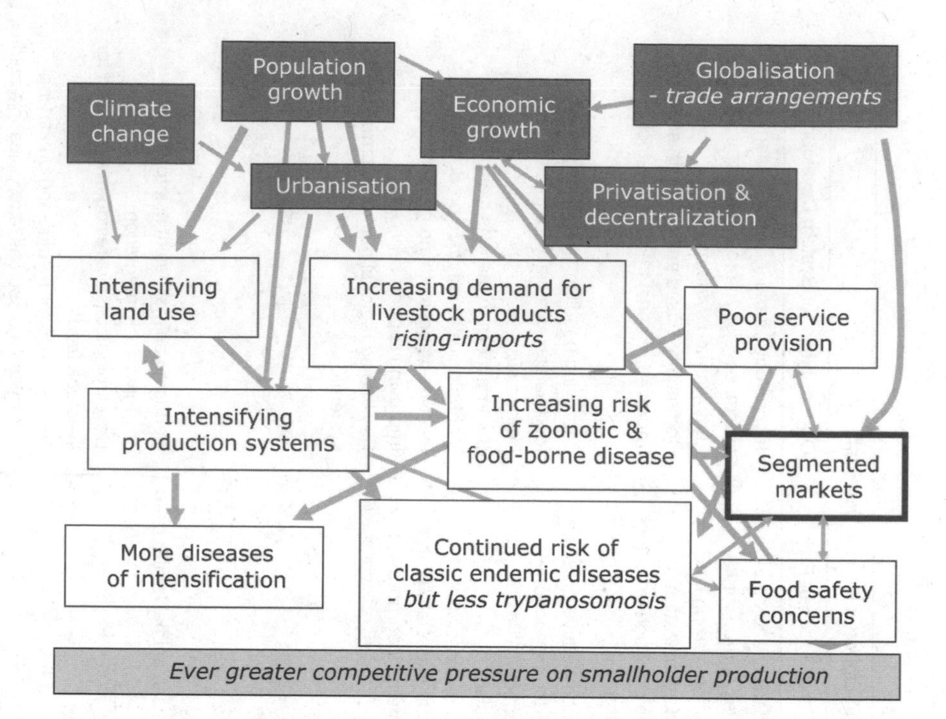

Figure 3.3. Trends and dynamics influencing livestock agriculture in West Africa.
Source: Randolph et al., 2006.

which will alter the types of crops that can be grown in a given area (McDermott et al., 2001, Born et al., 2004, UNEP 2008). If the Sahel becomes drier, a major reduction in the LGP will also be seen in a large area of the subhumid and humid zones of West Africa, stretching from Guinea to Benin. Climate variability is likely to increase, along with the frequency and intensity of severe weather events. Climate change per se is raising concerns as in other parts of the world. But here the variability of rains within the rainy seasons and between years has been a key factor in the dynamics of the production systems. Therefore one can expect that pastoral systems under appropriate grassland management could be better able to cope with climate change than other livestock systems (Neely and Bunning 2008).

Population

United Nations medium predictions of human population growth for Africa continue to exceed 2% annually until at least 2025 (United Nations 2009). An ever-greater proportion of this population will be concentrated in urban areas.

The West Africa region has one of the most rapidly growing human populations in the world. Its population, currently estimated at 300 million, will reach 383 million by 2020 (United Nations 2009). It will also be much more urbanized. More than half of the 2020 population will be living in settlements of more than 5000 inhabitants, instead of the current level of 39%. The number of urban areas will increase from 2500 in 1990 to 6000 by 2020. Some 300 are expected to have more than 100,000 inhabitants (WALTPS 1996, Cour 2001, Thornton et al., 2002). By 2050, 60% of West Africans will be found in urban areas if growth and massive urbanization continue as projected. Figures 3.4 and 3.5 in the color well show human population density for West Africa in 2000 and the projected density in 2050.

The trends are considerably affected by Nigeria's size and that of the coastal zones in general. In Nigeria, urbanization is even more pronounced. What is visually striking (Figure 3.4) is that human population density in Nigeria appears to be 50 years ahead of that of much of West Africa. By 2050 the rest of West Africa will look like the Nigeria of 2000 (Figure 3.5). In absolute terms, Nigeria's population of 130 million accounts for almost half the population of West Africa.

Economic Environment

ECONOMIC TRENDS AFTER STRUCTURAL ADJUSTMENT PROGRAMS

The dismal performance of the livestock sector in West and Central Africa over recent decades has been attributed in part to the policy and macroeconomic environment that prevailed in the 1970s and 1980s. The main features included price control (emanating from marketing boards and other parastatals), the imposition of import and export taxes, an array of tariff and nontariff barriers that provided disincentives to intraregional and international trade (quotas and bans), overvalued currencies, and restraints on private sector involvement in processing (Table 3.4).

In the 1970s and even the 1980s, governments in sub-Saharan Africa were skeptical of the notion that the

Table 3.3. West Africa—rural/urban population (1000) 1980–2015

Country	1980		1990		2000		2015	
Population	rural	urban	rural	urban	rural	urban	rural	urban
Nigeria	47,041	17,284	55,833	30,120	63,687	50,175	73,623	91,691
West Africa Total	94,916	32,783	115,824	55,328	136,125	88,058	168,078	161,477

Source: Renard et al., 2004.

market could be used to improve livestock production; they were concerned that equity distortions resulting from market imperfections would outweigh any market-based efficiency gains. Consequently, the livestock subsector was heavily assisted through government-led programs and projects. In the 1980s, policy reforms were implemented to counter the consequences of the inappropriate prior macroeconomic and sectoral policies. Their impact on income levels and distribution seriously affected the livestock sector. Interventions and actions were intended to achieve macroeconomic stabilization and structural adjustment. They included removal of inappropriate price policies, structural adjustment programs, currency alignment or devaluation, abolition of marketing boards, lifting of controls on livestock markets, and reduction of trade taxes. These changes initially altered the structure of incentives but also promoted expansion of livestock intraregional trade (Kamuanga et al., 2006a). Starting in the mid-1990s, with the collapse of traditional government interventions and public veterinary health and production systems, new challenges arose within the livestock subsector (Cheneau 1985, Leonard 2000, Ly 2002, Sandford 1983, Umali et al., 1992, Vétérinaires Sans Frontières 1994).

The three major coastal countries—Nigeria, Côte d'Ivoire, and Ghana—massively dominate the regional economy with jointly 70% of West Africa's Gross Regional Product (GRP). The 1991 West Africa Long Term Perspectives Study (WALTPS 1996) identified a narrow strip of economic activity along the Gulf of Guinea from Abidjan to Yaounde that accounted for over 80% of the regional GRP. WALTPS further notes that this strip constitutes the major part of the regional market and represents the greatest potential for regional trade, with Nigeria truly the heart of the regional economy. West Africa's coastal cities and the immediate hinterland, which account for about 70% of the region's human population, can also be seen as the region's food demand basin.

Table 3.4. Selected policy reforms affecting the livestock sector in West Africa

	The Gambia	Senegal	Guinea
Exchange rate adjustment	Flexible exchange rate in 1986, foreign currency exchange dealers authorized in 1990	Devaluation of the CFA in 1994	Flexible, market-determined exchange rates since 1983–86
• Increased producer prices • Liberalization of markets and prices • Liberalization of external trade and payments	• Economic recovery program (ERP) in 1985; SAP focused on privatization of public livestock projects • Marketing boards abolished • Sustainable development (PSD) in 1990, follow-up to ERP: injecting expansionary growth into productive sectors	• Restructuring and liberalization of the agriculture sector as 1985 (medium term) • Economic plan • Sectoral policies to liberalize prices • Reforms in the public sector and livestock marketing boards	• Most of the changes in macro and sectoral policies in effect in 1987–90 • Trade regulation for agricultural tradables introduced • Price incentives (producer), reduced role of parastatals (marketing boards) in livestock procurements • Monetary policy (devaluation)

Source: Kamuanga et al., 2006a.

GLOBALIZATION AND IMPORTS OF LIVESTOCK PRODUCTS FROM INTERNATIONAL MARKETS

Livestock trade constitutes an important economic activity in West Africa, and livestock is the highest valued agricultural commodity in intraregional trade (Akakpo et al., 1999, Williams et al., 2003). This trade has historically linked the Sahelian countries in the arid and semiarid parts of the region (e.g., Burkina Faso, Mali, and Niger) as exporters of livestock to the humid coastal countries in the south (e.g., Côte d'Ivoire, Ghana, and Nigeria). Differences in the biophysical production environment and in average incomes between the Sahel and coastal areas have divided the areas of livestock production from those of consumption, thus promoting this thriving intraregional trade in live animals. In cattle alone the trade increased in real value terms from $13 million in 1970 to $150 million in 2000 (Williams et al., 2003).

Major flows of livestock arrive at the urban terminal markets of the coast. The routes can be grouped in three zones:

1. The central corridor—Animals leave Mali and Burkina to supply Ivory Coast, Ghana, Togo, and Benin.
2. The Nigeria zone—Coming from Chad, Niger, Sudan, Republic of Central Africa, and sometimes Mali and Burkina, animals are directed to Cameroun, Nigeria, Benin, and Togo.
3. The West side—Animals from Mauritania and Mali are supplied to Senegal, The Gambia, and Guinea-Bissau.

Between 1970 and 1994 three major factors influenced the pattern of livestock marketing and trade in West Africa. Firstly, the severe droughts of the early 1970s and early 1980s disrupted the flow of animals from the Sahel to the coast and opened up the regional market to substantial extraregional imports of frozen meat from Argentina and the European Union. Subsequently the rapid declines in incomes of importing countries related to inappropriate macroeconomic policies caused falls in meat demand. In Nigeria, consumption fell from 12.2 kg per capita at the beginning of the 1980s to 11 kg by the end of the decade, and in Côte d'Ivoire from 8.4 kg to 4.2 kg due to the combination of population growth and relatively inelastic supply. Thirdly, globalization became a key factor in the livestock trade of West Africa, specifically the impact of significant quantities of subsidized imports of meat and dairy products from the European Union, which ate into the region's competitive advantage. In the early 1980s Sahelian beef cost about half as much as imports from Europe; by the end of the decade it cost about double, as import prices for European Union beef fell by about 29%. As a result, livestock exports from Sahelian countries to coastal countries, particularly to Côte d'Ivoire, dropped significantly. The share of frozen beef imports from countries outside the region, mainly from the European Union, increased threefold from 16% of total imports in the mid-1970s to 44% by the end of the 1980s (Williams et al., 2003). In recent years, imports of poultry meat have risen dramatically in West Africa (imports rose from 20,000 to 120,000 tonnes between 1998 and 2002) (Duteurtre et al., 2005).

The increasing flows and strengthening links of regional, national, and international markets and cultures are creating new dynamics, bringing new pressures to bear on the livestock sector. Markets are also breaking into previously closed environments, inducing shocks but also offering new opportunities for local producers, where they have market access (IFPRI 2006). Growing international markets are bringing competition from imports of livestock products (poultry, milk). International prices are playing a determinant role in the livelihoods of poor local herders, and animal transboundary diseases and food safety risks have a big impact on export possibilities. The prices for regionally produced livestock products are not competitive compared to international prices for similar products. In West Africa, the livestock trade is more competitive and functions better within countries than between countries. This is mainly due to high capital outlay, unavailability of credit, and the increased risks of losing animals associated with cross-border trade in livestock.

In all respects, West Africa's trade deficit in livestock products is growing, especially in milk and poultry since 2000 (Figure 3.6). The gap between demand and production is therefore increasing drastically (Figure 3.7).

Declining Interest from Donors and Government

The evidence of the widening technological gap and underinvestment in research targeting problems of tropical societies is there for all to see. These trends are perpetuating poverty and imply the case for greater funding of research that benefits poor countries and in particular for research on health as well as development initiatives. Apart from the difficulties and budget constraints that are widespread in overall national agricultural research in almost all the West African countries, research and development programs in livestock production are the least well endowed with financial, material, and human resources (Beintema and Stads 2006, Kamuanga et al., 2008).

In animal health similar disparities are evident. Many important diseases that constrain livestock production and affect millions of poor livestock keepers are unique to developing countries (e.g., trypanosomiasis, tickborne diseases, African swine fever). However, the pharmaceutical industry invests (largely) in products for the more lucrative pet market. In the United States, the main targets for this research were treatments for cancer, arthritis, and heart disease in domestic pets and a vaccine for West Nile virus (Maudlin and Shaw 2004).

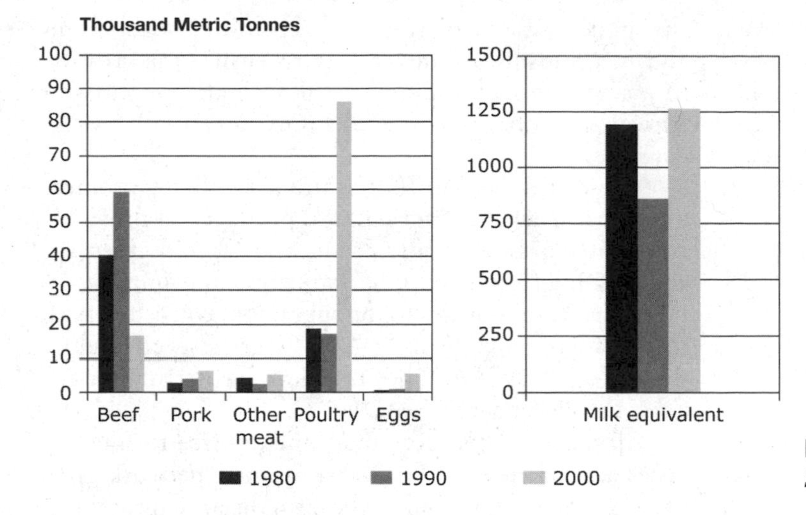

Figure 3.6. Imports in West Africa.
Source: FAO 2005.

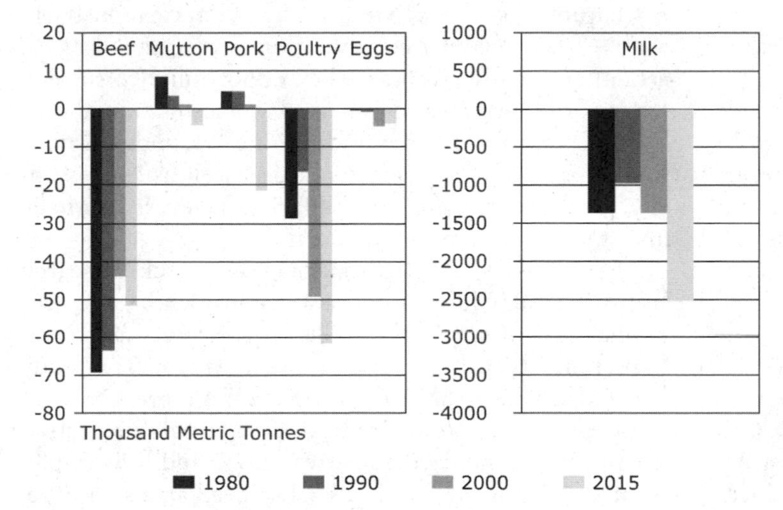

Figure 3.7. Gap (Demand-Production) of livestock products in West Africa.
Source: FAO 2005.

Changes in Livestock Sector Institutions

Institutions are important drivers because they can facilitate or hamper adaptive changes.

In the past, the livestock sector in West Africa was a key concern for colonial authorities because their political control was never certain, especially over nomadic and transhumant populations crossing borders and dwelling in remote areas. In recent decades, pastoral populations and livestock owners have been more and more incorporated in the political and social systems and superstructures at the national and regional levels. Over time, the livestock sector has been deeply affected by specific but also general institutional changes.

Institutions within Countries and Production Systems

Despite its importance in the economy, livestock production has never been the focus of major political attention, whereas huge amounts of development aid have been invested in modern schemes to increase crop production, often without the expected outcomes.

The failure to design coherent and integrated livestock production policies is linked to the sectoral or thematic organization of the relevant services. Livestock production has always been the preserve of veterinarians or animal scientists, who are usually not trained to take a holistic view of the entire sector. As a result, efforts to modernize livestock production have focused mainly on the animals and their performance (animal health, water supply, etc.) but have largely neglected rangeland improvement and management. With the exception of sanitary programs, most of the investments in livestock production (stock reduction programs, ranching, etc.) failed, partly because of a top-down approach that did not involve the herders themselves (Leonard 2004).

By encouraging privatization, structural adjustment in the livestock sector has been a leading factor in reshaping the supply of veterinary services. Prior to privatization, governments were overstretched in servicing major pastoral areas. Financial rationalization led to disengagement and divestment in public veterinary service delivery systems. Private sector development has been restructuring the economy of the livestock subsector, with new schemes to promote the opening or extension of private veterinary practice while maintaining the paraveterinarians. Today, a new breed of private veterinary professionals is present in the livestock sector. They are becoming more and more involved in curative and preventive services, but also in veterinary pharmaceutical sales. They are also involved in public contracting of mandatory immunizations in Senegal, Mali, Burkina, Niger, and so forth. (Leonard 1993, Vétérinaires Sans Frontières 1994, Leonard 2000, Ly 2000).

Institutions at the Regional Level

West Africa has three major subregional organizations engaged in regional agricultural policy making:

- The Economic Community of West African States (ECOWAS)
- Union Economique et Monétaire Ouest-Africaine (UEMOA) and
- Comité Permanent Inter-Etats de Lutte Contre la Sécheresse dans le Sahel (CILSS).

For more than three decades, these institutions have been facing considerable difficulties in coordinating policies in West Africa to better integrate divided markets and fragmented peoples, especially in the agricultural and livestock sectors. New challenges in the livestock subsector include the process of improving stakeholder consultation processes and ensuring coherence of the policies of the various overlapping regional bodies. At the nongovernmental organization (NGO) level, the Regional Network of Peasant and Agricultural Producer Organizations (ROPPA) has been working since 2000 to increase the value of smallholder agriculture in West Africa and to give a stronger voice to the region's producer groups. These groups have increasingly proved to be a major lobbying force, taking advantage of the waves of democracy, decentralization, civil societies, and stakeholders' participation.

The Conférence des Ministres de l'Agriculture de l'Afrique de l'Ouest et du Centre (CMAC/AOC), composed of agricultural and livestock ministers from 20 countries in the West and Central African subregions, is another regional intergovernmental organization. Since 2001, CMAC/AOC liaises with the Réseau des Chambres d'Agriculture de l'Afrique de l'Ouest in promoting the concept of decentralized decision making and supporting agriculture as a profession.

In addition to institutions, key initiatives are starting to shape the livestock sector. Since 2001, UEMOA (which groups the West African states using the CFA franc) has adopted the Politique Agricole de l'Union (PAU) as the common agricultural policy for its member states. The PAU is going to be financed through a special regional fund called Regional Agricultural Development Fund (FRDA). Since 2005, ECOWAS (which groups *all* West African countries) has also launched its own Agricultural Policy of the ECOWAS (ECOWAP). Because the two policies might be seen as a repetition, negotiations and mechanisms are under way to ensure the creation of a single agricultural policy valid for the whole West African Region, including UEMOA and non-UEMOA countries.

These regional institutions are designed to promote a regional market for agricultural products, improve the competitiveness of export products, and strengthen the capacity for formulating, harmonizing, and implementing agricultural policies in the subregion, consistent with the broader Comprehensive Africa Agriculture Development Program (CAADP) of the New Partnership for Africa's Development (NEPAD).

All these efforts show that, unlike three to four decades ago, a new regional institutional network is in place to implement public policies in the agricultural sector and the livestock subsector. Due to the transboundary nature of many activities in the livestock industry in West Africa, regional collaboration is considered as essential and is also seen as a key opportunity to remedy the lack of financial and human resources in animal production and health. Strong decisions are under way to harmonize biosafety frameworks, as well as food security, phytosanitary and zoosanitary policies, regulations, and trade conditions.

In 2000–01, the International Livestock Research Institute (ILRI) carried out a study to identify the economic, institutional, and policy constraints to cross-border livestock marketing in West Africa. The study concluded that UEMOA and ECOWAS protocols on intraregional livestock trade and regional integration need to be harmonized and streamlined and fully implemented (Williams et al., 2004). The need arises because, in theory, established institutions like UEMOA and ECOWAS should be able to undertake action on the issues of policy alignment for trade liberalization, facilitation, and exchange and payment systems. However, in reality, progress in implementation has lagged behind, and the focus so far has been on macroeconomic convergence; for example, in early 2000 UEMOA members adopted a customs union and common external tariff and have harmonized indirect taxation regulations (e.g., value added tax). The focus now needs to be extended to sectoral and trade policies that influence trade within the region, including livestock. UEMOA's progress needs to be extended to the additional six countries of ECOWAS in order to promote regional livestock trade.

Consequences

On the one hand population growth in the region is increasing competition for land and land use intensification and is therefore forcing adaptation of pastoral systems. On the other hand, increasing urbanization is creating the need for new supply chains, periurban production, and imports of frozen products. This trend is strengthened by the low capacity of pastoral systems, as well as the traditional mixed crop–livestock systems, to supply urban consumers. The more intensive periurban systems are not yet able to fill the gaps because they are just starting slowly to expand.

Environmental Consequences

Land Issues

The focus placed on crop production and food security since the drought of the 1970s has profoundly altered the structural basis of production systems in the Sahel region, particularly pastoralism, which has been comparatively neglected. This situation has led to further degradation of pastoral ecosystems, and ultimately to a true pastoral crisis. Climatic uncertainties, rapid demographic growth, and management practices have worsened the process by increasing cultivated area, instead of seeking sustainable benefits from the enhanced productivity of the different ecosystems. These factors combined have led to widespread erosion and degradation of natural resources. They have also accelerated the migration of population to urban centers; to the major wetlands of the Senegal, Niger, and Chad basins; and to the more southern and coastal parts of the region.

In arid and semiarid lands, species distribution and primary production are heterogeneous and patchy. Sudden variations in the availability of rain and of temporary water bodies are unpredictable. This results in significant production swings and a very fragile environment composed of an assemblage of biomes characterized by a large series of opportunistic species.

Sahelian pastoral ecosystems are now seriously threatened. There is a drop in biological productivity in many areas due to reduction of the vegetation cover, including grass and tree species, concentration of trees in limited areas with favorable water and soil conditions, reduction and sometimes disappearance of perennial grass species, reduction of plant diversity, extension of species with poor pastoral value, acceleration of wind and water erosive processes leading to the extension of glacis areas, moving dunes, and so forth; and loss of animal biodiversity. The conventional assumption is that livestock are responsible for rangeland degradation in the Sahel. However, Ellis (1992) and Hiernaux (1993) have challenged this thesis and provided evidence that climate is the main determinant of changes in the arid/semiarid environments and that the rangelands are resilient and capable of recovery. According to Ellis, "The strong seasonality of rangeland production in the Sahel limits the risk of overgrazing damaging the environment to short periods and consequently to confined areas."

Expanding Cropland versus Pastoral Land

The ever-increasing need to expand agricultural land, due to population growth and a depletion of resources linked to overexploitation of the soils, leads to a loss of pasture areas and water holes for pastoral herds. Sylvopastoral resources in rangelands have often been considered as potential agricultural areas to be cleared, rather than for their own native products (timber and fodder, etc.), services (soil and water conservation, etc.), and other values (biodiversity preservation, future option value, etc.). In addition, despite their importance and impact on production systems, transhumance routes are increasingly disrupted by the land allocation dynamics prevalent in most Sahel countries, which usually give no land tenure rights to pastoralists (Figure 3.8, FAO 2005, IFPRI 2006, Kamuanga et al., 2008).

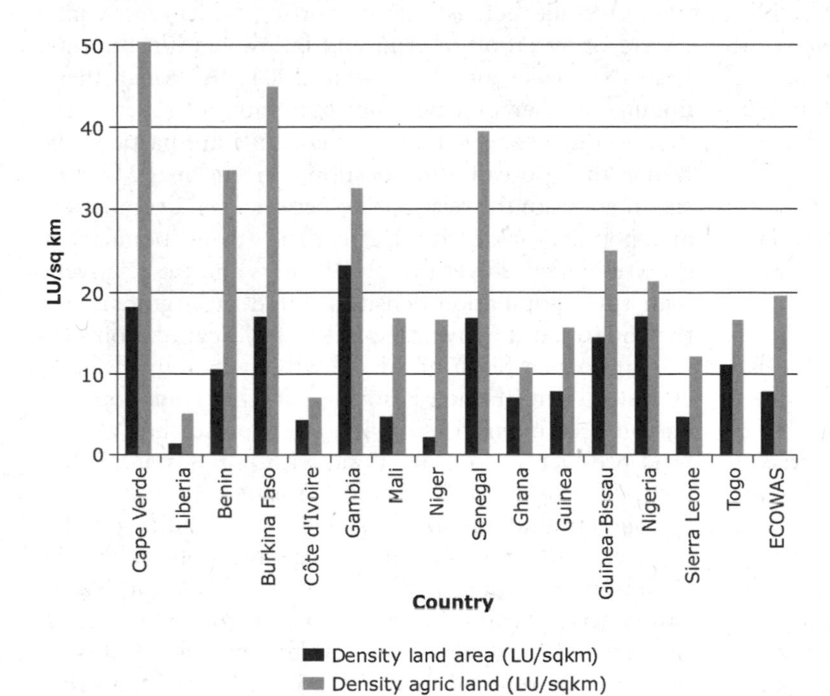

Figure 3.8. Livestock units in relation to available land, 2002.
Source: FAO 2005.
Note: Agricultural land does not cover permanent pasture.

Development workers and managers have operated with little respect, and often with contempt, for indigenous knowledge and practices. Since colonial times, administrations have had difficulties coming to terms with the fact that nomadic and transhumant herding is highly adapted to the ecological characteristics of the region. For example, in many countries burning of pastures, grazing in wooded areas, and shifting cultivation were discouraged. Topdown strategies have also led to inconsistencies, such as the promotion of intensified livestock production programs associated with networks of boreholes, which have accelerated erosion not only around the sources of water but also kilometers away, and at the same time the promotion of cash crops that may be detrimental to subsistence farming, as well as to nomadic and transhumant herding.

Processes of Land Degradation

Land degradation here is defined as the loss of actual or potential productivity or utility as a result of natural or anthropic factors; it is a decline in land quality and/or a reduction in its productivity. Mechanisms that initiate land degradation include physical, chemical, and biological processes (Lal 1994). The main forms of environmental degradation seen in West Africa include loss of natural vegetative cover, including deforestation; rangeland deterioration; reduction in floral and faunal diversity; and soil deterioration, including erosion, leaching, salinization, and decline in fertility.

Ecosystem degradation is well advanced. It has been estimated for Guinea, Mali, Burkina Faso, Niger, Côte d'Ivoire, and Ghana that 35 to 70% of rainfed croplands in each country are in a state of degradation (UNEP 2008). More than 70% of rangelands are degraded in Senegal, Mali, Burkina Faso, Niger, and Nigeria. In Niger, only 19% of the country is nondesert, and most of this is highly vulnerable to desertification. However, the extent and rate of ecosystem degradation are disputed because of conflicting interpretations of the available information and disagreements over the definition of the term *vegetation degradation*. It may imply reduction in biomass, decrease in species diversity, or decline in quality in terms of the nutritional value for livestock and wildlife.

In the Sahel area of western Africa, there is a growing risk of erosion and desertification. Since the 1960s, the area suffered a drastic decline in rainfall regimes, culminating in the drought years of the 1970s and '80s, with unprecedented food shortages and hunger crises. Combined with climate change, population growth has led over the years to adverse impacts, the most significant of which include unsustainable cropping practices, misuse of soil and water resources, soil compaction, sedentarization, and migration to cities (Thomas and Middleton 1994, Ozer 2000). However, a trend of increasing rainfall regimes is being observed in some West African countries, and since the mid-1980s in the Sahel rainfall trend, though this is still below the 1950–79 average (Hulme 2001, l'Hôte et al., 2002, Ozer et al., 2003, JISAO 2008).

Vegetation cover is a key indicator for the assessment of land degradation. In general, qualitative observations show a dominant picture of a significant reduction of vegetation cover, with bare soils reaching up hills and slopes and the concentration of vegetation in more humid areas of valley bottoms. Time series data (1981 to 1999) from Niger indicate a progressive diminution of the vegetation productivity as highlighted by the lowering of the maximum normalized difference vegetative index/rainfall ratio ($NDVI_{Max}/RR$). This situation suggests that consistent environmental degradation and desertification processes continued during the last two decades over most of the Sahelian belt of Niger. Data from a transect in Mali (Mainguet 1991) showed that land degradation in the 600 to 800 mm rainfall areas was significantly greater than in the 350 to 450 mm rainfall areas. In the higher rainfall areas, the percentage of bare soil increased from 0 to 10% over the period 1950 to 1990, whereas in the more arid areas there was no significant change.

However, there seem to be insufficient data on actual soil conditions and/or productivity to support findings of inexorable and inevitable degradation and serious nutrient depletion. Following Malthus's theory, the trends of strongly rising population densities, growing herds, deepening poverty, and limited agricultural intensification have been said to cause land degradation as well as declining soil productivity. The validity of this paradigm has been tested (Mortimore 2005) in West and East African countries experiencing demographic and economic trends that are usually associated with soil degradation. Overall, there is little supporting evidence of widespread degradation of crop and fallow land in Burkina Faso (Niemeijer and Mazzucato 2001). Although these findings neither preclude localized spots of severe degradation nor suggest that Sahelian soils are particularly fertile, they do call into question conventional Malthusian theories of the relationship between soil degradation and population density. Despite the strong population growth observed over the last 40 years and the relatively high rural population densities found in large parts of the country, a downward spiral of soil degradation and starvation as a result of this growth seems unlikely. The evidence from Burkina Faso suggests that some form of agricultural intensification is taking place that allows food production to grow along with population, as the theories of Boserup (1965, 1981) suggest.

Similar findings also resulted from studies on the long-term effects of climate change and policies on the evolution of farming systems in semiarid Senegal, Niger, and Nigeria. Analysis of long-term data gave evidence of significant achievements in ecosystems management (stabilization or reversal of degradation), land investment, and productivity (maintenance or increase) that runs counter to some current perceptions (Mortimore 2005). Land-use data from aerial photographs and satellite data from the two departments of Bambey and Diourbel in Senegal, showed that "saturation" was reached before 1960. Land under cultivation occupied over 82% of the surface in 1954 and 93% in 1999. Fallows virtually disappeared. The reduction and, in some places, near-elimination of fallows made a strong a priori case for the decline of fertility. A comparison of four soils that were sampled and analyzed in 1966 and resampled in 1999 showed that the topsoils (0–10 mm) had deteriorated during the intervening period. They had become more acidic and sandier and contained less carbon (Badiane et al., 2000). These soils had been cultivated without significant fertilization, being far from animal and human compounds.

However, the indicators show that management of plots close to compounds receiving manure and domestic waste may be sustainable (though at low levels of fertility) on sandy soils, whereas that of plots far from compounds is not. Furthermore, the number of TLUs per hectare rose from 0.12 in 1960 to 0.46 in 1995 despite the transfer of part of the livestock population to other areas. This suggests that livestock productivity expressed as output of meat, milk, and manure per hectare has substantially increased since 1960. In addition, the relatively new cattle functions of transport and traction energy for cropping have substantially increased the value of their output of services (Faye and Fall 2000).

Threats to Biodiversity: Rangelands and Protected Areas

A side effect of the reduction in pastoral land availability is increased pressure on and hostility to protected areas, and hence rising threats to biodiversity. The whole process is aggravated by the alienation of pastoralists and their exclusion from the decision-making process relating to natural resources access and management. This is partly because of a deficiency of true herder organizations, compounded by the multiplicity of institutional settings with various backgrounds and goals, such as various donor and government projects, NGOs, and the private sector. The specific dynamics of power redistribution and/or sharing need to be reformed so as to include all stakeholders and to reinforce the path toward sustainability. This institutional dimension needs to be integrated in all activities aiming at pastoral development in the Sahel.

The difficulty of finding practical solutions to the pastoralist crisis disrupts and brings distress to local communities. Social tensions, both within and between ethnic groups, conflicts between pastoralists and crop producers, and the disorganization of pastoralist social structures and populations eventually contribute to accelerate the crisis, to aggravate food insecurity and poverty, and to threaten even further the fragile stability of the region.

These changes affect the biodiversity of endemic livestock breeds as well as wildlife. Due to human population increase, the habitat for endemic livestock is being increasingly converted to cropland, and deforestation is rampant due to high demand for fuelwood. In southern Mali, for example, the land under cultivation increased from 5 to 18% of the total area between 1977 and 1994, due in large part to the continuous flow of humans and their livestock herds from drought-stricken Sahel areas to the south. Similar trends can be seen in southeastern Senegal, where decreasing fertility in the so-called Peanut Basin pushes farmers to migrate into virgin land to the south. In eastern Gambia, the surface area of cultivated land has doubled in the last 15 years (Fall et al., 2003).

These pressures are transforming indigenous woodlands into croplands, open savannas, and fallows. In addition, population growth has led to pronounced increases in demand for crops (in particular cereals), livestock and livestock products, and forest resources, prompting rural inhabitants to seek out higher productivity livestock breeds, and to engage in more intensive and often unsustainable resource use.

There has also been a breakdown in traditional rules and practices for use or control of common resources related to endemic ruminant livestock herds and rangelands. This has been precipitated by the influx of significant numbers of people and nonendemic animals into areas that support endemic ruminant livestock, as well as by changing patterns of resource use and demand, exacerbated by government policies and subsidies.

As traditional mechanisms have declined, state-sponsored resource management systems have not materialized to fill the need for coordinated control and use of common resources. Existing laws, regulations, and enforcement mechanisms for pastoral management, land tenure, and conflict resolution remain piecemeal and inadequate. In particular, unclear land tenure, combined with increasing competition for land and water resources, has led to increased conflict between farmers and herders and to overgrazing of communal pastures. Also, the lack of cross-border agreements or coordinated management of pastureland and livestock herds, despite increasing patterns of cross-border transhumance on the part of livestock herders, has made the sustainable management of communal grazing areas increasingly rare.

Threats to Biodiversity: Endemic Livestock Breeds

Investigations have shown that many of the endemic ruminant livestock breeds in West Africa are currently threatened with significant population decline, including possibly extinction, as well as the dilution of their unique genetic traits. The sources of the threats to these populations are varied and complex, but they can be broadly grouped into three primary categories:

1. Destruction and degradation of habitat critical for endemic ruminant livestock (discussed earlier)
2. Cross-breeding between endemic ruminant livestock and exotic livestock breeds
3. Declining interest among livestock producers in raising endemic ruminant livestock stemming from constraints on production levels and limited marketing opportunities.

The original African cattle are humpless cattle (*Bos taurus*) that evolved under high trypanosomiasis pressure. Other breeds (exotic taurine and *Bos indicus* cattle) that were imported in Africa later are unable to survive without veterinary intervention. There is an overall trend of the spread of *B. indicus* cattle, and of cross-breeding.

The humpless *B. taurus* cattle present in West Africa are trypanotolerant—able to survive and produce under trypanosomiasis pressure. They are also adapted to hot and humid areas and are resistant to ticks and tickborne diseases or associated diseases such as dermatophilosis. *B. taurus* breeds are divided into two main categories: the West African longhorn (N'dama) and the West African shorthorn (WAS), which is further divided into a savanna type (SWAS) (Baoule, Somba, Namchi) and dwarf type (DWAS) (Lagoon). The geographic distribution of trypanotolerant *B. taurus* cattle is similar to the geographic distribution of tsetse flies, the vectors of trypanosomiases.

The absolute number of *B. taurus* cattle increased from 6.92 million in 1985 to 8.03 million in 1998 (Agyemang 2005). In 1998, N'dama cattle made up 66.5% of pure taurine cattle, SWAS 31.5%, and DWAS 2%. However, the share of these breeds in the overall cattle population decreased: N'dama fell from 13.1% in 1985, to only 10.5% in 1998, whereas SWAS decreased from 5.3 to 4.2%. Even when cross-bred cattle are also considered (e.g., Borgou), trypanotolerant cattle decreased from 26.5 to 19.2%.

In addition to these general data, some specific cases give cause for concern: although the absolute numbers of trypanotolerant taurine cattle seem to be high, some breeds have very small populations. For instance, it is considered that the Pabli breed in Benin (SWAS) has disappeared. The Lagoon breed population also seems very small: in Côte d'Ivoire, 1000 individuals were censused in 1985, and no recent data are available (Agyemang 2000). In Togo, the Lagoon population is smaller than 1000 individuals, and only Benin has a large population of more than 20,000 individuals. The Baoule breed was considered to represent 73% of the cattle population in Côte d'Ivoire in 1973, but only 39% in 1998.

Other genetic threats derive from crossbreeding. Public actions during the last decade have focused on crossing local breeds by artificial insemination with imported semen from Europe or South America, mostly in periurban systems. Improvements in production are recorded for short or medium term, but follow-up and sustainability are not guaranteed. Rigorous breeding programs on local breeds are still not generalized and remain without continuous assessment (Boly et al., 2006, Fall 2006).

The low-input livestock systems of southern Burkina Faso, northern Côte d'Ivoire, northern Benin, and Togo and Guinea are characterized by mixtures of trypanotolerant, trypanosensitive, and stabilized crosses. Crossbreeding of tolerant and susceptible breeds in West Africa has occurred for decades due to the following:

- Dry season transhumance into the subhumid zones
- Recent droughts and the resulting southward movement of the Zebu herds
- Changing climatic patterns, land and bush clearing, and specialized tsetse control programs that reduced trypanosomiasis pressure and hence caused the number of trypanosensitive livestock to increase in the subhumid zone.

However, the determining factor has been the change in the often sophisticated breeding practices of indigenous farmers, through a combined process of passive and deliberate selection practices (Kamuanga et al., 2006b).

Health Consequences

Diseases remain major hindrances to livestock production, particularly among poor livestock keepers with few animals (Akakpo 1994, Perry et al., 2001, Sidibé 2001). Major diseases are not yet fully controlled and the following are still of concern:

- Among small ruminants: Peste des petits ruminants and contagious caprine pleuropneumonia (CCPP).
- Among village poultry: Newcastle disease.
- In intensive poultry production: Gumboro disease, colibacillosis, salmonellosis, Newcastle disease, and coccidiosis are major concerns due to their impact on the profitability of periurban livestock production. Every year, Newcastle disease is the number one killer of rural poultry, with 30 to 80% mortality levels.
- Among cattle: Trypanosomiasis, contagious bovine pleuropneumonia (CBPP), brucellosis, anthrax, and foot and mouth disease (FMD).

Botulism is still present in semiarid areas around the end of the dry seasons. However, the zoosanitary situation is relatively stable as far as epizootics are concerned.

Shifts in the Animal Health Paradigm

Some progress has been made in animal health. Since the mid-1990s, most of the West African countries have been declared partially or entirely rinderpest-free, and the disease has now probably been entirely eradicated. National and regional networks have been created for the epidemiological surveillance of animal diseases under the supervision of the World Animal Health Organization (OIE). The key objective is to monitor rinderpest, but other diseases are under passive surveillance (OIE 2004).

In addition, some diseases that are closely linked to natural habitats have been brought under control in many parts of West and Central Africa. For example, the effects of increasing human population, expansion of settlements and roads, and conversion of natural habitats to farmland have led to a reduction in the numbers and distribution of tsetse (the vector that transmits trypanosomiasis) in the semiarid and subhumid zones of West Africa.

Changing climate and land use patterns such as human occupation are likely to continue to benefit the region by gradually relaxing the pressure from trypanosomiasis, which has long limited the development of livestock. Simulations done by McDermott et al. (2001) show declining tsetse fly populations across much of West Africa. The net impacts are characterized in Figure 3.9. A band across the Sahelian zone would become essentially tsetse-free, but also will be experiencing declining agricultural potential. In the subhumid transition zone, however, where more of the crop production will be concentrated, significant reductions in tsetse pressure should enhance the ability of farmers to practice integrated crop–livestock systems.

Climate change in the region will also change the epidemiology of animal diseases, increasing the risk of some diseases but reducing that of others. The distributions of certain species of tsetse, for example, are quite sensitive to the length of growing period, and some may disappear from the drier areas, reducing the pressure of trypanosomiasis (Bourn et al., 2001).

However, low availability and use of veterinary services have allowed the classic endemic infectious diseases to persist across the region, including African swine fever, brucellosis, dermatophilosis, and bovine tuberculosis. As livestock systems intensify, especially in periurban areas, the incidence of associated production diseases and syndromes such as pneumonia, mastitis, and foot problems can be expected to become a significant veterinary problem. Indeed, the reduced role of the state in the provision of veterinary services—and the inability of the private sector to fill the void—have led to a resurgence of endemic animal diseases and reduced livestock productivity in many parts of the Sahel. Moreover, increasing concentrations of animals kept under poor conditions, often in close quarters within or near the households that keep them, together with poor access to veterinary and human medical care, will raise the risk of zoonotic diseases, especially among the poor.

This situation will also pose a growing threat of outbreaks of new emerging diseases, such as Nipah, bovine spongiform encephalopathy (BSE), and Ebola (McDermott et al., 2004). Because of market segmentation based on food safety concerns, larger volumes of low-quality livestock products will pass through informal marketing channels, further compounding the risks of zoonoses and

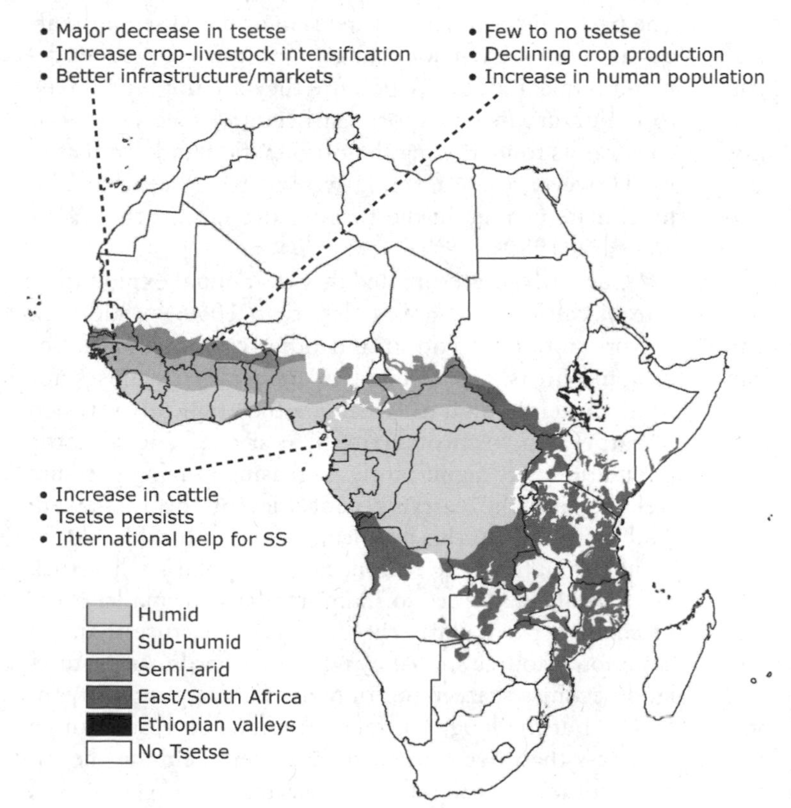

Figure 3.9. Expected impacts of declining tsetse populations and geographical distribution. *Source:* McDermott et al., 2004.

foodborne diseases for low-income consumers (e.g., brucellosis, salmonellosis; *E. coli* O157:H7, cysticercosis).

These trends pose a serious threat to incomes and livelihoods of poor livestock keepers. Not only do diseases limit production but in the case of FMD, CBPP, and CCPP they also prevent smallholders from participating in rapidly expanding external markets for livestock products.

The classification of diseases by Perry et al. (2001) suggests the need for a shift in the animal health paradigm (Tables 3.5, 3.6). The appearance of avian influenza in 2006 in Nigeria, Niger, and Burkina is of great concern and represents a key factor in the shifts of animal health paradigms.

Deterioration of Animal Health Support Systems

The structural adjustment policies of the 1980s led to state withdrawal from production operations. Privatization and declining public resources have left their mark on the livestock sector, as governments across the region have withdrawn significantly from the provision of veterinary and other input services.

Although more and more professionals are entering the private veterinary business and are playing active roles in immunization schemes in partnership with government and producer organizations, the anticipated emergence of private-sector provision of the full range of veterinary and advisory goods and services has not been as successful as hoped. It will continue to develop only gradually as long as production systems remain largely based on low levels of inputs.

Thus livestock keepers have found it more difficult to have access to veterinary and extension services, and to information and innovations needed to improve farm productivity. Poor access to products, services, and information contribute to poor performance, profitability, and competitiveness, and will continue to limit the ability of livestock keepers to address disease and other production constraints.

Zoonotic Diseases

Zoonotic diseases are of economical significance and constitute a public health threat in most sub-Saharan African countries. Access to markets for livestock products in West African countries requires provision of safer products of high quality—hence the prevalence of zoonoses can constitute serious obstacles for poor livestock producers. Losses have occurred in high prevalence areas through abortion, loss of weight, abattoir inspections, and carcass condemnations.

Accurate assessment of the economic impact on productive agriculture and public health is difficult because of inadequate information on the prevalence and clinical significance of the diseases. A few studies provide some sporadic information on the economical importance of zoonotic diseases. Prevalence of brucellosis as reported in livestock-raising areas of Mali in West Africa varies between 7% and 30%. Studies in the same country report prevalence figures for tuberculosis varying between 3% and 20%, affecting cattle, sheep and goats, and camels. Rabies is endemic in urban areas, where most of the samples submitted to the laboratory analysis are found to be positive. Dogs play the role of reservoir and main vector. Meatborne helminth infections such as cysticercosis/taeniasis cause medical and food safety problems and are becoming more of a public health concern (Tembely 2006).

Effective control strategies require a holistic approach with collaboration of biological, social, and medical scientists. Public awareness, improved diagnostic tools, and increased knowledge of the epidemiology of zoonotic diseases will play a major role in the implementation of any control strategy (Tembely 2006).

Reliable, up-to-date information on the prevalence of zoonoses and their zoonotic significance for farmers is very limited. Because locally produced milk in the region is mostly consumed either raw or fermented (Hempen et al., 2004), this commodity is an ideal medium for the spread of these milkborne zoonotic infections.

Social and Economic Consequences

One of the primary underlying threats to the long-term viability of rangeland ecosystems in West Africa is an evolving, unsustainable agropastoral system characterized by a low rate of cattle destocking.

Pressure on Pastoralism and Livestock Production

The pastoral production systems in arid and semiarid areas have not seen major changes in response to higher demand in meat and milk because they are not well linked to major urban or export markets and face enormous constraints to increasing their production and productivity. However, these farming systems have been hard hit by climate change in the form of drought in the 1970s and early 1980s.

Currently, it is estimated that the annual exploitation rate of cattle in the area is less than 10%. In addition, the promotion of cotton and other cash-earning crops in some areas has resulted in monetary surpluses for some rural inhabitants, which are then typically invested in cattle as a form of savings. As a result, local cattle populations are significantly increasing grazing pressure well beyond the carrying capacities of the rangeland. Adding further to this problem, as agricultural lands expand throughout the region, larger and larger livestock herds are being forced to share smaller and smaller areas of pasture, particularly the dry season pasture that is a common resource shared by migratory herds. As pastureland becomes scarcer, not only does grazing intensity increase, but the length of fallow periods decreases (often now less than five years), further overwhelming the capacity of the rangeland to regenerate.

Table 3.5. Major animal diseases with an impact on the poor, by species

Rank	Cattle	Sheep/Goats	Poultry
1	Foot and mouth disease	Helminthosis	Newcastle disease virus
2	Nutritional/micronutr def.	PPR	Helminthosis
3	Reproductive disorders	Haemonchosis	Coccidiosis
4	Hemorrhagic septicemia	Neonatal mortality	Ectoparasites
5	*Brucella abortus*	Respiratory complexes	Neonatal mortality
6	Trypanosomosis	Sheep and goat pox	Fowl cholera
7	Liver fluke	Ectoparasites	Infectious coryza
8	Anthrax	Anthrax	Fowl pox
9	CBPP	Liver fluke	DVE
10	*Toxocara vitulorum*	Heartwater	Nutritional/micronutr def
11	Mastitis	CCPP	Gumboro
12	Helminthosis	Foot problems	Mycoplasmosis
13	Babesiosis	Rift Valley fever	Salmonella
14	Neonatal mortality	Foot and mouth disease	DVH
15	Diarrheal diseases	Trypanosomosis	
16	*Theileria annulata*	Clostridial diseases	
17	Rinderpest	Para-tb	
18	Dermatophilosis	*Brucella melitensis*	
19	Blackleg	Orf	
20	IBR	Blue tongue	

CCPP = Contagious caprine pleuropneumonia

DVE = Duck viral enteritis

DVH = Duck viral hepatitis

IBR = Infectious bursitis rhinotracheitis

Nutritional/micronutr def = Nutritional micronutrients deficiency

Para-tb = Paratuberculosis

PPR = Peste des petits ruminants

Source: Perry et al., 2001.

Table 3.6. Ranking of major zoonoses with an impact on the poor in West Africa

	Global Ranking		Regional Ranking	Production System Ranking		
			WA	Pastoral	Agropast	Periurban
Disease or Organism	Rank	Index	Rank			
Brucella abortus	1	100	1	3	1	2
Brucella melitensis	2	29	2	7	2	3
Trypanosomiasis	3	16	n/a	1	3	n/a
Bovine tuberculosis	4	15	4	6	4	7
Leptospirosis	5	13	n/a	12	5	n/a
Anthrax	6	12	5	5	6	10
Cysticercosis	7	11	n/a	10	7	1
Buffalo pox	8	8	n/a	n/a	8	11
Rift Valley fever	9	8	3	2	9	n/a
Botulism	10	1	6	9	14	13

n/a = not available.

Source: Perry et al., 2001.

The expansion of crop agriculture into marginal grazing lands has implications for land degradation and the shrinking of grazing resources, which led Nori et al. (2005) to infer picturesquely that pastoralists are herding on the brink. Mortimore (2001) agrees that pastoralism may well survive for several more decades in the low population density parts of the Sahel where it is the logical response to large land resources and few marketing points. Foreseeing that slowly growing rural populations are likely to continue to nibble away at any land that has potential for crop production, Mortimore (2001) postulates that "the future of livestock producing systems rests with enabling closer forms of integration with farming, rather than attempting to stop the inevitable." Does this imply, for example, that the very foundation of the livelihood of pastoralists is vulnerable to change? It can be imagined that conflicting arguments could be advanced regarding what could or could not be considered the "inevitable" evolutionary pathway of livestock production, but this question goes beyond the scope of this chapter.

Changing Animal Production Systems and Marketing Channels

Major changes have been brought about in the livestock production systems in West Africa by climate change, increased demand for livestock products driven by rapid population growth and urbanization, and policy changes.

At the regional level, the devaluation of the CFA franc in 1994 led to short-term improved competitiveness of livestock exports from the West African Sahel zone to the Gulf of Guinea coast. Livestock exports expanded as a result of this policy change, but only briefly, during the first half of 1994. After this surge, flows returned to normal levels.

For three livestock exporting countries (Burkina, Mali, and Niger) between 1990 and 2000, there were three peak cattle export years (1992, 1995, and 2000) and two low cattle export years (1994 and 1998) (Williams et al., 2003). During peak cattle export years, the corresponding values in US dollars were higher, whereas during the low export years the reverse was the case. In other words, cattle prices and CFA exchange rates have been a major determinant of the volume of exported cattle from Burkina Faso, Mali, and Niger (Figure 3.10).

In the cotton belt of Mali and Burkina, Hamadou et al. (2006) observe that the predominant traditional herding faces important constraints to dairy production, including low income of herders, low adoption of productivity technologies, poorly performing breeds and poor breeding management practices, poor feeding and watering, and animal diseases (Hamadou et al., 2006). Liehoun and Sidibé (2006) find two strategies in subhumid savanna areas:

- An adaptive strategy involving better handling of forages and increasing storage and use of crop residues, forage production, and diversification of livestock activities (different species, fattening, etc.)
- A conservative strategy based on intensifying mobility to solve feeding constraints and land overstocking.

Which of these strategies is chosen depends on the evolution of household animal distribution. The result is that livestock systems are evolving toward sedentary systems integrated with other household activities—small herds with draught animals. These sedentary systems coexist alongside agropastoralists who have larger herds and a higher mobility.

The patterns of pastoral mobility are also changing. Dry-season transhumance is becoming a common practice even with mixed farmers. A major change is the now regular practice of wet season transhumance. The intensity and speed of change depend on the level of cropland expansion and saturation of agricultural land. Extensive

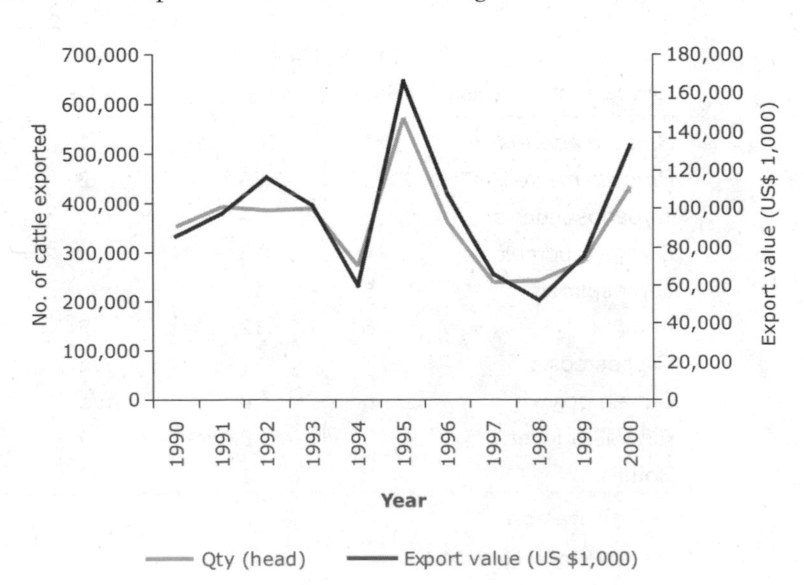

Figure 3.10. Number of cattle exported from Burkina Faso, Mali, and Niger and their corresponding export values (US$1,000), 1990–2000.
Source: Williams et al., 2003.

practices are increasing, based on more mobility of the herds and their dispersal in various locations. In the zones of origin, these practices are adding to the difficulties of traditional management of soil fertility based on manure. Ethnic and cultural drivers are less and less significant, whereas technical and socioeconomic constraints are dominating more and more in explaining pastoral mobility in West Africa.

Among the few studies based on long-term data analysis, Tiffen (2002, 2003) and Faye et al. (2000) have tried to capture the structural trends in semiarid zones of West Africa. In the case of semiarid areas of Senegal (Diourbel region), since 1960, the populations of cattle, small ruminants, and equines have maintained an upward trend, and meat production has risen. However, the trend in marketed milk production has become unstable because more cattle are now involved in transhumance.

The most rapid increase has been in the numbers of small ruminants. The real price of sheep and goat meat increased after 1980, unlike the prices of beef, grain, and groundnuts. In one Sereer rural community, near the south of Diourbel Region, small stock increased from 3042 in 1954 to 8836 in 1990 (Faye et al., 2000). However, fattening has become widespread for both cattle and small livestock, which probably increases their productivity in terms of weight gain, although in drought years there is a fall in the average weight of carcasses at abattoirs. The number of TLUs per hectare rose from 0.12 in 1960 to 0.46 in 1995. If we ignore transhumance outside the region, this suggests that output of meat, milk, and manure per hectare has increased substantially since 1960. The value of manure has always been recognized, especially by the Sereer, but also by the Wolof, who traditionally paid Fulani herdsmen to manure their fields. In addition, livestock's relatively new functions of transport and traction energy for cropping have substantially increased the value of their output of services.

Where transport infrastructure exists in rural areas close to cities, mixed crop livestock production systems have evolved into more intensified patterns. Pilot fattening schemes and small-scale dairy systems with small-scale milk processing units have been scaled up in a number of farming systems in West Africa.

The rapid development of industrial poultry and dairy production systems in urban and periurban areas is a major change that has taken place over the last 20 years as a result of the increased demand for livestock products in large cities.

The presence of horses has been very much restricted to the Sahelian zone due to the African animal trypanosomiasis. However, in some of these areas the role of equines in animal traction is increasing. Detailed studies are available for the groundnut basin of Senegal. Some changes that took place in the use of work animals include (1) the rapid spread of the use of equines for transport into eastern and southern zones even in areas where health hazards previously constrained the use of these animals, and (2) the stagnation of the use of work animals because of changes in policies that reduced the capacity of farmers to have access to credit to get implements. Studies conducted in the Senegalese groundnut basin on the dynamics of work animals on farms showed that (1) the vast majority of farms keep equines, especially male equines; the proportion of farms with equines was 69% in 1975 and 81% in 1996, (2) the number of farms owning draft oxen was 51% in 1975 and 46% in 1996, and (3) draft cows were present in 77% of farms in 1996, up from 53% in 1975 (Fall et al., 2003).

Substitution of Red Meat by White, Short-Cycle Meat (Pork, Poultry)

Structural changes in marketing systems for livestock and their products are also occurring in response to increasingly segmented markets (Weatherspoon and Reardon 2003).

Average per capita demand for livestock products has decreased in West Africa during recent decades. Local supply of milk products available was around 40 kg per capita per year in the 1980s, with imports filling gaps. In 1997, local supply is estimated at 33 kg per capita—below the proper level for child growth (though there are important geographical variations) (FAO 2007). Meat consumption per person fell by 8% from 1960 to 1990. Fish consumption rose by 9% during the same period, but not enough in absolute terms to compensate. Meat consumption actually increased modestly between 1961 and 1982, from 9 kg to 11 kg per capita/year. However, since then it fell to 10 kg per capita/year in 2003. The explanation for this trend is twofold: lower red meat consumption (beef and mutton-goat meat: 4.9 kg per capita/year in 1961 to 4.7 kg per capita/year in 2003) along with higher consumption of white meat (pork and poultry: 1.4 kg per capita/year in 1961 to 4 kg in 2003) (Table 3.7, Figure 3.11).

Importance of the Livestock Subsector in the Agricultural and Overall Economy

Despite a significant role in the overall macroeconomy and potential market niches, the performance of the livestock sector in West Africa remains poor. The distribution of population in poverty tends to match the distribution of livestock density, especially for large animals.

Countries can be grouped into two main categories (Sahelian and coastal) with respect to the contribution of livestock to the economy. In addition to Mauritania where livestock accounted for 85% of agricultural GDP in 2000, the Sahel countries are showing contribution around 30 to 50% (Burkina, Mali, Niger, Senegal). The coastal countries show a lower importance of livestock. The case of Cape Verde is atypical because small animals play an important role in the islands (Table 3.8).

Projections suggest a general increase in each type

Table 3.7. Per capita consumption of livestock products and annual growth rate (1992–2002)

	Meat		Milk		Eggs	
Country	kg/capita	Growth rate (%)	kg/capita	Growth rate (%)	kg/capita	Growth rate (%)
Benin	18.0	4.1	11.1	7.3	0.9	0.0
Burkina Faso	11.3	0.4	18.0	2.5	0.9	–2.0
Cape Verde	27.7	–0.6	85.6	0.4	4.0	3.3
Côte d'Ivoire	11.3	–1.0	5.9	5.7	1.6	4.8
Gambia	5.2	–3.2	18.8	3.6	1.2	2.9
Ghana	9.9	–1.2	3.3	6.9	0.9	4.1
Guinea	6.5	2.9	10.9	–1.0	1.5	4.1
Guinea-Bissau	13.0	–1.1	13.7	–0.8	0.6	1.8
Liberia	8.1	–3.3	1.7	1.3	1.1	–5.3
Mali	19.0	0.8	44.5	0.7	0.4	–6.7
Niger	11.3	–1.5	10.2	–0.8	0.6	–1.5
Nigeria	8.6	0.4	5.4	5.4	3.3	–0.9
Senegal	17.7	0.7	21.7	3.8	2.6	1.7
Sierra Leone	6.2	1.8	4.9	0.4	1.6	2.1
Togo	8.5	0.4	5.5	0.0	1.0	–1.8
ECOWAS	10.3	0.2	9.5	2.7	2.2	–0.5

Source: FAO 2005.

of animal production. However, unfortunately the strong human population growth and stagnant or decreasing low trends in production per capita are not favorable (Tables 3.9, 3.10; Figure 3.12).

Contribution of the Livestock Sector to the Millennium Development Goals

The general prospects for economic recovery in West Africa are not very favorable. The International Monetary Fund suggests that at least 40% of countries in sub-Saharan Africa are "off-track" or "seriously off-track" on reaching most millennium development goals (MDGs) (IMF 2006). Only one West African country (Senegal) is considered well positioned to meet the income poverty goal in addition to Cameroun, Ethiopia, South Africa, and Swaziland. The trend of net trade per capita has been stagnating for the last decade with low to negative growth rate (Table 3.11, Figure 3.13). Low productivity and trade deficit in livestock products (Figure 3.14) can explain part of the negative forecast for the MDGs (FAO 2005, IFPRI 2006, Kamuanga et al., 2008).

The changes that have occurred over the last 30 years

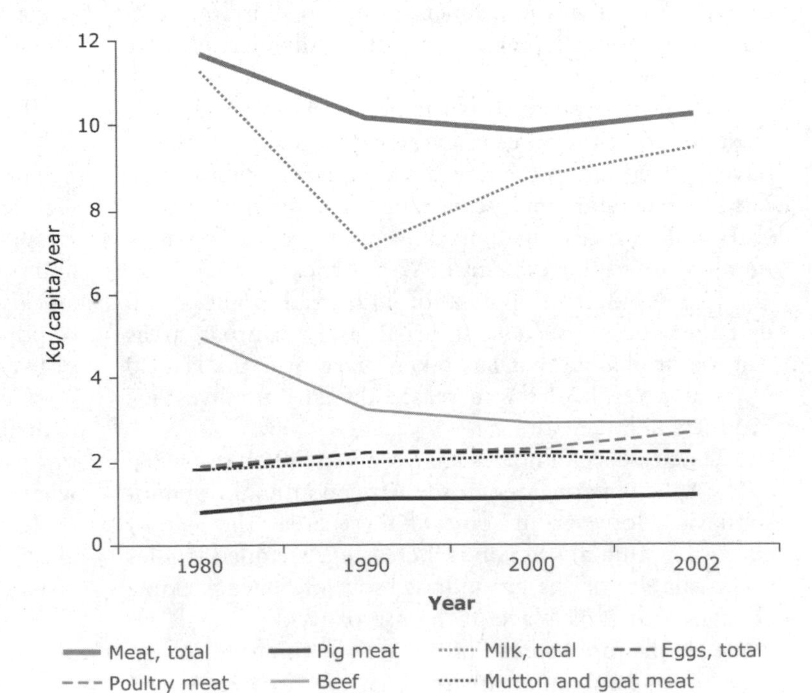

Figure 3.11. West Africa trends in annual per capita consumption of meat, milk, and eggs.
Source: FAO 2005.

Table 3.8. West Africa—contribution of livestock to agricultural GDP 1980–2000

Country	1980	1990	2000
Burkina Faso	33.7	34.1	34.7
Mali	59.2	45.7	48.8
Mauritania	91.2	89.9	85.1
Niger	42.5	37.9	37.4
Benin	23.6	14.9	8.9
Cape Verde	30.7	50.7	61.2
Côte d'Ivoire	8.2	7.8	7.3
Gambia	24.2	21.3	13.8
Ghana	15.0	15.8	9.4
Guinea	14.9	14.9	17.3
Guinea-Bissau	30.1	25.2	23.2
Liberia	11.1	17.3	14.4
Nigeria	23.6	15.7	13.8
Senegal	30.9	29.7	30.9
Sierra Leone	13.3	12.9	20.3
Togo	11.6	15.3	13.0
West Africa average	25.2	19.7	17.2

Source: Renard et al., 2004.

Table 3.9. West Africa—trends in production (1000 Mt)

Production	1980	1990	2000	2015
Beef	541.4	492.5	647.1	1204.2
Mutton and goat meat	257.7	356.7	470.4	798.2
Pig meat	85.5	169.6	216.1	205.1
Poultry meat	218.5	368.2	465.0	877.7
Milk	1481.4	1625.5	2056.4	3461.4
Eggs	274.8	433.3	598.1	1025.9

Source: FAO 2005.

Table 3.10. West Africa—trends in production per capita (kg/person/year)

Production	1980	1990	2000	2015
Beef	4.6	3.2	3.3	4.0
Mutton and goat meat	2.1	2.1	2.2	2.5
Pig meat	0.6	1.0	0.9	0.7
Poultry meat	1.7	2.1	2.1	2.6
Milk	12.4	10.1	10.0	11.3
Eggs	2.1	2.5	2.7	3.0

Source: FAO 2005.

in West Africa would have required an important first wave of socioeconomic investment to redress the initial situation, and a second more significant wave to address the basic needs of a population in rapid expansion. In reality, the region has not benefited from any significant

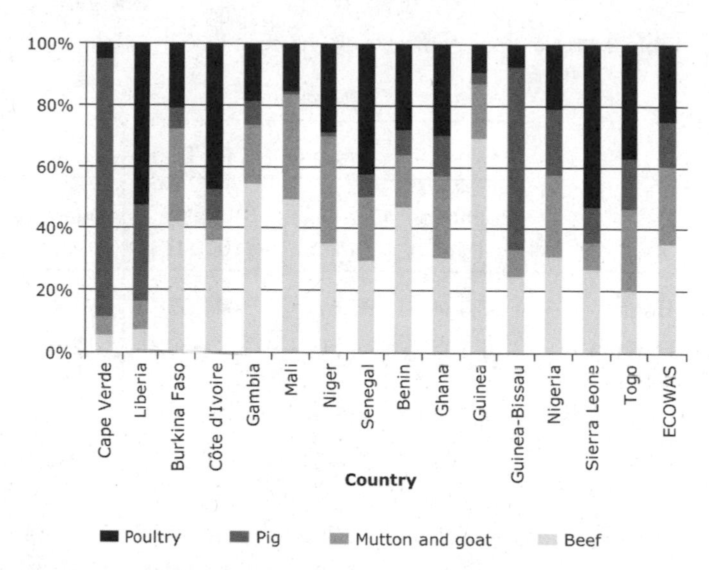

Figure 3.12. Composition of total meat production, 2002. *Source:* FAO 2005.

amount of foreign direct investment. Its share of world trade is still extremely low, and despite many efforts to cancel public debt, debt has remained in many countries disproportionately high in relation to reimbursement capacity. In addition, the region has not received a level of Official Development Assistance (ODA) commensurate with its needs, nor has it been able to mobilize internal savings. Given the very low level of financial resources, it is not surprising that most of the countries in the region will not be able to reach the MDGs. Commitment and determination will be required from all development players to correct the current situation.

Responses

Decision makers at the farm, community, national, and regional levels have developed strategies and taken action in response to changes in the livestock sector in order to take advantage of new opportunities, to better adapt to the consequences of changes, or to mitigate negative impacts of changes. Adjustment in production systems and development of new supply chains combined with traditional ones are among the most notable responses to significant changes in the West African livestock industry, such as the consequences of climate change, increasing population density, and changing economic environment. At the macro level, key responses from institutional and government stakeholders include institutional and policy adaptations. Here we focus on the major responses.

Adjustments of Production Systems and Development of Supply Chains

Many West African livestock producers have responded to the increased demand of livestock products brought about by dietary changes due to urbanization and

Table 3.11. Net trade of meat, milk and eggs (2003)

	Meat			Milk			Eggs		
	Export	Import	Net Trade	Export	Import	Net Trade	Export	Import	Net Trade
Country	Quantities (Mt)	Quantities (Mt)	Value (1000 US$)	Quantities (Mt)	Quantities (Mt)	Value (1000 US$)	Quantities (Mt)	Quantities (Mt)	Value (1000 US$)
Benin	28,484	95,670	–44,658	426	38,335	–16,629		0	–2
Burkina Faso	105	503	–502	2,406	64,780	–25,847	4	2	–13
Cape Verde	0	5,334	–809	0	27,478	–11,495	0	54	–213
Côte d'Ivoire	101	31,421	–25,892	59,435	163,351	–26,502	14	17	–73
Gambia	0	9,283	–6,836	84	36,462	–12,730		997	–1,062
Ghana	788	51,131	–33,564	8,161	129,545	–32,056	196	167	–679
Guinea	0	4,858	–4,283	0	26,615	–11,459		72	–74
Guinea-Bissau	0	670	–1,084	0	3,772	–1,331			
Liberia	32	3,938	–3,012	0	5,120	–2,429		917	–876
Mali	0	257	–442	0	40,710	–14,954		27	–94
Niger	13	107	–87	232	49,751	–8,813	0	237	–104
Nigeria	0	4,892	–8,176	1,165	671,928	–249,009		230	–809
Senegal	509	14,564	–22,719	22,372	182,023	–48,570	34	284	–1,005
Sierra Leone	0	6,113	–5,636		16,491	–7,954		1,671	–1,057
Togo	494	9,910	–4,246	22,146	27,533	4,050	10	192	–25
ECOWAS	30,526	238,651	–161,946	116,427	1,483,894	–465,728	258	4,867	–6,086

Source: FAO 2007.

growing city populations. New supply chains have emerged and agricultural intensification is allowing food production to grow along with population. Agricultural production in West Africa has been forced to intensify and expand, and increasingly to integrate previous stand-alone crop and livestock production enterprises into mixed crop–livestock enterprises. Underlying this integration are livestock-for-land and land-for-livestock exchanges between erstwhile "sole" livestock and crop farmers (McIntire et al., 1992, Jagtap and Amissah-Arthur 1999, Okike et al., 2004, Thys 2006).

In densely populated semiarid areas, farmers have increased livestock output per unit of land. They have changed the composition of their livestock holdings, with a significant shift from cattle to small ruminants (Kamuanga et al., 2008) or camels. They have also developed fattening schemes for both cattle and small ruminants. Whereas real beef prices have remained steady over time, mutton prices have escalated since 1980. As a result, fattening of sheep to meet the increasing demand of sheep during the Muslim Tobaski celebration has become a lucrative business for many farmers in rural and urban areas, especially women. Advances are taking place in crop–livestock integration in West African cities where the livestock sector has been buoyant, leading to readiness to invest in fattening by buying in animals and fodder (Akinbamijo et al., 2002).

In subhumid areas, the expansion of cotton production has led to further integration of crop and livestock activities through increased use of animals for draft. The availability of cotton seeds laid the foundation of a more intensified cattle production system (Ly and Diaw 1996). Cattle fattening schemes have been developed to supply slaughter cattle in good condition when cattle from the northern Sahel are affected by early feed shortages. Pilot fattening schemes and small-scale dairy systems with small-scale milk processing units have been scaled up in a number of farming systems in West Africa (Dièye 2006, Hamadou 2006).

The increasing demand for animal protein due to increasing urbanization and growing human populations has encouraged short-cycle private sector meat, egg, and

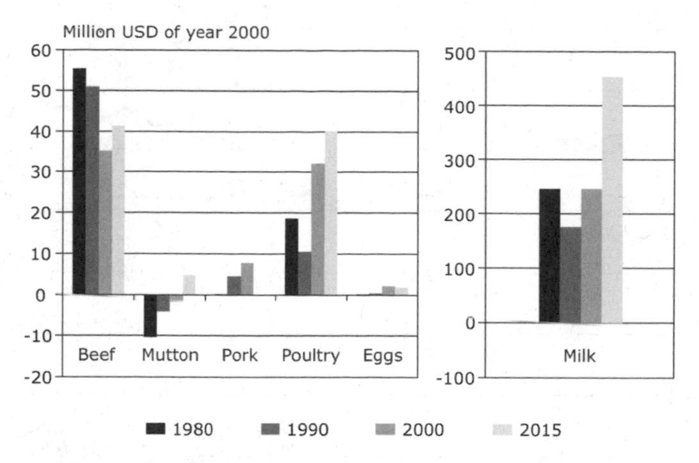

Figure 3.13. Trade deficit of West Africa.
Source: FAO 2005.

milk production for strictly commercial purposes. In West Africa—exemplified in this case by Nigeria before highly pathogenic avian influenza (HPAI)—favorable government policies led to the emergence of a number of industrial-scale integrated poultry enterprises. Some even came to the point of exporting day-old chicks (parent stock and commercial), broilers, hatchable eggs, and so forth, to other West African countries (Okike 2002). After the outbreak of HPAI was reported in Nigeria in 2006, imports of poultry products from Nigeria were banned by neighboring countries. The immediate short-term economic impact has been a decline in commercial poultry production. It is notable that previously ignored legislation banning commercial livestock production within residential areas is gradually coming into force as people take into account the environmental pollution caused in the vicinity of such enterprises, as well as the health hazards that may arise in the case of zoonoses.

However, despite positive response from farmers and communities to meet the growing demand in animal products in a context of a shrinking resource base, there has been a widening gap between demand and domestic supply of animal products. This is indicated by the surge of milk and poultry meat imports. In recent years, imports of poultry meat have risen dramatically in West Africa. Several import surges have been reported in various countries. In Côte d'Ivoire, poultry imports went from 2840 tonnes in 2000 to 15,400 tonnes in 2003. In Senegal, the situation is very similar with a rise from 506 tonnes in 1996 to 16,900 tonnes in 2002 (Duteurtre et al., 2005).

Policy and Institutional Adaptations

Livestock policies are responses that are influenced by historical circumstances and emergencies, as well as by the spectrum of means and resources available and by the stakeholders' respective political strengths and interactions. Policies are the outcomes of competing forces, negotiation capacities, and opportunities of various actors, and have been shaped by the access to policy design mechanisms and political will of the government and livestock owners.

From the end of colonial rule in the 1960s to the droughts in the 1970s, the livestock policy approach of national administrations was focused on livestock products as exports or food supply for the new growing cities. There were also relative successes in vaccination programs against major diseases, especially rinderpest. Livestock projects funded by donors were directed at mitigating the social and economic impact of drought on pastoral populations and production systems.

Strategies were based on government interventions in production and national markets through administrative price-setting, the creation of livestock and meat marketing boards, trade components in projects, and government-centred rules and regulations. These characteristics led to mechanisms of government control aiming at the modernization of marketing networks and systems. Despite partially meeting urban demand and partially inserting traditional herders in the market economy, projects did not succeed in creating a self-driven development of pastoral systems. Those systems remain heavily dependent on foreign aid and the vagaries of climate (Ancey and Monas, 2005).

In the 1980s, various policies seeking economic recovery and structural adjustment became dominant despite their exogenous origin and blueprint approaches. Adjustment policies—the framework for all macroeconomic and subsector policies—profoundly affected explicit and implicit livestock policies, especially in West Africa (Table 3.4). Policy reforms were implemented to counter the consequences of the inappropriate prior macroeconomic and sectoral policies, whose impact on income levels and distribution seriously affected the livestock sector. Interventions and actions were intended to achieve macroeconomic stabilization and structural adjustment. They included removal of inappropriate price policies, structural adjustment programs, currency alignment or devaluation, abolition of marketing boards, lifting of controls on livestock markets, and reduction of trade taxes. The structural adjustment policies of the 1980s led to state withdrawal from production operations. Privatization and declining public resources have left their mark on the livestock sector as governments across the region have withdrawn significantly from the provision of veterinary and other input services. These changes initially altered the structure of incentives and promoted expansion of intraregional trade in livestock (Pica-Ciamara 2005, Kamuanga et al., 2006a). Today, national policies still follow the same patterns, despite new constraints deriving from external causes such as world market issues or trade negotiations outcomes (WTO, SPS agreement, EU policies, emerging diseases such as avian influenza).

Starting in the 1990s, poverty reduction has come to the forefront in the design of livestock policies.

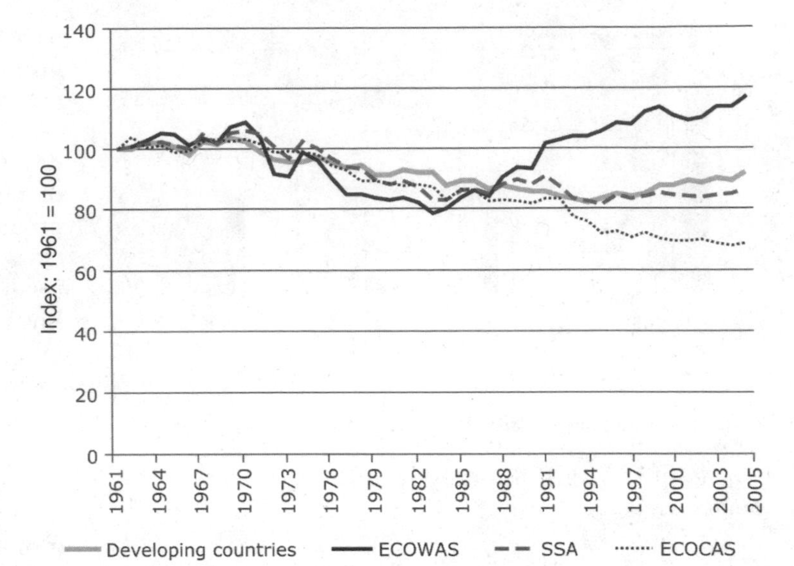

Figure 3.14. Net per capita agricultural production, 1961–2005.
Source: IFPRI 2006.
Note: SSA (sub-Saharan Africa); ECOWAS (Economic Community of West African States); ECOCAS (Economic Community of Central African States)

Interventions have been made so that monetary incomes could be increased and assets improved; they have included access to inputs, pastoral land management, and investment (water points, rangeland management, trekking improvement) (Barret 2003, Duteurtre and Faye 2003). When necessary, safety nets have been used to tackle temporary risks and constraints, relying on emergency aid schemes such as emergency livestock feed supply, free vaccination campaigns, veterinary pharmaceuticals distribution, and so forth. In parallel, withdrawal of government services, privatization, markets free of government interventions and deregulation, were all meant to remove economic and social inefficiencies as well as management constraints from public services. Unfortunately, national and local markets became suddenly and entirely open to stronger international competitors without benefiting from the safeguard and mitigation mechanisms provided by the World Trade Organization.

The important potential of the livestock subsector is recognized as a powerful factor in favor of its economic integration, but its positioning as an engine for economic development is still lagging. A thorough assessment of livestock policies in West Africa for the last three decades is still lacking and urgently needed. Ex-post and comparative analyses are required for an objective assessment to derive lessons for development policies and livestock policies. Indeed there is still a need to define and to agree on a regional livestock policy—although there has been some recent progress toward a regional harmonization of zoosanitation and food safety within UEMOA, leading to an agreement signed in 2007.

Outcomes and Lessons

The livestock subsector remains very weak in West Africa as far as allocation of resources and effort are concerned. Despite the potential of livestock for reducing poverty and enhancing the economy, appropriate strategies for livestock development are still lagging (Leonard 1993 and 2000, Ly and Duteurtre 2004, Pica-Ciamarra 2005). To cope with this situation, institutional and policy innovations are still much needed. Promising approaches include involving producers' groups in policy design, market monitoring, institutional arrangements to smooth up market flows, financing mechanisms adapted to livestock production duration and cycles, capacity building for producers organizations, and so on (Duteurtre and Faye 2003).

There is a growing concern to foster changes in livestock policies and institutions that will benefit both the economy and poor livestock owners. A balance is needed so that subsidies and policies that favor crop production over grazing will not lead to a widespread conversion of grazing lands to agricultural production. This is particularly true for cereals and for cash crops such as cotton. A balance is also needed with respect to subsidies; for example, policies that promote and subsidize exotic livestock breeds over endemic breeds are widespread and distort the real cost of production of the different races, which without subsidies would often favor local breeds.

The history of agricultural development in Europe and North America shows that it was not severely hampered by financial constraints, and that farmers were willing and able to adopt available technologies once an enabling policy and institutional environment were in place. Such an enabling environment is influenced by economic and institutional factors that are beyond the individual farmer's immediate control. Consequently, in order to strengthen positive developments for livestock development in West Africa, major institutional and policy reforms are required at national and regional levels targeting the livestock subsector.

Conclusions

Unlike in other parts of the world, the Livestock Revolution is hardly taking place in West Africa. Constraints to production and productivity are lasting features in the regional livestock subsector, along with high zoosanitary risks, which are difficult to reduce by technology investments because of narrow marketing margins in the local value chains, and high transactions costs.

Livestock production and marketing are, however, undergoing change in West Africa. Economic strategies are changing. Diversification is under way, with increased mixed farming, fattening, and market-oriented destocking for local and regional markets. New risks faced in the sector also bring changes in capitalization strategies; cattle will decreasingly be kept for savings and insurance, as market access improves and favors novel activities and opportunities.

The future of the livestock subsector in West Africa lies in properly managing a threefold approach based on the following:

- Sustainable management of rangeland, shared pastoral resources, and transhumance
- Enhanced livestock productivity and competitiveness in the ECOWAS member states
- Improved market access (local, regional, and international).

Development of the livestock subsector in West Africa benefits considerably from regional harmonization and integration because most issues affecting the subsector can be adequately addressed only at the regional level. Issues with strong regional aspects include natural resource management, market integration, and improving the competitiveness of the livestock subsector.

In individual countries as well as at the regional level of the common ECOWAS and UEMOA agricultural policy, specific implementation strategies and programs still await formulation and negotiation. However, the capacity for formulating, negotiating, and implementing sustainable livestock development strategies and institutional changes focused on growth and poverty reduction is weak, both at the regional institutional level of ECOWAS as well as in the member countries. Therefore, there is a critical need to strengthen the regional capacities in order to combine stakeholder engagement with research and analysis and information dissemination for livestock development strategy. In addition, program formulation, implementation, and coordination of actions need to be organized within the framework of maximizing the potential contribution of the livestock subsector to food security, poverty reduction, and economic development.

In the future, participatory institution building and monitoring through local networks, producer organizations, and regional integration will be key elements for better access to innovations, knowledge and information, and better access to input and output markets. Policies are needed to support all livestock stakeholders at the national and regional level to enter competitive food chains to supply urban consumers in local products at affordable prices and good quality.

References

Agyemang, K. 2005. *Trypanotolerant livestock in the context of trypanosomiasis intervention strategies.* PAAT Technical and Scientific Series 7. Rome: FAO.

Akakpo, A. J. 1994. Mode d'élevage, épidémiologie des maladies infectieuses animales et Santé publique en Afrique au Sud du Sahara. *Cahiers Agricultur,* 1994(3): 361–368.

Akakpo, J. A., Ch. Ly, and R. Bada-Alambedji. 1999. Le commerce du bétail et de la viande en Afrique de l'Ouest et du Centre, facteur d'intégration économique en Afrique tropicale. *Revue de Médecine Vétérinaire* 150(5): 453–462.

Akinbamijo, O. O., S. T. Fall, and O. B. Smith (eds). 2002. *Advances in crop–livestock integration in West African cities.* Dakar: ITC-ISRA-IDRC.

Ancey, V. and G. Monas. 2005. Le pastoralisme au Sénégal entre politique "moderne" et gestion des risques par les pasteurs. *Revue Tiers Monde* XLVI(184): 761–783.

Badiane, A. N., M. Khouma, and M. Sène. 2000. Région de Diourbel: gestion des sols. *Atelier sur les rapports entre politiques gouvernementales et investissements paysans dans les régions semi-arides,* Bambey and Dakar, 12–14 January 2000. Drylands Research Working Paper 15. Crewkerne, UK: Presstige Print.

Barret, C. B. 2003. Rural poverty dynamics: Development policy implications. In *Proceedings of the 25th International Conference of Agricultural Economists,* 16–22 August 2003, Durban, South Africa.

Beintema, N. M. and G.-J. Stads. 2006. *Agricultural R&D in Sub-Saharan Africa: An era of stagnation.* Agricultural Science & Technology Indicators (ASTI) Initiative. Washington DC: IFPRI.

Boly, H., M. Boundaogo, W. Sanogo, L. Sawadogo, and P. Leroy. 2006. Evolution des Programmes d'Amélioration Génétiques de Production Laitière Bovine au Burkina Faso. In *International Conference on Livestock Agriculture in West and Central Africa: Achievements in the past 25 years, Challenges ahead and the way forward,* ed. A. Schoenefeld et al. Banjul: ITC-CIRDES -CTA.

Born, van den G., H. Leemans, and M. Schaeffer. 2004. Climate change scenarios for dryland West Africa, 1990–2050. In *The Impact of Climate Change on Drylands,* ed. A. J. Dietz, R. Ruben, and A. Verhagen. Netherlands: Springer.

Boserup, E. 1965. *The Conditions of Agricultural Growth: The Economics of Agrarian Change Under Population Pressure.* New York: Aldine Publishing.

Boserup, E. 1981. *Population and Technological Change: A Study of Long Term Trends.* Chicago: University of Chicago Press.

Bourn, D., R. Reid, D. Rogers, W. Snow, and W. Wint. 2001. *Environmental Changes and the Autonomous Control of Tsetse and Trypanosomosis in Sub-Saharan Africa: Case Histories from Ethiopia, The Gambia, Kenya, Nigeria and Zimbabwe.* Oxford: Environmental Research Group.

Cheneau, Y. 1985. L'Organisation des services vétérinaires en Afrique. *Revue Scientifique et Technique—Office International des Epizooties* 5(1): 57–105.

Cour, J. M. 2001. The Sahel in West Africa: countries in transition to a full market economy. *Global Environmental Change* 11: 31–47.

Dennis, K. C., I. Niang-Diop, R. J. Nicolls. 1995. Sea level rise and Senegal: potential impacts and consequences. *J Coastal Res* 14: 243–261.

DFID. 2004. Climate change in Africa. In *Climate Change in Africa,* ed. D. Carney. London: Department for International Development.

Dièye, P. N. 2006: *Arrangements contractuels et performances des marchés du lait local au sud du Sénégal. Les petites entreprises de transformation face aux incertitudes de l'approvisionnement.* Thèse Doctorat en agroéconomie. Montpellier, France: ENSA.

Dixon, J., A. Gulliver, and D. Gibbon. 2001. *Global Farming Systems Study: Challenges and Priorities to 2030. Synthesis and Global Overview.* Rome: FAO.

Duteurtre, G. and B. Faye (eds). 2003. *Elevage et pauvreté. Actes de l'Atelier Cirad Montpellier,* 11 et 12 Septembre 2003.

Duteurtre, G., P. N. Dièye, and D. Dia. 2005. *Ouverture des frontières et développement agricole dans les pays de l'UEMOA: l'impact des importations de volaille et de produits laitiers sur la production locale au Sénégal.* Etudes et Documents, 8(1): 78. Dakar: Institut Sénégalais de Recherches Agricoles.

Ellis, L. 1992. *ILCA's Rangeland Research Program in the Arid and Semiarid Zones: Review and Recommendations.* Addis Ababa: International Livestock Center for Africa.

Fall, A. 2006. Towards sustainable cattle genetic improvement programmes in West Africa: The contribution of PROCORDEL. In *International Conference on Livestock Agriculture in West and Central Africa: Achievements in the Past 25 years, Challenges Ahead and the Way Forward,* ed. A. Schoenefeld et al. Banjul: ITC-CIRDES-CTA.

Fall, A., A. Diack, and F. Dia. 2003. The role of work animals in semi-arid West Africa: Current use and their potential for future contributions. In *Working Animals in Agriculture and Transport: A Collection of Some Current Research and Development Observation,* ed. R. A. Pearson, P. Lhoste, M. Saastamoinen, and W. Martin-Rosset. EAAP Technical Series No. 6. Wageningen: Wageningen Academic Publishers.

FAO. 2005. *Economic Community of West African States: Trends in Livestock Production.* AGAL. Rome: FAO.

FAO. 2007. *Global Livestock Production and Health Atlas.* Animal Production and Health Division. Rome: FAO. Cited November 2007. Available at: www.fao.org/ag/aga/glipha/index.jsp

Faye, A. and A. Fall. 2000. *Région de Diourbel: Diversification des revenus et son incidence sur l'investissement agricole.* Drylands Research Working Paper 22. Crewkerne, UK: Drylands Research.

Faye, A., A. Fall, and D. Coulibaly. 2000. *Région de Diourbel: Evolution de la production agricole.* Drylands Research Working Paper 16. Crewkerne, UK: Drylands Research.

Faye, S., I. Niang-Diop, S. Cisse-Faye, D. G. Evans, M. Pfister, P. Maloszewski, and K. P. Seiler. 2001. Seawater intrusion in the Dakar (Senegal) confined aquifer: calibration and testing of a 3D finite element model. In *New Approaches Characterizing Groundwater Flow. Proceedings of the XXXI IAH Congress,* ed. K. P. Seiler and S. Wohnlich. Munich, 10–14 Sept. 2001, Vol. 2: 1183–1186. Lisse, Netherlands: A.A. Balkema.

Fernández-Rivera S., I. Okike, V. Manyong, T. O. Williams, R. L. Kruska, and S. A. Tarawli. 2004. Classification and description of the major farming systems incorporating ruminant livestock in West Africa. In *Sustainable Crop–Livestock Production for Improved Livelihoods and Natural Resource Management in West Africa.* Proceedings of an international conference held at the International Institute of Tropical Agriculture (IITA), Ibadan, Nigeria, 19–22 November 2001, T. O. Williams, S. A. Tarawali, P. Hiernaux, and S. Fernández-Rivera, 89–122. Wageningen: Technical Centre for Agricultural and Rural Cooperation. Nairobi: International Livestock Research Institute.

Gabre-Madhin, E. Z. and S. Haggblade. 2004. Successes in African agriculture. MSSD Discussion Papers 53. Washington DC: International Food Policy Research Institute.

Hamadou, S. 2006. Lutte contre la Pauvreté par une Amélioration de la Production Laitière en Zone Périurbaine: Cas de Bobo-Dioulasso. In *Agriculture, Elevage et Pauvreté en Afrique de l'Ouest,* ed. A. A. Mbaye, D. Roland-Holst, and J. Otte, 17–24. Dakar: CREA-FAO.

Hamadou, S., H. Marichatou, M. Kamuanga, B. A. Kanwé, A. G. Sidibé, J. Paré, H. Djouara, M. I. Sangaré, and O. Sanogo. 2006. Diagnostic des Systèmes de Production Laitière en Afrique de l'Ouest: Typologie des Elevages Périurbains. In *International Conference on Livestock Agriculture in West and Central Africa: Achievements in the Past 25 years, Challenges Ahead and the Way Forward,* ed. A. Schoenefeld et al. Banjul: ITC-CIRDES-CTA.

Hempen, M., F. Unger, S. Muntermann, M. T. Seck, and V. Niamy. 2004. The hygienic status of raw and sour milk from smallholder dairy farms and local markets and potential risk for public health in The Gambia, Senegal and Guinea. Animal Health Research Working Paper 3. Banjul: International Trypanotolerance Centre.

Hiernaux, P. 1993. *The Crisis of Sahelian Pastoralism: Ecological or Economic?* Addis Ababa: International Livestock Center for Africa.

Hillel, D. and C. Rosenzweig. 2002. Desertification in relation to climate variability and change. *Advances in Agronomy* 77: 1–38.

Hulme, M. 2001. Climatic perspectives on Sahelian dessication: 1973–1998. *Global Environmental Change* 11: 19–29.

IFPRI. 2006. *Regional Strategic Alternatives for Agriculture-Led Growth and Poverty Reduction in West Africa.* Vols. 1 and 2—annexes. Dakar: IFPRI-IITA-ECOWAS, CORAF-WECARD.

IMF. 2006. World Economic and Financial Surveys, World Economic Outlook Database.

Jagtap, S. and A. Amissah-Arthur. 1999. Stratification and synthesis of crop–livestock production systems using GIS. *GeoJourna* 47: 573–582.

Jahnke, Hans E. 1982. *Livestock Production Systems and Livestock Development in Tropical Africa.* Kiel: Wissenschaftsverlag Vauk Kiel.

JISAO. 2008. *Sahel Rainfall Index.* Seattle: Joint Institute for the Study of the Atmosphere and Ocean, University of Washington. Cited 27 July 2008. Available at: http://www.jisao.washington.edu/data_sets/sahel/

Kamuanga, M., J. Somda, E. Tollens, and T. Williams. 2006a. Policy Reforms and Performance of the Livestock Sub-Sector in West Africa. Case Studies: The Gambia, Guinea and Senegal. In *International Conference on Livestock Agriculture in West and Central Africa: Achievements in the Past 25 Years, Challenges Ahead and the Way Forward,* ed. A. Schoenefeld et al. Banjul: ITC-CIRDES-CTA.

Kamuanga, M., K. Tano, K. Pokou, G. d'Ieteren, C. Mugalla, and J. Somda. 2006b. Farmers' preferences for cattle breeds and prospects for improvement in West Africa. In *International Conference on Livestock Agriculture in West and Central Africa: Achievements in the past 25 Years, Challenges Ahead and the Way Forward,* ed. A. Schoenefeld et al. Banjul: ITC-CIRDES-CTA.

Kamuanga, M., J. Somda, Y. Sanon, and H. Kagoné. 2008. *Livestock and Regional Market in the Sahel and West Africa, Potentials and Challenges.* Paris: SWAC/OECD/ECOWAS.

Kristjanson, P. M., P. K. Thornton, R. L. Kruska, R. S. Reid, N. Henninger, T. O. Williams, S. A. Tarawali, J. Niezen, and P. Hiernaux. 2004. Mapping livestock systems and changes to 2050: Implications for West Africa. In *Sustainable Crop–Livestock Production for Improved Livelihoods and Natural Resource Management in West Africa.* Proceedings of an international conference held at the International Institute of Tropical Agriculture (IITA), Ibadan, Nigeria, 19–22 November 2001, ed. T. O. Williams, S. A. Tarawali, P. Hiernaux and S. Fernández-Rivera, 28–84. Wageningen: Technical Centre for Agricultural and Rural Cooperation. Nairobi: International Livestock Research Institute.

Kruska, R. L., R. S. Reid, P. K. Thornton, N. Henninger, and P. M. Kristjanson. 2003. Mapping livestock-oriented agricultural production systems for the developing world. *Agricultural Systems* 77: 39–63.

Lal, R. 1994. Sustainable land use systems and soil resilience. In *Soil Resilience and Sustainable Land Use*, ed. D. J. Greenland and I. Szabolcs, 41–67. Wallingford, UK: CAB-International.

Leonard, D. K. 1993. Structural reform of the veterinary profession in Africa and the new institutional economics. *Development and Change* 24: 227–267.

Leonard, D. K. (ed.). 2000. *Africa's Changing Markets for Human and Animal Health Services: The New Institutional Issues.* London and New York: Macmillan Press/St. Martin's Press.

Leonard, D. K. 2004. *The Political Economy of International Development and Pro-Poor Livestock Policies: A Comparative Assessment.* PPLPI Working paper 12. Pro-Poor Livestock Policy Initiative. Rome: FAO.

L'Hôte, T., G. Mahé, B. Somé, and J. P. Triboulet. 2002. Analysis of a Sahelian annual rainfall index from 1896 to 2000: the drought continues. *Hydrological Sciences Journal* 47: 563–572.

Liehoun, B. E. and I. Sidibé. 2006. Dynamiques Agraires et Evolution des Pratiques Pastorales dans les Zones de Savane Sub-humides: Etude de Cas au Burkina Faso. In *International Conference on Livestock Agriculture in West and Central Africa: Achievements in the Past 25 Years, Challenges Ahead and the Way Forward,* ed. A. Schoenefeld et al. Banjul: ITC-CIRDES-CTA.

Ly, C. 2000. Veterinary professionals in Senegal: Allocation of priorities and working behavior. In *Africa's Changing Markets for Health and Veterinary Services: The New Institutional Issues,* ed. D. K. Leonard, 119–145. London: Macmillan.

Ly, C. 2002. The economics of community-based animal health workers, Parts 1 and 2. Economic theory and community-based animal health workers. In: *Community-Based Animal Health Workers: Threat or Opportunity*, ed. The IDL Group, 91–96, 99–107. Crewkerne: The IDL Group.

Ly, C. and A. Diaw. 1996. La rentabilité de la production laitière en étable fumière: Cas de la zone cotonnière du Sénégal. *Revue de Médecine Vétérinaire* 7(3): 203–210.

Ly, C. and G. Duteurtre (eds). 2004. Pour des politiques d'Elevage "Partagées." *Atelier Régional sur les Politiques d'Elevage,* Dakar, 17–18 November 2004. ISRA, CIRAD, DIREL, ODVS, Initiative PPLPI, FAO. Rome: FAO.

Mainguet, M. 1991. *Desertification: Natural Background and Human Mismanagement.* Berlin: Springer-Verlag.

Manyong, V. M. 2002. Economic research at IITA for the improvement of agriculture in the sub-humid and humid zones of West Africa. In *Economic Analyses of Agricultural Technologies and Rural Institutions in West Africa: Achievements, Challenges, and Application to Rice Farming Research*, ed. T. Sakurai, J. Furuya, and H. Takagi, 37–58. Proceedings of JIRCAS International Workshop 12–13 July 2001. JIRCAS Working Paper Report No. 25. Tsukuba, Japan: JIRCAS.

Maudlin, I. and A. Shaw. 2004. Donor and Private Sector Participation in Support of Livestock Development. Issues in the Field of Animal Health. In *International Conference on Livestock Agriculture in West and Central Africa: Achievements in the Past 25 Years, Challenges Ahead and the Way Forward,* ed. A. Schoenefeld et al. Banjul: ITC-CIRDES-CTA.

McDermott, J. J., P. M. Kristjanson, R. L. Kruska, R. S. Reid, T. P. Robinson, P. G. Coleman, P. G. Jones, and P. K. Thornton. 2001. Effects of climate, human population and socio-economic changes on tsetse-transmitted trypanosomiasis to 2050. In *World Class Parasites*, ed. R. Seed and S. Black. Boston: Kluwer.

McDermott, J., D. Richard, and T. Randolph. 2004. Incidence and impacts of animal diseases on the development of crop–livestock systems in West Africa. In *Sustainable Crop-Livestock Production for Improved Livelihoods and Natural Resource Management in West Africa.* Proceedings of an international workshop held at the International Institute of Tropical Agriculture (IITA), Ibadan, Nigeria, 19–21 November 2001, ed. T. O. Williams, S. Tarawali, P. Hiernaux and S. Fernandez-Rivera. Nairobi: International Livestock Research Institute, and Wageningen: CTA.

McIntire, J., D. Bourzat, and P. Pingali. 1992. *Crop–Livestock Interactions in Sub-Sahara Africa.* Washington DC: World Bank.

Mortimore, M. 2001. Overcoming variability and productivity constraints in Sahelian agriculture. In *Politics, Property and Production in the West African Sahel: Understanding Natural Resources Management,* ed. T. A. Benjaminsen and C. Lund, 233–255. Uppsala: Nordiska Afrikainstitutet.

Mortimore, M. 2005. *Why Invest in Drylands?* Rome: Global Mechanism of the UNCCD.

Neely, C. and S. Bunning. 2008. *Review of Evidence on Dryland Pastoral Systems and Climate Change: Implications and Opportunities for Mitigation and Adaptation.* FAO–NRL Working Paper. Rome: FAO.

Niasse, M., A. Afouda, and A. Amani. 2004. *Reducing West Africa's Vulnerability to Climate Impacts on Water Resources, Wetlands and Desertification. Elements for a Regional Strategy for Preparedness and Adaptation.* Gland: IUCN, The World Conservation Union.

Niemeijer, D. and V. Mazzucato. 2001. Productivity of Soil Resources in Sahelian Villages. In *Agro-Silvo-Pastoral Land Use in Sahelian Villages,* ed. L. Stroosnijder and T. van Rheenen, 145–156. Reiskirchen: Catena Verlag.

Nori, M., J. Switzer, and A. Crawford. 2005. Herding on the Brink: Towards a Global Survey of Pastoral Communities and Conflict. An Occassional Paper from the IUCN Commission on Environmental, Economic, and Social Policy. IUCN: Gland, Switzerland. Available at: http://www.iisd.org/publications/pub.aspx?id=705

OIE. 2004. *Handistatus II* [online]. Cited October 2004. Paris: Office International des Epizooties. Available at: http://www.oie.int/hs2/

Okike, I. 2002. Transregional analysis of crop–livestock systems: understanding intensification and evolution across three continents—the case of West Africa. Consultant's report submitted to International Institute of Tropical Agriculture, Ibadan, Nigeria, and International Livestock Research Institute, Nairobi, Kenya.

Okike, I., T. O. Williams, and I. Baltenweck. 2004. *Promoting Livestock Marketing and Intra-regional Trade in West Africa.*

West Africa Livestock Marketing: Brief 4. Nairobi: International Livestock Research Institute.

Ozer, P. 2000. Les lithométéores en région sahélienne. *Geo-Eco-Trop* 24: 1–317.

Ozer, P., M. Erpicum, G. Demaree, and M. Vandiepenbeeck. 2003. The Sahelian drought may have ended during the 1990s. *Hydrological Sciences Journal* 48: 489–492.

Parry, M. L. 2002. Turning up the heat: how will agriculture weather global climate change? In *Sustainable Food Security for All by 2020*. Proceedings of International Conference, 4–6 September 2001, 117–123. Washington DC: IFPRI.

Perry, B. D., J. J. McDermott, T. F. Randolph, K. R. Sones, and P. K. Thornton. 2001. *Investing in Animal Health Research to Alleviate Poverty.* Nairobi: International Livestock Research Institute.

Pica-Ciamarra, U. 2005. Livestock policies for poverty alleviation: theory and practical evidence from Asia and Latin America. Working paper No. 27. Pro-Poor Livestock Policy Programme. Rome: FAO.

Randolph, T. F., O. Diall, M. Kamuanga, J. van Binsbergen, and P. Thornton. 2006. Mega-trends shaping livestock agricultural development in West and Central Africa. In *International Conference on Livestock Agriculture in West and Central Africa: Achievements in the Past 25 years, Challenges Ahead and the Way Forward,* ed. A. Schoenefeld et al. Banjul: ITC-CIRDES-CTA.

Renard J.-F, C. Ly, and V. Knips. 2004. *L'élevage et l'intégration régionale en Afrique de l'Ouest.* Montpellier: Ministère des Affaires Etrangères-France-FAO-CIRAD.

Sandford, S. 1983. *Management of Pastoral Development in the Third World.* New York: John Wiley & Sons.

Schoenefeld, A., K. Agyemang, A. Gouro, and I. Sidibé (eds). 2006. *International Conference on Livestock Agriculture in West and Central Africa: Achievements in the Past 25 Years, Challenges Ahead and the Way Forward.* Banjul: ITC-CIRDES-CTA.

Sene, S. and P. Ozer. 2002. Evolution pluviométrique et relation inondations—événements pluvieux au Sénégal. *Bulletin de la Société Géographique de Liège* 42: 27–33.

Sére, C., H. Steinfeld, and J. Gronewold. 1996. *World Livestock Production Systems: Current Status, Issues and Trends.* FAO Animal Production and Health Paper 127. Rome: FAO.

Sidibé, M. 2001. Impact économique des maladies animales sur l'élevage en Afrique subsaharienne. In *Séminaire sur l'utilisation des trypanocides en Afrique subsaharienne*, EISMV, Dakar, 6–9 February 2001. Dakar: EISMV.

Tarawali, G., E. Dembélé, B. N'Guessan, and A. Youri. 1998. Smallholders' use of Stylosanthes for sustainable food production in subhumid West Africa. *International Workshop on Green-Manure Cover Crop Systems for Smallholders in Tropical and Subtropical Regions*, 6–12 Apr 1997, Chapeco, Brazil.

Tembely, S. 2006. Zoonotic Diseases in sub-Saharan Africa: Impact on animal production and public health. In *International Conference on Livestock Agriculture in West and Central Africa: Achievements in the Past 25 Years, Challenges Ahead and the Way Forward,* ed. A. Schoenefeld et al. Banjul: ITC-CIRDES-CTA.

Thomas, D. S. G. and N. J. Middleton. 1994. *Desertification: Exploding the Myth.* Chichester: Wiley.

Thornton, P. K., R. L. Kruska, N. Henninger, P. M. Kristjanson, R. S. Reid, F. Atieno, A. N. Odero, and T. Ndegwa. 2002. *Mapping Poverty and Livestock in the Developing World.* Nairobi: International Livestock Research Institute.

Thys, E. 2006. *Role of Urban and Peri-urban Livestock Production in Poverty Alleviation and Food Security in Africa.* Brussels: Académie Royale des Sciences d'Outre-Mer.

Tiffen, M. 2002. The evolution of agro-ecological methods and the influence of markets: case studies from Kenya and Nigeria. In *Agroecological Innovations,* ed. N. Uphoff, 95–108. London: Earthscan.

Tiffen, M. 2003. Transition in sub-Saharan Africa: Agriculture, urbanization and income growth. *World Development* 31: 1343–1366.

Tiffen, M. 2004. Population pressure, migration and urbanization. In *Sustainable Crop–Livestock Production for Improved Livelihoods and Natural Resource Management in West Africa.* Proceedings of an international workshop held at the International Institute of Tropical Agriculture (IITA), Ibadan, Nigeria, 19–21 November 2001, ed. T. O. Williams, S. Tarawali, P. Hiernaux, and S. Fernandez-Rivera. Nairobi: International Livestock Research Institute, and Wageningen: CTA.

Umali, D. L., G. Feder, and C. de Haan. 1992. *The Balance between Public and Private Sector Activities in the Delivery of Livestock Services.* World Bank Discussion Papers 163. Washington, DC: World Bank.

UNEP. 2008. *Africa: Atlas of Our Changing Environment.* Division of Early Warning and Assessment, United Nations Environment Programme. Malta: Progress Press.

United Nations. 2009. *World Population Prospects 2006.* New York: United Nations Population Division.

Vétérinaires Sans Frontières (eds). 1994. *Actes du Colloque de Bamako sur la privatisation des services vétérinaires.* Lyon: VSF.

WALTPS. 1996. *Preparing for the Future: A Vision of West Africa in the Year 2020. West African Long Term Perspective Study,* ed. J.-M. Cour and S. Snrech. Paris: OECD.

Weatherspoon, D. D. and T. Reardon. 2003. The rise of supermarkets in Africa: Implications for agri-food systems and the rural poor. *Development Policy Review* 21(3): 333–355.

Williams, T. O., P. Hiernaux, and S. Fernández-Rivera. 2000. Crop–livestock systems in sub-Sahara Africa: Determinants and intensification pathways. In *Property Rights, Risks, and Livestock Development in Africa,* ed. N. McCarthy, B. Swallow, M. Kirk, and P. Hazell, 132–149. Washington, DC: IFPRI, and Nairobi: ILRI.

Williams, T. O., B. Spycher, and I. Okike. 2003. *Final Report for Component 2: The Determination of Appropriate Economic Incentives and Policy Framework to Improve Livestock and Intra-regional Trade.* CFC Project CFC/FIGM/06. Nairobi: ILRI.

Williams, T. O., I. Okike, I. Baltenweck, and C. Delgado. 2004. *Marketing Livestock in West Africa: Opportunities and Constraints. ILRI/CFC.* West Africa Livestock Marketing: Brief 1. N.P.: ILRI/CFC/CILSS.

Wint, W., J. Slingenberg, and D. Rogers. 1999. *Agro-ecological Zones, Farming Systems and Land Pressure in Africa and Asia.* Report prepared for FAO by the Environmental Research Group. FAO Animal Production and Health Division. Rome: FAO.

4

India

Growth, Efficiency Gains, and Social Concerns

C. T. Chacko, Gopikrishna, Padmakumar, Sheilendra Tiwari, and Vidya Ramesh

Abstract

Until quite recently, the environmental, social, and health impacts of livestock production in India have generally been considered to have more positive implications than negative ones because the production system is still numerically dominated by rural-based, integrated, smallholder crop–livestock mixed farming systems.

However, there is some justification for environmental, social, and health concerns about ongoing changes in livestock production. In the context of rapidly increasing demand for livestock products, large-scale industrial production units are likely to grow in number in the future, displacing, at least partially, the smallholder production system. Similarly there is a risk of overuse of natural resources leading to environmental damage.

This chapter examines livestock production trends in India and the factors driving them over the last two and a half decades; the beneficial and adverse environmental, social, and human health consequences of production changes; and the public and private responses and mechanisms to address the consequences. Five case studies highlight specific situations resulting from changing livestock-related systems in different parts of the country, which have both positive and negative impacts on the environment.

Introduction

India is the world's largest democracy, embracing countless cultures, languages, and religions. The population exceeds one billion. With a GDP annual growth rate of 8% (driven mainly by industrial growth of 9% and service sector growth of 9.8%) and inflation around 5%, India ranks today as the world's fourth largest economy.

Agriculture is the mainstay of the Indian economy. Agriculture and allied sectors contribute nearly 18% of the gross domestic product (GDP), whereas about 65 to 70% of the population is dependent on agriculture for their livelihood.

The livestock sector plays an important role in the Indian economy. Estimates (Government of India 2006), show that the GDP from the livestock sector at current prices was about Rs. 935 billion in 1999–2000 (about 22.5% of agriculture and allied GDP). This rose to Rs. 1239 billion in 2004–05—24.7% share in agriculture and allied GDP. The sector also plays an important role in providing nutritive food (rich in animal protein) and generating employment in the rural sector, particularly among the landless, small, marginal farmers and women. Because the distribution of livestock wealth in India is more egalitarian than that of land, from the equity and livelihood perspectives, livestock are considered an important component in poverty alleviation programs (Government of India 2006).

Agriculture, including livestock, is handled mainly by the state governments (provinces). The Government of India deals in livestock issues on a policy level and controls the import and export of livestock and their products. Though the National Agricultural Policy (Government of India 2000) targets a 4% annual growth rate of agriculture by 2020, currently the sector seems trapped in a low growth regime of below 2% per annum. However, the contribution of livestock to the agriculture sector GDP has been steadily increasing, mainly thanks to the dairy and poultry sectors. Because the demand for livestock products is on an increasing trend, driven by sustained economic growth, rising incomes, and urbanization, it is likely that increasing numbers of organized, larger, industrial livestock production units will sooner or later emerge to meet the growing demand. Smallholders own about three fourths of the country's livestock wealth and predominantly follow mixed crop–livestock farming systems. Hence pro-poor policies need to be put in place, to support small stock keepers, of whom 22% are landless and 63% have farms of less than 2 ha.

Drivers of Changes and Trends in the Livestock Sector in India

The Drivers of Change

With more than a billion people, India's population is still growing at the rate of 1.6% per year, currently adding 17 million people per year (UN 2008). The UN medium projection expects India's population growth to continue until at least 2050, before stabilizing above 1.6 billion, by which time India will have overtaken China as the world's most populous country. The UN's medium projection estimates that there will be 1447 million people in India by 2025, and 1658 million by 2050.

India is urbanizing at a rapid rate of 2.5% per year. The number of cities over one million is expected to double from 35 in 2001 to 70 by 2025. Between 1981 and 2001 India's urban population grew at an annual rate of 3% compared to 1.7% for the rural population (Jones Lang LaSalle 2005).

According to the World Bank, India's per capita income is growing fast and is estimated to be over $800 in 2006. With a booming economy, real annual personal disposable incomes are set to increase by 8 to 10% per year over the period 2006–10, providing a significant boost for the demand for lifestyle products and services (Jones Lang LaSalle 2005). Median household incomes are expected to grow from $2250 in 2005 to $3600 by 2010. A large middle class has emerged, currently estimated at 120 million. India's National Council of Applied Economic Research expects a further 180 million to join the middle class category by 2010.

Increasing population, urbanization, and sustained income growth are causing significant changes in food consumption patterns in India. Kumar and Birthal (2004) report that between 1977 and 1999 the per capita cereal consumption declined by 20%, whereas there was a significant increase in the consumption of fruits (up 553%), vegetables (up 167%), milk products (up 105%), and meat, eggs, and fish (up 85%). The demand for animal food products is more income-elastic than for staples: low income groups spend more on such high-value foods as their incomes rise.

The consumption of milk and meat during the last few years shows an impressive growth of 2.3% and 1.3%, respectively (Table 4.1).

Although small ruminant meat consumption has not changed much (1.1 kg to 1 kg), beef and buffalo meat have increased by 50% (0.6 kg to 0.9 kg) and poultry meat by 133% (0.3 kg to 0.7 kg) during the same period.

Interestingly, the consumption of milk and eggs in rural areas increased faster than in urban areas, and the difference between the urban and rural consumption of animal products is narrowing over the years. Birthal and Taneja (2006) report that "the income elasticity of demand for animal food products for the very poor households is 0.70 and that for the very rich is 0.39. This implies that the demand for animal food products will grow faster if there is a rapid increase in the purchasing power of poor people. Between 1983 to 2000, retail prices of meats and eggs (except mutton and goat meat) declined at between 0.2 and 3.6 percent per year. With real prices going down, growth in demand is expected to accelerate." Government trade liberalization policy also helped to drive fast growth in India's livestock sector. As part of the economic reforms in 1991, the Government of India introduced a number of trade reforms such as reduction in tariffs, removal of quantitative restrictions, and demonopolization of imports and exports. Import tariffs were reduced significantly. The government of India also took policy initiatives to boost exports of livestock products, especially buffalo meat. The minimum export price for meat was abolished in 1993, whereas exports of milk, cream, and butter were liberalized. Export-oriented units and companies in export processing zones are allowed duty-free import of goods for manufacturing and processing. They also enjoy tax holidays and other benefits such as concessional rents, lower sales tax, excise duty, corporate taxes, etc."

All these factors suggest that the demand for livestock products will keep on increasing in the years to come. Projections to 2020 indicate that demand for milk is expected to double compared to 2000, to a range of 132 to 140 million tonnes, whereas demand for meat would triple to 8 to 9 million tonnes (Parthasarathy Rao et al., 2004). The current changes in the sector such as production trends, population dynamics, species shifts, and so forth, already signal a booming livestock sector future for India.

Table 4.1. Per capita consumption of livestock products (grammes/day/person)

	1996	1997	1998	1999	2000	2001	2002	2003	2004	Average Growth %/yr
Milk	193	200	206	214	218	225	228	234	233	2.3
Meat	14	13	14	14	14	15	15	15	15	1.3

Source: FAOSTAT 2006.

Current Changes in the Livestock Sector

Trends in Livestock Production and Trade

Between 1980–81 and 2003–04, livestock production increased at an annual rate of 4.3%, much faster than the agricultural sector as a whole (2.8%). Notable growth occurred in the dairy and poultry sectors (Table 4.2). In the case of poultry meat production, the increase was tenfold.

Milk production increased from 44 million tonnes in 1985 to 91.9 million tonnes in 2005. This sustained growth, due to technological change and improved market access, has made the country self-sufficient in milk. Milk markets are still largely informal. Dairy cooperatives make up an important segment of organized milk markets and their number has expanded considerably since 1970.

Meat production increased from 2.67 million tonnes in 1985 to 5.66 million tonnes in 2005. Birthal and Taneja (2006) report that "in the early 1980s small ruminants were the major sources of meat, followed by large ruminants and poultry. The meat production structure underwent a drastic shift in recent years, with poultry emerging as one of the major contributors."

Changes in Livestock Population

As of 2003, India's livestock population was made up of 185 million cattle, 98 million buffaloes, 61.5 million sheep, 124.4 million goats, 13.5 million pigs, and 489 million poultry (Table 4.3). Cattle numbers actually declined over the last 10 years, and the growth of the buffalo, goat, and swine population decelerated during the same period vis-à-vis the previous decade. Sheep showed a faster annual growth rate compared to the previous decade, whereas poultry grew almost 60%.

Species Shift

Table 4.3 shows that monogastrics, mainly poultry, are gaining importance. Between 1992 and 2003 the poultry population increased by 59%, whereas pig and ruminant populations showed only marginal increases (except for cattle, which showed a decline).

Table 4.2. India's livestock production, 1985–2005 (million tonnes)

Year	1985	1990	1995	2000	2005
Milk	44.02	53.68	65.25	80.83	91.94
Beef and buffalo meat	1.95	2.40	2.72	2.86	2.98
Sheep and goat meat	0.53	0.61	0.66	0.70	0.71
Poultry meat	0.19	0.37	0.62	1.14	1.97

Source: FAOSTAT 2006.

Table 4.3. Livestock population 1992–2003 (million)

	1992	2003	Change (1992–2003)
Cattle	204.58	185.2	–09.47%
Buffalo	84.21	97.9	16.26%
Sheep	50.78	61.47	21.05%
Goat	115.28	124.36	07.88%
Total Ruminants	454.85	468.93	3.0%
Pigs	12.79	13.52	05.71%
Poultry	307.07	489.01	59.25%

Source: Government of India 2005.

Poultry is one of the fastest growing segments of the agricultural sector in India today. While the production of crops has been rising at an annual rate of 1.5–2%, the production of eggs and broilers has been rising at a rate of 8–10% (Mehta et al., 2002). The growth of the poultry sector in India has also been marked by an increase in the size of the poultry farm. For example, in earlier years broiler farms used to produce a few hundred birds per cycle on an average. Now, units with less than 5000 birds are becoming rare, and units with 5000 to 50,000 birds per cycle are common.

Geographical Shift

Livestock production is largely a rural activity. About 95% of ruminants, 84% of pigs, and 92% of poultry are still raised in rural areas (Table 4.4). No significant geographical shift has been noticed from rural to urban areas during the last 10 years, though proportionate increases in population have been observed in both urban and rural areas (except for cattle). Urban livestock production is small, but specialized dairy and poultry enterprises may emerge in the future in response to rising demand for animal foods by urban populations.

Although there was a reduction in the rural cattle population by about 20 million, the urban cattle population, though numerically small, showed a marginal increase. The increase in the buffalo population was noticed in both urban and rural areas, with the ratio between rural and urban populations tilting slightly in favor of the urban population. A noticeable shift to urban can also be seen in the pig population.

Changes in Draft Animal Populations

Male zebu cattle and buffalo have been the main draft animals in India. There has been a shift away from draft animals, facilitated by rising mechanization of agriculture. The population of draft animals declined from 80.8 million in 1971–72 to about 62.2 million in 2003

Table 4.4. Livestock population trends (urban and rural) in million

Species	1992		2003		Change	Change
	Rural	Urban	Rural	Urban	Rural	Urban
Cattle	195.88 (96%)	8.69 (4%)	175.65 (95%)	9.53 (5%)	–20.23 million	+0.84 million
Buffalo	79.92 (95%)	4.29 (5%)	91.93 (94%)	5.99 (6%)	+12.01 million	+1.70 million
Sheep	48.86 (96%)	1.91 (4%)	57.99 (94%)	3.48 (6%)	+9.13 million	+1.57 million
Goat	109.36 (95%)	5.92 (5%)	117.48 (94%)	6.88 (6%)	+8.12 million	+0.96 million
Pigs	11.25 (88%)	1.54 (12%)	11.41 (84%)	2.10 (16%)	-0.16 million	+0.56 million
Poultry	282.67 (92%)	24.40 (8%)	449.14 (92%)	39.87 (8%)	+166.47 million	+15.47 million

Source: Government of India 2004.

(Government of India 2005). During the same period the number of tractors increased rapidly from 150,000 to 1,820,000 (Birthal and Parthasarathy 2002). The use of male buffaloes for meat purposes has been increasing over the years.

Shifts in Producer Categories

Table 4.5 shows that changes in the scale of livestock holding by different classes of land holding vary widely. On large farms, the average number of cattle rose by 58% between 1991–92 and 2002–03, and on medium farms by 17%, whereas on small farms there was a decline in the range of 6 to 20%. For the landless, the size of cattle holding increased marginally. The average buffalo holding increased in all landholding categories except small and medium range. The average number of small ruminants declined by just over half in the landless households, remained almost stable on small and medium farms, and increased by 25% on large, 13% on marginal, and 9% on submarginal farms. The scale of pig production increased on submarginal and marginal farms. Elsewhere it declined in the range of 12 to 48%, the maximum being in the landless households. For poultry there was a decline of 43% in the landless and 24% in the medium land class,

Table 4.5. Average number of animals per 100 households in India

	Landless (<0.002ha)	Sub-marginal (0.002–0.5 ha)	Marginal (0.5–1.0 ha)	Small (1.0–2.0 ha)	Medium (2.0–4.0 ha)	Large (>4.0 ha)	All
Cattle							
1991–92	196	281	335	340	306	274	305
2002–03	200	226	293	318	357	433	295
Buffalo							
1991–92	151	190	211	259	287	352	246
2002–03	153	197	225	256	286	366	245
Small ruminants							
1991–92	335	339	378	427	513	800	419
2002–03	153	371	428	443	523	998	433
Pigs							
1991–92	337	266	262	267	298	486	285
2002–03	177	319	283	233	261	311	304
Poultry							
1991–92	641	701	783	816	1138	1029	790
2002–03	366	794	876	1025	867	3311	888

Source: Birthal, Jha, and Joseph 2006 (based on Government of India data).

whereas for others there was an increase, the maximum being for the large landholders.

Changes in Feed/Grazing Resources
Estimates by the Planning Commission's working group on animal husbandry and dairying show that for 2002–03 there was a deficit (based on the difference between optimum animal feed requirements and actual consumption) of 157 million tons of green fodder, 44 million tons of dry fodder, and 25 million tons of concentrates. This shortage is expected to persist and even worsen in the future because of threats to some of the major fodder sources. On average about 35% of livestock-keeping households use common land for grazing and about 23% for fodder collection (Government of India 1999). The area under permanent pastures and grazing lands represents a mere 3.3% of the total area and has been declining steadily, down from 12 million ha in 1981–82 to 10.5 million ha in 2001–02 (FAI 1982, 2002).

Changes in the Production System
A large proportion of cattle and buffaloes in India are either nondescript or belong to draught breeds that have a poor milk production potential. This has been a major constraint in raising livestock productivity, along with feed and fodder scarcity. Scientific genetic selection has yet to be put in place for many parts of the country and for most of the species. Exceptions are certain states like Kerala for cattle, and certain species like poultry in the organized private sector. Despite the aforementioned constraints, it is encouraging to note that there was increased production and reproduction per animal (Table 4.6).

Economic reforms have paved the way for increased participation by the private sector in the livestock products market. The markets are now transforming from open to vertically coordinated structures like cooperatives, producers' associations, and contract farming. The private sector has been increasingly relying on contracts to source a sustained supply of raw material. Much of the poultry in major producing states is now produced under contract farming, which provides an assured market and returns to the producers.

In India mixed rainfed farming systems are practiced on 46% of agricultural land and mixed irrigated systems on 37%. In these systems, cattle or buffalo are the second or third largest economic activity (Parthasarathy Rao et al., 2004). However, there is increasing pressure on livestock to produce more to meet the growing food demand (Birthal and Taneja 2006). As a result mixed farming systems are undergoing a steady transformation, and the integration of crop and livestock production is likely to weaken, giving way to commercial production systems based on high-producing animals and externally purchased inputs. For instance, the bulk of poultry production in India has been transformed from a backyard activity to a commercial activity.

Contribution of Livestock to National Income
The livestock sector in India has grown impressively during the last 20 years and now contributes about 24.7% of the agricultural GDP in 2004–05—up from about 22.5% in 1999–2000 (Birthal et al., 2003). The contribution of agriculture as a whole to national GDP fell from 23.2% to 17.6% during the same period.

It is expected that India's livestock sector will emerge in the immediate future as an engine of growth of the agricultural economy, driven mainly by urbanization, increased purchasing power, and changing consumption patterns. However, it will be essential to mitigate the possible adverse social and environmental impacts of this economic upsurge, which we will examine in the following section.

Consequences of Changes in the Livestock Sector in India

Social Consequences
From the perspective of the poor, small animals like sheep, goats, pigs, and (backyard) poultry are considered important because of their low initial investment, zero/low input requirements, and quick returns to investment on a continuous basis (Birthal et al., 2006). Because the majority of livestock wealth is concentrated among marginal and small landholders in India, in theory it might be expected that growth in the livestock sector would

Table 4.6. Milk and egg yield and proportion of producing livestock

	Milk/egg yield (kg/animal/year)				Percentage milked/laying			
	1990	1995	2000	2005	1990	1995	2000	2005
Cattle	732	806	944	1000	15.0	18.6	19.9	21.2
Buffalo	1122	1294	1423	1450	32.1	33.2	33.5	34.9
Chicken	10.1	11.7	11.6	11.6	39.1	40.3	41.6	42.7

Source: FAOSTAT 2006.

bring prosperity to the smallholders and assist in poverty reduction.

However, trends in India's livestock sector show that sector growth does not go hand in hand with poverty reduction. Except in the case of pigs, the most rapid growth in livestock numbers has been among larger landholders (4 ha or more) and the really rapid growth is mainly seen among big industrial poultry production units and large cattle farms. The landless poor are becoming increasingly marginalized with respect to small ruminants, pigs, and poultry: the numbers of stock they own, and their share of the total stock, is declining dramatically (Table 4.5). There is an increasing exodus of landless households out of livestock production, mainly because of reduced access to grazing resources, and lack of access to nonexploitative markets, credit, and services.

Given the rapid increase in demand for quality meat and milk products, it seems likely that smallholder livestock producers may be displaced by large industrial producers, who are better able to invest in food quality and safety and to sell the products through well organized outlets such as supermarket chains.

Environmental Consequences

Increasing Grazing Pressure in Arid, Semiarid Dry Lands

Grazing intensity in India is already very high. In rainfed areas, the present stocking rate is 1 to 5 adult cattle units (ACUs) per hectare—compared with the rate of 1 ACU/ha allowed by government norms. In arid zones overstocking is even more pronounced: the actual stocking rates are 1 to 4 ACU/ha as against norms of 0.2 to 0.4 ACU/ha (Shankar and Gupta 1992). It is estimated that about 100 million adult cattle units graze in forests with a sustainable capacity of only 31 million.

More than 80% of resource-poor households depend on common property resources for the fodder requirements of their livestock. Several studies (Jodha 1992, FAI 2002) show that there has been a decline in the area under permanent pastures and grazing land from 1950 onward because of privatization, encroachment, government distribution of land to the poor, increase in built-up areas, and expansion of national parks and wildlife sanctuaries.

During the period from 1997 to 2003, overall livestock numbers have been static at around 485 million adult cow units. Within this total, there has been a shift from large ruminants to small ruminants. The large ruminant population reduced from 289 million in 1992 to 283 million in 2003, whereas the small ruminant population increased from 166 million to 186 million during the same period.

The increase in small ruminant populations, especially sheep (which are exclusive grazers) is causing alarm. It is further aggravated by a steady decline in common grazing areas, leading to increasing levels of overstocking. The quality and productivity of grazing lands are also showing a declining trend due to improper management, unregulated land use, overgrazing, and lack of reseeding of pastures.

The pastoral system is putting more pressure on the limited land available to it. It is argued that one of the reasons for deforestation is uncontrolled grazing of livestock in forestland. Further, the food function of livestock is nowadays becoming more important than draft and manure, so soils are receiving less organic matter fertilization.

All these factors contribute to land degradation, particularly in open grazing areas in the arid and semiarid ecosystems. The Livestock, Environment and Development (LEAD) study conducted in five semiarid watersheds in India revealed that common grazing lands in most of the villages studied are under various stages of degradation (CALPI-IWMI 2005). In most villages, grazing lands are used as open access resources without any control on the intensity of use. Moreover, insecurity of user rights deters villagers from investment in biomass development. Encouragingly, the study also indicated that when management systems are in place, land quality is improved even in places with high aridity.

Involution of Mixed Farming in High Input Intensive Areas

Integrated crop–livestock mixed farming is generally considered as a sustainable system because it promotes enhancement of soil and vegetation resources, as well as reducing resource losses, pollution, and degradation.

However, current trends indicate that in some parts of India like the Indo-Gangetic river basin, where high-input farming is practiced, livestock are not properly integrated with crops such as paddy, mainly due to replacement of draft animals by mechanized harvesting. In the western part of the Indo-Gangetic region large amounts of straw are left in the fields and need to be removed for agronomic or management reasons. Farmers normally burn the straw in the field as an easy solution (Parthasarathy Rao 2003). Over 70% of rice straw and 50% of wheat straw produced in the region is burned. This results in loss of valuable organic carbon necessary to maintain soil health and also increases greenhouse gases and air pollution.

The decline in livestock–crop integration also leads to a decline in recycling of farmyard manure. This necessitates increased use of inorganic fertilizers, in soils already overdosed with chemical fertilizers. This reduces soil quality, soil health, water holding capacity, and infiltration.

Another important consequence of the shift from draft animals to meat and reduced crop–livestock integration is the impact on water use efficiency. A much-discussed study to estimate the water productivity of dairy animals in Gujarat (Singh et al., 2004) found that between 1900 and 4600 liters of water were used to produce one liter of milk. Milk and meat production requires 10 to 50 times more water than crop production, particularly if based on intensive grain feeds and irrigated forages (Onyekakeyah 2006). Thus, for efficient use of water, especially in water deficient areas in India, it is essential to promote the mixed crop–livestock farming system through appropriate policies and incentive mechanisms.

Industrial Poultry and Dairy Production Units

Poultry is one of the fastest growing segments of the agricultural sector in India. Whereas the population of other livestock species showed only slight changes between 1992 and 2003, the poultry population has shown a massive increase of 59% (Table 4.3).

Fast growth in the commercial poultry sector has serious environmental, social, and health implications. The main feed ingredients for poultry production are grains. Maize constitutes 50 to 55% of broiler feed. Increasing demand for feed grains will create greater pressure on land to cultivate them, which will increase competition with grain production for human consumption. Currently India produces only 11 million tonnes of maize, of which 5 million tonnes are used in the poultry sector.

In terms of environmental impacts, the grain-based intensive system, though efficient in terms of output per unit of input, is less efficient in terms of energy and greenhouse gas emissions (IFPRI-FAO 2002). Large amounts of fossil fuels are used to produce meat and eggs under the intensive system.

In the poultry industry, almost 85% of the feed nitrogen is unutilized and excreted. Air pollution results as the nitrogen in poultry manure is converted to ammonia. Soil toxicity occurs as nitrogen and phosphorous from manure build up in the soil.

There are also human environmental health concerns. Rampant use of antibiotics leads to antibiotic resistance among bacteria. Consumers' preference for live and fresh chicken forces retailers to slaughter birds in their shops—in many cases in a very unhygienic manner.

A study conducted by Mehta et al. (2002) shows that there are biosecurity issues associated with industrial poultry production in India. They include air and water pollution, soil toxicity, waste disposal, and health hazards, especially when production units are located too close to densely populated areas. It is reported that 100 chickens produce about 54 kg of nitrogen and 38 kg of phosphorous per year. Water pollution may occur if nutrients from manure enter the water course, especially when there is rain. Farms close to population centers and watercourses create human health problems and ecological damage due to overconcentration of nutrients. The same thing will happen in rural industrial units if wastes are not properly managed.

Industrial dairy and piggery production units cause similar threats to the environment. Here the issues are all the more serious because manure is produced in liquid form and can more easily enter water bodies unless strict precautionary measures are taken. The case study (discussed later) on periurban dairy colonies in Mumbai (Maharashtra) provides a vivid account of the gravity of the issue.

Greenhouse Gas Production

The important greenhouse gases associated with livestock are methane (CH_4), nitrous oxide (N_2O), and carbon dioxide (CO_2). Methane and other gases are produced from enteric fermentation in ruminants as well as from dung. Released into the environment, they join methane produced from other sources such as rice fields, coal burning, biomass burning, transport, solid waste treatment, coal beds, mines, and so forth. N_2O production in the livestock industry is mostly from manure.

India has the highest density of cattle and buffaloes, as well as of small ruminants reared under the extensive system—large numbers of small herds dispersed over a vast area. Livestock is fed poorly under this type of rearing—with inadequate rations based on less digestible crop by-products and grazing on poor quality rangelands. These conditions are most conducive for release of high levels of methane from enteric fermentation into the atmosphere.

India's initial national communication to UNFCC (NATCOM, 2004) indicates that the methane contribution by livestock in India toward global warming is significant. A cow emits around 100 kg of methane every year. Over a 100-year period, methane gas is 21 times more aggressive than CO_2 in contributing to climate change. NATCOM has estimated that in 1994 around 300 million bovines plus 180 million small ruminants produced around 10 million tonnes of methane in India, which is 15% of global methane production from livestock. Because CH_4 loss by livestock means some 8 to 10% loss of energy to the animal, any steps taken to reduce enteric methane emission will also improve animal condition and production.

Thus livestock production makes a significant contribution to climate change. The potential effects of climate change on agriculture are as yet uncertain. At the regional level, changes in precipitation and temperature patterns could jeopardize current agricultural practices. The frequency of extreme weather phenomena like floods, droughts, severe storms, and so forth, may threaten the livelihood security of small and marginal farmers, particularly in the rainfed regions (NATCOM 2004).

Health Consequences

Another issue is the possibility of epidemic zoonoses emerging from livestock production units. This was dramatically illustrated when the outbreak of avian influenza in India during 2005 triggered widespread concern and fear. The disease was detected at several commercial farms in the western state of Maharashtra. Large numbers of poultry deaths were noted at more than 50 farms in the area. The disease can be transmitted to humans by direct or indirect contact with infected poultry and wild ducks. Timely steps like isolation of the area, mass killing of birds in affected farms, and a ban on importing of live chickens and other poultry products from countries affected with avian flu were taken by state and central governments.

In general, inadequate slaughtering facilities and techniques cause meat losses on a wide scale, and also result in food poisoning by bacterial toxins. The economic value and marketability of pastoral products are often reduced due to hygienic problems.

Responses

The global food market is undergoing major changes, especially in the developing world. Driven by increasing income levels of large numbers of city dwellers, the per capita consumption of food of animal origin has increased dramatically. As the economy grows and drives social changes, the relative importance of livestock for nonfood functions like draft power, status symbols, insurance against income shocks, and so forth, is growing less and less important, and the food functions are strengthening.

With the rapid growth of milk and poultry production in India between the 1980s and 2004, the critical question for economic managers and planners is no longer whether there is a livestock revolution, but to what extent poor people and smallholders can play a significant part in this enterprise. There is a risk that the livestock revolution, similar to the green revolution, will deepen the inequality between rich and poor. Decisive action needs to be taken to ensure that the poor benefit from such developments (Khan and Bidabadi 2004).

It is anticipated that world demand for production of milk, meat, and poultry products could double by 2020 and that the bulk of production would shift from temperate to humid and warm regions of the developing world (Delgado et al., 1999). There could be three scenarios in this change:

- Demand will be met by large-scale industrialized units.
- Small-scale producers will develop livestock production that can satisfy the demand.
- A harmonized combination of the first two scenarios could develop.

Private Sector

With the reduction in European subsidies under World Trade Organization (WTO) agreements, India's exports of dairy products are likely to expand because of its price competitiveness. The private sector, which is already handling more than 75% of poultry production in the country, anticipates growth in the dairy, poultry, and meat industries. Although large-scale livestock production units are in a position to cope with the increasing demand, they can be a threat to the environment if not properly regulated.

The private sector is playing a proactive role in the marketing of livestock products. It has a vital role in strengthening forward linkages and value-addition, particularly in areas that have remained neglected. There are however, some constraints that hinder their entry in these marginal areas. The much-needed interface between public and private sectors is sadly missing. Investment in the livestock sector is mainly in production systems and processing. Good examples include Nestlé, Cadbury's, and Dumex in dairying/dairy products, Venketeswara hatcheries, and Shanti group in the poultry industry, and Alkabeer in the meat industry. Large farms are mostly located near large urban centers and have relatively better awareness of the environmental issues of livestock production. However, the system of contract-rearing of poultry, and milk collection by private dairies from large numbers of small farms, have the advantage of shifting the major production units to the villages (cf. chapter on Nestlé in this volume). Such systems allow better use of wastes.

The private sector and the dairy cooperative federations dominate the livestock feed manufacturing industry of the country. More than 80% of compound feed for livestock, including poultry, is manufactured by these sectors. Since the late 1990s the private sector has also provided breeding services for large ruminants. The Gopal Mitras of Andhra Pradesh, the Para-vets of Uttar Pradesh, and the Bharatiya Agro Industries Foundation (an NGO) are providing breeding and health care for cattle and buffaloes on a large scale (Ahuja et al., 2008).

Government Policies

Livestock Related

NATIONAL AGRICULTURAL POLICY

Agriculture is accorded a high priority in central and state government policies. The National Policy on Agriculture seeks to actualize the vast untapped growth potential of Indian agriculture; strengthen rural infrastructure to support faster agricultural development; promote value addition; accelerate the growth of agribusiness; create employment in rural areas; secure a fair standard of living for farmers; discourage migration to urban areas; and face the challenges of economic liberalization and globalization. The policy emphasizes the watershed approach to managing land resources, which helps to develop rainfed agriculture while protecting the inhabitants of fragile ecosystems through technology, credit, market and roads, and a remunerative price environment.

NATIONAL LIVESTOCK POLICY

India's National Livestock Policy aims to improve the quality of livestock and livestock products. The policy changes identified for the livestock sector in the twenty-first century include the following:

- Improvement of livestock breeds through genetic upgrading
- Eradication of livestock diseases
- Constitution of the Indian Council of Veterinary Research
- Intensification of fodder development on wastelands and degraded lands
- Development of poultry, small ruminants, and swine
- Preservation of endangered indigenous livestock breeds
- Production-linked livestock insurance.

NATIONAL PROJECT FOR CATTLE AND BUFFALO BREEDING

Operated by the Government of India, the National Project for Cattle and Buffalo Breeding has the major objectives of supporting conservation of genetic diversity among cattle and buffaloes, providing quality artificial insemination (AI) services on payment, and increasing the coverage of AI from the present 12% to 40% in the next 10 years. Many states have not yet succeeded in implementing the breeding policy, in involving farmers in drawing up programs and policies, or in developing a long-term plan to increase the coverage of AI.

NATIONAL DAIRY DEVELOPMENT BOARD

India dairy cooperative movement is the biggest in the world. The contribution of the cooperatives to handling milk produced by millions of small farmers is a positive example of dairy production with few environmental hazards and with the involvement of smallholders. The National Dairy Development Board establishes and supports dairy cooperatives throughout the country. There are more than 10 million farmers in dairy cooperatives in more than 80,000 villages. The cooperatives handle 16% of India's marketable milk surplus and reach out to 15% of the milk animal households (Indian Dairyman 2006).

Environmental Policy

The Government of India has recognized reduction of land degradation and improved natural resource management as key priorities in a number of key policies, strategies, and action plans.[1] India thus has a large number of environmental acts and regulations. However, opinions differ on the effectiveness of implementation. Pollution limits for various industries have been prescribed in the Environmental Protection Rules. Environmental clearance from the Union Ministry of Environment and Forests is mandatory for setting up new industries in many sectors. All major industry associations have a climate change division and have taken initiatives to conduct training and generate awareness in key areas, such as energy efficiency and other environmentally friendly projects.

The central government has undertaken a broad spectrum of initiatives on climate and related issues, such as the diffusion of renewable energy technologies, joint forest management, water resource management, petroleum conservation research and consumer awareness, energy parks for demonstration of clean energy technologies, and so forth.

Increasing demographic pressures and the transition of livestock production from subsistence to market driven have been causing accelerated degradation. The absence of appropriate pollution control measures, such as manure management, waste disposal, and so forth, render the environment more vulnerable to irreparable damage. A clear understanding of the interactions between livestock and environment is a prerequisite in designing programs and projects to mitigate negative interactions and enhance positive ones, so that the livelihoods of livestock keepers can be substantially improved.

An example can be seen in the new Scheduled Tribes (Recognition of Forest Rights) Bill, which gives rights to hold and live in forestland to those who have been living in the forests for the last three generations (forest dwellers). Forest-dependent communities living near and around forests, and scheduled caste pastoralists are authorized to use forestland. The government has also regulated encroachments that occurred before 13 December 2005. Similarly the land ceiling limit has been removed (now there is no limit for the land to be allotted). Though this has been appreciated by many people, there are also critics who suggest that this amounts to privatization, turning common property resources into a commodity.

---

1. Including: the National Water Policy (1987); National Land Use Policy Outline (1988); National Forest Policy (1988); National Agricultural Policy (2000); and National Policy for Common Property Resource Lands. Some of the major environmental acts and rules in India are the Water (Prevention and Control of Pollution) Act (1977); the Air (Prevention and Control of Pollution) Act (1981); the Environment (Protection) Act (1986); the Hazardous Wastes (Management and Handling) Rules (1989); the Public Liability Insurance Act (1991); the Environmental (Protection) Rules—"Standards" (1993); and the National Environment Tribunal Act (1995).

Watershed Development

The increasing pressure of human and livestock populations on natural resources in the semiarid zones of India has impacted agroecosystems. Many efforts are being made to reverse these trends and to promote sustainable natural and land management practices. The largest effort made by the Government of India in addressing this issue is through the implementation of the Watershed Development Program. However, issues have been raised over the sustainability of the interventions carried out so far. The Government is now searching for effective project completion protocols to ensure that the outcomes can be sustained.

FAO's LEAD research on livestock–environment interactions in the watershed context created greater awareness among policy makers, planners, implementers, and researchers about the importance of sustainable land and water management in livestock development projects. It could also influence the Government of India to reformulate watershed guidelines, with an emphasis on livestock integration and common land management in watershed development programs.

Because of the impact created by the LEAD study, different organizations with no direct livestock or environment linkage started discussing and integrating "livestock–environment" themes in their agenda (e.g., IWMI, ICRISAT, TERI). The Government of India, for the first time, included "Livestock and Environment" as one of the working group themes in the proposal for the 11th Five Year Plan.

Case Studies

The case studies focus on five distinctive livestock-linked systems, located in different parts of the country (Figure 4.1), that have environmental, social, and health implications in the context of change. They include the following:

1. Periurban dairy colonies in Mumbai (Maharashtra)
2. Periurban industrial poultry production in Chhattisgarh
3. Slaughterhouses
   a. Organized slaughterhouses in Bangalore city
   b. Village slaughterhouses in Kerala
4. Grazing land in Kalyanpur watershed (Rajasthan)
5. Tanneries in Kanpur (Uttar Pradesh).

It is to be appreciated that these cases are location-specific and hence cannot be generalizable for India as a whole.

Case Study: Periurban Dairy Colonies in Mumbai (Maharashtra, India)

Background

To meet the liquid milk requirement of urban populations, big dairy colonies of milking buffaloes have operated around the large urban areas. Though their numbers have been decreasing over the years, in 2006 there are still some 1000 such colonies keeping around 100,000 milking buffaloes in and around Mumbai, the second biggest city of India.

The authors visited two colonies in the busy area of Mumbai, one keeping around 350 milking buffaloes in an area of less than 1200 m^2, and the other 25 milking buffaloes in an area of 50 m^2. Only milking animals are kept here, whereas unproductive animals are either sold out or sent for contract rearing.

The sheds are semitemporary structures. The floor is made of concrete and is kept dry and clean. There is a narrow and shallow drain at the side for draining out urine, dung, and shed washings. The drain comes out as an open channel leading to the bigger open sewage canal. Waste management varies according to season. During the summer months the dung is stored within the farm or on adjacent land belonging to another person until it is sold to villagers. During the rainy season the dung is pushed into the drains.

The animals are stall-kept throughout the year. Because there are difficulties in getting sufficient water, only half the required quantity of water is used per animal per day. Most animal diseases are treated by the supervisor of the dairy colony, and the use of antibiotics and hormones promoting milk flow seems high. Except for rare cases of mastitis and calcium deficiency, the animals are reported to be in good health. All animals are vaccinated routinely against foot and mouth disease. The animals are stall fed with paddy straw and very little green and concentrate mix based on yield potential.

Drivers

The relevance of large dairy colonies in big urban centers has been decreasing in recent times. However, a small group of traditional urbanites still want to get fresh buffalo milk for their domestic use. Since the land was obtained on long-term lease at nominal cost, land costs are low, which allows the dairy owners to continue the dairies despite many threats and problems.

Consequences

There are serious constraints for the viability and effective functioning of these dairies. They include competition from pasteurized milk in sachets; mounting pressure from the Mumbai city authorities; resistance from the public to smells, flies, and wastes; rising urban demand for new buildings, and resource shortages in terms of water, waste disposal facility and so forth.

Stationing a large number of milking animals in a small area in the middle of a big city causes environmental problems on a big scale. The land available is far too small to allow a satisfactory dairy farm management system. This adversely affects the waste management system, which in turn contributes to environmental pollution by

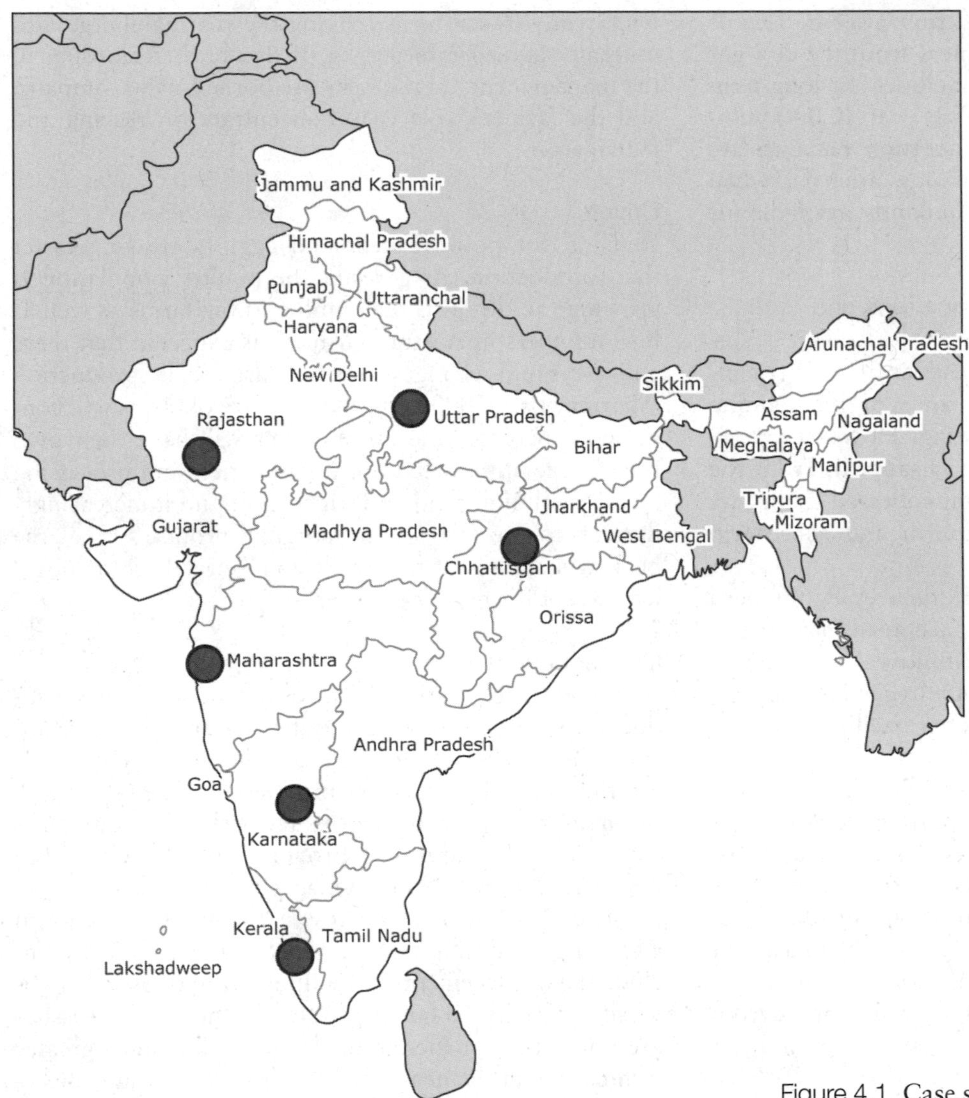

Figure 4.1. Case study locations.

way of soil nutrient overloading, water and air pollution, and problems associated with flies and mosquitoes. The animals also lack the required standing space. The owners report hoof problems and very high calf mortality on account of lack of exercise for the stock. Waste management and keeping the surroundings clean and hygienic is all the more difficult when water sources are limited and tank loads of water must be bought at exorbitant cost.

Neighboring residents often make big issues out of the problems created by the dairy colonies, and complain about increasing the risk of human diseases. The owners are willing to move dairy colonies to village areas if they are given sufficient land and required support. It was also reported that many dairy colonies were closed down during the last decade on account of pressure from society, government, municipality, builders, and so forth, as well the younger generation's lack of interest in continuing in this profession.

The working and residential facilities available for the employees are rather minimal. In the center of the animal sheds, a platform is made at a height of 2 meters from the ground and used as a temporary quarters for the laborers. However, the employees look healthy and contented even in such a setting.

Public and Private Responses

In 2005, the dairy colonies faced public interest litigation on the grounds that they caused a general nuisance from health hazards, traffic jams, burden on sewers and drains, eyesores, flies, noise, disease risk, and so forth. This resulted in a court order to implement environmental norms for dairies in urban areas. Mumbai's Pollution Control Board issued guidelines for prevention of pollution by urban dairy colonies. These guidelines include relocating dairies 1 km away from public residential areas, 1 km away from rivers and lakes, 15 meters away from existing wells, and 100 meters away from state and national highways. However, so far there have been no concrete actions to enforce the guidelines. A large public sector dairy colony, the Array Milk Colony, was restructured with private participation after it became nonviable. Around 100 dairy owners from the city got the shed and fodder cultivation facilities on long-term lease and relocated their dairies with over 40,000 milking animals. However, the basic question remains: are cities the right place to keep milking animals, when milk produced in rural areas is abundantly available for cities?

Options to Mitigate Negative Implications and Strengthen Positive Ones

With the expansion of the city and availability of alternate sources of milk the relevance of maintaining urban dairy colonies is fast declining. However, it will take many more years before they disappear. From the primary and secondary information collected it appears that there could be several long-term and short-term options.

Options for the short term include keeping the herd size optimum with regard to the space available; awareness creation and training of the employees and owners for proper waste management and hygienic practices; providing sufficient water; and better facilities for employees' accommodation.

Government authorities have already initiated action to relocate the dairy colonies to appropriate places for the benefit of the people involved as well as the animals. The restructuring of the Array Milk Colony with private participation is said to be a good move in this direction. Plans must also be developed to gradually phase out all the dairies or move them to appropriate locations, with adequate infrastructure and financial support from government.

Case Study: Periurban Industrial Poultry Production in Chhattisgarh

At present the poultry industry in the state of Chhattisgarh has emerged as the most dynamic and fastest expanding segment in the animal husbandry sector within the state. The state has a poultry population of 8 million, of which 75% is in the organized commercial poultry industry. The broiler poultry population has been growing at an annual rate of 13.3%, and layers at 11.1%. The state's annual production of chicken meat is 42 million tonnes, whereas consumption is 44 million tonnes, leaving a deficit of 2 million tonnes. Similarly there was a deficit of 176 million eggs in 2005–06.

The current case study focuses on the Shanti group of industries, where nearly 1 million birds are reared per cycle by 1200 contract farmers, and the manure is sold to nearby crop farmers. Besides supervision and technical advice, Shanti provide chicks mainly from their hatchery, feed from their factory, and vaccines and medicines to farmers. The broiler farmers' contribution comes in the form of a shed, equipment, litter material, water, electricity, labor, and management. The company takes back ready-to-sell birds, paying the rearing charge at a mutually agreed rate per kg of live bird. According to the management, all the risks are borne by the company, and the farmer's role is to concentrate on rearing and management.

Drivers

Demand for poultry meat and eggs is growing faster than production. As a result the poultry population is growing rapidly, as is the number of big farms as well as bird numbers in existing farms. It is expected that there will be rapid changes toward large-scale production (Sharma et al., 2003). The average size of farms will continue to increase, and larger farms will have their own breeding facilities, feed mills, hatcheries, and processing units. Small independent farmers will find it increasingly difficult to run farms with marginal profits, so that the backyard system of poultry keeping may become obsolete except for household use.

Consequences

At present this contract-farmed system of poultry production seems to have no significant negative externalities. The hatchery is located away from the city, birds are then reared by farmers in a decentralized way, and the manure is sold to nearby farmers. The company's contract farming approach provides a good livelihood to many small farmers in the villages.

The rapid wider changes in the poultry sector in Chhattisgarh do have some negative impacts. These include the displacement of smallholders and loss of biodiversity in backyard farming systems. Though these issues are not critical at the moment, they will assume greater significance in the near future unless appropriate corrective mechanisms are put in place.

Public and Private Responses

At the moment there seems to be no awareness among the stakeholders and the public about the environmental or health issues arising out of commercial poultry production in the state. Regulations are weak and not strictly enforced, making it difficult to check potential hazards in the future.

Options and Models to Mitigate Negative and Strengthen Positive Implications

Although the Chhattisgarh case study does not currently present any serious environmental or social problems, it may do so in the future, and many other locations already do present such problems. Hence we offer the following suggestions, which may be applicable to periurban poultry operations in general.

Relocating Commercial Production Units

Commercial units near cities and rivers may be relocated to rural areas with the help of policy regulations and incentive/disincentive systems. Zoning policies can be developed (as in China), where large scale production can be restricted to preidentified suitable areas.

Regulating the Industrial Production System
There should be strict regulations and technical support for pollution-neutralizing mechanisms. Regulatory mechanisms to control pollution can take a variety of forms. Negative environmental costs can be internalized into consumer prices. To achieve this, a wide variety of financial instruments can be used such as levies on waste discharges, taxes on excess animals or phosphate loads, and removal of subsidies favoring concentrate-based intensive production. Provision should be made for subsidies to encourage investment in emission control technologies, and import restrictions on materials and equipment that improve feed efficiency should be removed (FAO 2006).

Effluent charges may be imposed, based on the amount of pollutants discharged. Limits may be fixed on the number of birds per hectare. Technical options for manure management should be specified, including improving feed conversion rates with enzymes. Synthetic amino acids may be advised and installation of biogas may be made compulsory. Use of antibiotics should be regulated and checked at frequent intervals.

Managing Emerging Disease Outbreaks
In view of the risk of emerging diseases like avian influenza, there is a need to strengthen biosecurity (hygiene, cleaning, and movement of birds). There should be mechanisms for compensating farmers in the event of outbreaks and consequent culling.

Promoting Ecologically Friendly Production Systems
Incentives and policy support should be provided to promote environment-friendly production systems. The required funds can be generated by applying the "polluter pays" principle (at present the negative environmental externalities are imposed on society as a whole). If poultry meat is produced by industrial methods that cause damage to the environment, high taxes should be levied to discourage such methods or to put in place pollution-neutralizing mechanisms such as waste treatment plants, biogas digesters, and so forth. The money thus generated should be used to provide incentives to adopt ecofriendly production processes. Thus the negative environmental externalities can be internalized. However, this requires political will and sustained effort. The priorities will depend on the importance assigned to the environment compared to other objectives such as livelihoods or cheap supply of animal products.

Knowledge Sharing
Awareness creation and knowledge sharing on appropriate technologies and practices of industrial production and the pollution pathways should be given high importance.

Case Study: Slaughterhouses

Nationwide there are more than 3600 authorized slaughterhouses in the government sector. Most of them are operated and maintained by municipal bodies. The capacity of these units varies from 100 to 500 large animals and 25 to 800 small ruminants per day. A large number of these slaughterhouses maintain poor standards of hygiene and sanitation. In states like Kerala where there is no taboo for slaughtering cattle, almost all villages have slaughterhouses run by the *panchayat* (village council) or by private people. This case study looks at two types of slaughterhouse: a large slaughterhouse in a metropolitan area of Karnataka, and the village-level slaughterhouses of Kerala.

Large Organized Slaughterhouses in Bangalore City (Karnataka)

The slaughterhouse, spread over an area of 4.5 acres owned and managed by the Karnataka Meat and Poultry Marketing Corporation (KMPMCL), is providing facilities to contractors for slaughtering and dressing of sheep, goats, and cattle. The slaughterhouse has a lairage (holding area), small ruminant slaughter area, large ruminant slaughter area, solid waste disposal yard, and office complex. There is an open well and bore well to meet the water requirements, and an effluent treatment plant with a 150,000 liters per day capacity. Animals are brought to the slaughterhouse a day before slaughter, rested in the lairage, and examined for health. Because the number of animals brought for slaughter often exceeds the capacity, animals are held in public parks and open spaces in adjoining localities, disturbing neighborhoods and generating conflicts between the contractors and local residents.

About 650 to 700 small and 100 to 150 large animals are slaughtered on normal days. The number rises to 1000 small ruminants and 200 large animals on Sundays and up to 2000 small animals and 300 large animals on festival days. The butchers employed by the contractor carry out all slaughter-related jobs. After slaughter/dressing, the veterinary surgeon appointed by the KMPMCL certifies the carcass as fit for human consumption. Generally a very small percentage of the carcasses or their parts are rejected. Stomach and intestinal contents together with inedible/nonsaleable portions form the solid waste. The estimated solid waste on a normal day is 14,200 kg, which goes up by 35% on Sundays and up to 100% on festival days. All the solid waste is collected and disposed of as landfill, for which trucks are engaged. There is no designated site for disposal of solid waste, and the contractors dump it in unauthorized suburbs. The effluent treatment plant installed to treat the wastewater from the slaughterhouse was not functioning during our visit. There are no high-pressure pumps and jets for proper floor cleaning of the slaughterhouse.

Village Slaughterhouses in Kerala

An estimated 1.5 to 1.7 million cattle and buffaloes are slaughtered annually in Kerala, where most people are reported to eat beef. With an estimated average meat yield of 60 kg per large animal, Kerala handles around 90,000 to 100,000 tonnes of beef per year. More than 45% of the large animals are slaughtered and marketed in villages that do not have even minimum facilities for these purposes.

Village-level slaughterhouses in Kerala are of two types, the *panchayat* slaughterhouses and the village slaughter places. There are more than 1000 *panchayats* in the state and each has at least one place to slaughter animals. In the *panchayat* slaughterhouses, slaughter is often done in the open. Usually inedible parts like bones, fat, and blood are not used. Visceral contents and effluent are disposed of on nearby land or into waterways. Butchers are not properly trained and practice below-average hygienic procedures. Village slaughter places are temporary thatched sheds where one or two large animals are slaughtered on weekends. Their surroundings are relatively clean. Slaughter and sale are done from the same place.

Drivers

The fast growth in demand for meat and meat products, changing food habits, and accelerated growth in urbanization are the driving forces that have increased the number of large and small ruminants slaughtered in the country. According to official statistics (FAOSTAT 2006) over the period from 1985 to 2005 India's beef production increased from 1.95 million tonnes to 2.98 million tonnes and sheep and goat meat production from 0.53 million tonnes to 0.71 million tonnes. However, due to nonreporting of animal slaughter at the village level and the prevalence of unauthorized slaughter in many municipalities and corporations, the actual quantity of meat production, especially from large ruminants, is far higher than the reported figures.

Consequences

The capacity of many of the major slaughterhouses is inadequate to handle the present demand for meat. In the absence of a fully equipped slaughterhouse, proper use of the by-products and proper disposal of effluents and waste from the slaughter is not possible, and there is no organized system for disposal of solid wastes from slaughterhouses. The capacity of the trucks used for solid waste transport in Bangalore is only 60% of requirements. As a result, the wastes and effluent from many village slaughterhouses are often left fully or partially in or near the slaughter place to leach into the soil. Such wastes often pollute waterways nearby and adjacent drinking water wells. Farmers refuse to allow wastes to be dumped in their farmlands because of the stench and protests from the local residents. Blood, urine, and water used for cleaning are sent to the open drains, which overflow during the rainy season, contaminating land and water sources.

Unhygienic slaughter and sale of meat poses serious hazards to the health of people handling meat as well as of consumers. Even if the animals themselves are not infected, meat kept at room temperatures in a hot, humid atmosphere serves as a potent breeding ground for bacteria and toxins. Animals are flayed on the ground and the carcass is eviscerated and cut into pieces without lifting it from the ground, providing ample opportunities for contamination from the soiled ground as well as from the visceral contents and dung. However, health problems from precooking contamination are comparatively less because meat is consumed only after it is fully cooked.

The places of meat sale in the municipalities and village towns are identical in appearance. They have different levels of display and storage. Some are well maintained and have cold storage/refrigeration facilities. The carcass is often hung by the legs in front of the shop, and customers get the required quantity of beef carved out from the hanging pieces of the carcass. It is customary that the skinned head with the horns is exhibited in front of the sale point to indicate whether it is cattle meat or buffalo meat. People in different parts of the state and among different communities have different preferences for these two types of meat.

Apart from pollution-related issues, nonutilization of slaughter by-products creates a considerable economic loss. It can be shown that the funds needed to create facilities for production and sale of good quality meat could easily be recovered by the sale of by-products within a period of five to seven years.

Often the very setting of some of these slaughterhouses and the sales counters for meat are appalling: blood everywhere, mutilated parts of the carcass lying around—all in an unclean surrounding with marshy areas formed by waste water and visceral contents. Solid wastes are generally left in the premises, emanating bad odors, and are scavenged by predator birds and stray dogs. Wastes invite vultures and other such birds in large numbers, a potential risk for aircraft and helicopters.

Public and Private Responses

Solid waste disposal and the transport of meat have become emerging issues, creating urban conflicts and communal disharmony in periurban areas. The respective local bodies are mainly responsible for day-to-day operation/maintenance of the slaughterhouses. Greater attention is being bestowed nowadays to issues related to pollution from livestock waste, thanks to the heightened awareness and efforts of the media. The importance of using disease-free and healthy animals, clean meat production, minimizing environmental pollution, and aesthetic marketing are at a lower plane in public awareness. Local bodies (*panchayats*, municipalities, and corporations) can do a lot more to improve this situation.

The involvement of Bangalore Municipal Corporation (BMC) for solid waste management and effluent treatment and KMPMCL for meat certification and outsourcing of slaughtering/dressing services causes confusion about the roles and responsibilities of respective organizations. In the absence of designated sites for waste disposal, contractors are dumping solid wastes in unauthorized private or municipal lands in city outskirts.

As a result of public pressures or court directives, laws have emerged regulating the health, manner, place, and number of animals that may be slaughtered for meat. However, the absence of adequate infrastructure, institutions, and enforcement mechanisms makes enforcement difficult.

State Pollution Control Boards have powers to take action against defaulting slaughterhouse owners. It is important that we adopt community-friendly, decentralized, and low-energy systems for management of meat production and marketing activities. Whenever modernization, expansion, or addition of new slaughterhouses is planned, there must be involvement of the different stakeholders concerned. The system should be developed in a planned manner, rather than thrust upon the authorities by crises and protests.

Options

Slaughterhouse issues, especially in village slaughterhouses, can be addressed through awareness creation, strict ante- and postmortem examination of animals, improving infrastructural facilities, training and orientation of butchers and others, effective supervision and inspection of slaughter-related processes, improving sales places, and maintaining cold chains.

A well planned public awareness campaign highlighting the environmental, health, social, and economic issues must be attempted with the participation of government agencies (animal husbandry, health, and local bodies and departments), NGOs, meat traders, media, schools, and colleges.

There must be strict ante- and postmortem examination before the meat is passed for human consumption. Local bodies and statutory organizations must ensure that the quality and regularity of these inspections are satisfactory. The government's plan to set up a meat board to oversee and regulate the meat industry in the country is a welcome move.

Slaughtering facilities are often inadequate and insufficient. Even though there are modern slaughterhouses in corporations and major municipalities, they are rarely fully used. It is felt that there is some kind of mismatch between what is offered in the form of facilities and what is acceptable to butchers. A possible solution could be to create facilities as discussed and decided in a forum with the participation of all concerned, looking into hygienic and environmental aspects, and making the operations user-friendly and acceptable to all stakeholders. Such a facility must be simple and moderately priced and must ensure hygienic meat production with an effective system to handle by-products, effluents, and solid waste.

Training and orientation of butchers with a view to make positive changes in their mindset has proved extremely difficult. An analysis of strengths, weaknesses, opportunities, and threats (SWOT analysis) to identify the reasons for butchers' resistance to the systems advocated could be helpful to orient the training efficiently. Rather than trying to totally change the operational style of the butchers the objective should be to make slow but steady incremental progress, allowing them time to feel the advantage of each successive change.

Case Study: Grazing Land in Kalyanpur Watershed (Rajasthan)

A very large number of livestock farmers in India still depend on common lands, known as common property resources (CPRs), for grazing their animals, including village pastures, revenue land, and forestland. Several studies show that there has been a decline in the area under CPRs from 1950 onward.

This case study was undertaken in Kalyanpur (Rajasthan) to understand the changing trends in CPR management. In Kalyanpur the common grazing area shrank from 1.85 million ha in 1983 to 1.70 million ha in 2005. A snapshot study conducted by an NGO (Sevamandir) showed that 100% of revenue lands (land belonging to the revenue department of the state government), 56% of *panchayat* lands, and 24% of forest lands are encroached.

There are 6000 ha of watershed area in Kalyanpur, half of which are public land. Marginal and small landholders keep higher numbers of small ruminants than large ruminants. Small ruminants, kept mainly by lower castes, largely depend on common or fallow lands for grazing.

Drivers

With increasing human populations and industrial development there is pressure on the CPRs for other purposes as well as grazing. Privatization, encroachment, distribution of land by government to the landless, establishment of national parks and sanctuaries are all forces that reduce the area under CPRs. The management of common lands used to be the responsibility of the village community who are the beneficiaries of the CPRs. Under the Land Settlement Act of 1956, the control and management of such lands changed hands from the community to the *gram panchayat,* the official local body, and thus community control of the land was lost. In most villages, grazing lands are now used as an open access resource with no control on the intensity of use.

Consequences

Because of shrinking grazing resources, poor biophysical conditions, and excessive livestock numbers, the grazing lands are subjected to degradation. A larger number of animals than the resource base can support, coupled with shrinking grazing areas and lack of regeneration efforts, have led to overuse of grazing areas and loss of vegetation cover. Most of the CPRs are now unavailable for grazing because they are contested, degraded, and encroached.

The lack of institutional mechanisms to regulate the use of CPRs is a big issue in their development and sustainability. It threatens rural livelihoods and the ecological security of the region, which is already in a depleted state. The people most affected by the reduction of the CPRs are small and marginal farmers whose major means for livelihood is livestock, especially small ruminants.

The grazing system, besides its economic contribution (milk, meat, wool) is very valuable in conserving animal biodiversity and improves the dryland ecology. Because of reduction in grazing areas, pastoralists are forced to migrate for longer distances and periods in search of grazing land. There has been increased tension between agricultural and pastoral communities over livestock grazing. Some members of pastoral groups have tried to cope by quitting the pastoral way of life, moving to cities to take up menial jobs, but often they are not successful.

The *gram panchayat*, having no direct involvement in these CPR lands, has shown less responsibility for their upkeep, resulting in their shrinkage and degradation. Lack of user rights also deters villagers from investment in biomass development.

Public and Private Responses

Public and private sector responses are now emerging to address the issue of land degradation, from local communities, NGOs, government, and so forth. The government of Rajasthan's Ministry of Rural Development has launched a watershed development program for reversing land degradation. The present project period of five years needs to be extended to achieve sustainability of the land development interventions.

The Joint Forest Management (JFM) program implemented by the Ministry of Environment and Forests is another attempt to promote the development and sustainable management of forest areas, including up to 25% of watershed areas. The forest department in association with the Kalyanpur area *panchayat* developed 45 ha of degraded forestland with community participation. The land was fenced and closed for five years for regeneration. It has now been opened for controlled livestock grazing.

Case Study: Tanneries in Kanpur (Uttar Pradesh)

In India, tanning industry establishments are located along rivers. It has been estimated that annually 0.9 million tonnes of hides and skins are processed in India. More than 90% of this involves chrome tanning. This case study examines the tanneries in Kanpur, a city in Uttar Pradesh. Tanning is a traditional occupation in the area, and most of the tanneries are family owned and managed.

Drivers

Production of hides and skin in India has increased from 0.7 million tonnes in the early 1970s to 1.1 million tonnes in 2000–01. The major source (85%) is buffaloes and cattle. The remaining 15% are obtained from small ruminants. Finished leather from India is highly rated in the international market and is a source of foreign exchange for the country. India's competitive advantage is helped by less stringent government regulations on environmental issues, cheap labor, and facilities to set up tanneries near riversides. These conditions are also conducive to the expansion of tanneries handling imported rawhides.

Consequences

Tanneries are a source of livelihood to families from the lower economic strata in the society. Dixit (1995) estimated that the Indian tanning industry employs 80,200 people. This number has probably increased by now. After the massive closing down of textile mills in Kanpur area during the 1980s and 1990s, many unemployed persons were engaged in the tanneries, which now represent the primary source of livelihood in Kanpur. Although tanneries are a source of employment for people from the economically weaker sections of society, they have several negative social, health, and environmental impacts.

During the tanning process, 68 to 80% of the hides processed and 90% of the water used end up as waste. Pollution from tanning is both organic and chemical in nature. It has been estimated that 35 to 40 liters of water is required per kg of hide/skin processed (from raw to finished stage). The solid waste generated by tanneries includes hair, trimmings, flesh, sludge, salts, shavings, and sources of vegetable tannins like bark and nuts (Yadav 1998). Some of these are used to make by-products like glue, dog bones, chicken feed, organic fertilizer, heel-caps for shoes, toy stuffing, and so forth, though demand for these is very small in comparison to the waste generated.

In chrome tanning, the most common method employed, chemicals are dissolved in water and are not absorbed by the hide. As a result, the effluent contains huge quantities of chrome and other fixing chemicals, which are discharged into water bodies. Annually the Kanpur tanneries discharge 1500 tonnes of chromium sulfate as waste (CLRI 1996).

No economically viable use for tannery sludge has been identified. The sludge therefore needs to be dumped on special grounds to prevent chemicals leaking into the groundwater. In reality such precautions are not taken, and there is a visibly adverse impact on the quality of groundwater, which contains high concentrations of chromium, pesticides, nitrates, and dyes.

The performance of the Central Effluent Treatment Plant (CETP) established by the government to treat the tannery effluents is often below average. The central pollution control board (CPCB) reported in 1997 that the treated water coming out of the CETP had chromium concentrations 124 to 258 times higher than the permissible limit (CLRI 1996). The posttreated water, meant for irrigation, is heavily laden with toxic pollutants such as arsenic, cadmium, mercury, nickel, and chrome VI. These pollutants not only cause grave damage to the soil but also pollute groundwater resources. Frequent breaches in sewage irrigation water channels allow the hazardous water to seep into the River Ganges. Due to the toxic concoction that is used to irrigate land, year after year agricultural crops are destroyed. Food chains, including milk, are contaminated, leading to several diseases affecting humans and livestock in the area. Aquatic life in the river has almost disappeared.

Public interest litigation and regulations to control the environmental impacts of the tannery waste during this period have had mixed results, but overall the standards for control of chemicals and organic matter in the effluents have not been met.

Chrome VI, a product transformed from chrome III used in the tanning process, is a known carcinogen (UNEP 1991). Although the polluted water is not fit even for irrigation, people continue to drink it because alternate supplies are not available. Air pollution in the tanneries is mainly from the release of dust from the buffing of the leather, and use of solvents and dyes that can be toxic when released.

Public and Private Responses

As a result of the ruling on public interest litigation, a central effluent treatment plant (CETP) was established in 1994 to treat effluents from all of the 354 tanneries in the area. Tanneries not connected to a CETP must have their own individual ETP that takes care of both primary and secondary treatment.

The state pollution control board (SPCB) has the authority to inspect any tannery at any time and can initiate action against those not conforming to standards. Because the CETP is run by the government, the control function of the SPCB is less toward CETP than toward the tanneries.

Tannery regulations are all related to water pollution. All tanneries are required to treat their effluent before releasing it into either the sewage system or a river and must meet certain standards for pH, total suspended solids, sulfides, and chrome. Tanneries are either connected to a CETP (and therefore have a primary treatment plant where sludge in the effluent can settle and where the pH is adjusted prior to going to the CETP) or they must have their own ETP. The cost of complying with environmental regulations in the tanning industry has been estimated to be around 5% of the production cost (Schjolden 2000).

The NGO Eco-friends is working to create awareness and networks among the local people and to shift the mindset of tannery owners and government officials toward less polluting and more community-friendly production systems. Litigation has also produced results. After social worker M. C. Mehta filed a court petition, the court ordered relocation or closure of tanneries that have not established an effluent treatment plant (ETP) or connected to a CETP. Several litigations that followed helped to strengthen the case for tighter environmental regulation of tanneries.

Development of cleaner process technology is another positive development that has occurred in the tanneries. One such technology is automatic feeding of chemicals, which considerably reduced the uptake of chemicals. Two tanneries in Kanpur adopted this technology as part of a collaboration project with the United Nations Industrial Development Organization (the cost of the equipment was partly covered by UNIDO). Another is addition of a chrome recovery unit that precipitates and separates the unabsorbed chrome, which can then be reused. This technology not only saves the cost of chromium but also drastically reduces the chrome content of effluents. Even though this technology can bring paybacks to the tannery, less than 25% of the tanneries using chrome have installed it due to the high cost. However, this system is likely to be installed by larger tanneries in due course.

UNIDO also introduced the use of enzymes into the tanning process, which increases the uptake of chrome and therefore reduces chrome in the effluent. By using enzymes and magnesium oxide for basification instead of soda, chrome uptake can rise from 40% to 80 to 85%. UNIDO covered six large and medium tanneries in Kanpur in this exercise. What needs to be looked at is an affordable working system for the small-scale tanneries that make up 80% of tanneries in India.

Options

In spite of court rulings and establishment of environmental regulations, these are not being implemented or enforced in an effective manner. The following are options to improve this situation and bring about the desired changes in the tanneries:

- Strengthening of awareness among tanners: There is a need to combine awareness generation with strict regulation application.

- Research into viable clean technology: Several new technologies are being tried in Kanpur. Unfortunately, the costs are way beyond affordable for small-scale tanneries. Government and other supports would be valuable to encourage adoption of the new environmentally friendly technologies.
- Reuse of water: Presently, the CETP managed by the Uttar Pradesh water and sewage authority releases its treated water into waterways or irrigation channels. Since this water is unhealthy efforts must be made to develop systems for tanneries to reuse this water.
- Stakeholder network: There is a need to set up a network of stakeholders involved in tannery activities and those impacted (both positively and negatively) by tannery activities. This will generate greater understanding of the way different communities are affected by the tannery activities and serve as a forum to develop collective understanding and identify solutions to address the negative impacts of the tanneries.

Conclusions

The current policy framework often favors the development of large-scale industrial production making the poor even more vulnerable. Promotion of technology and input-intensive industrial production may lead to faster growth of the livestock sector in the country. But such an approach could further marginalize small-scale producers in India. Industrialization has a high risk of changing livestock production from an important factor in the rural economy to an activity with limited effect on poverty and equity. This is particularly so in a developing country like India, where the vast majority of livestock keepers are smallholders, keeping livestock for their livelihood or sustenance and depending on traditional natural-resource-based production systems: pastoral, agro-pastoral, or integrated mixed crop–livestock systems.

Geographically, most large-scale industrial production takes place in and around major cities. This leads to substantial pollution, particularly of surface and groundwater. In smallholder production systems, particularly in the extensive system, the increasing market for livestock products might lead to overuse of existing land resources in the absence of mechanisms for development and regulated use of natural resources.

Industrialized large-scale livestock production is expanding due to economies of scale, vertical integration, and high demand for quality products. Industrialized production is often favored by more or less hidden subsidies, but it entails environmental hazards and the risk of enzootic diseases. It also results in changing consumer preferences because of cheaper prices. It tends to marginalize small-scale producers and can have a negative impact on economic growth and employment in rural areas.

The development of semi-industrialized livestock production systems is already familiar in India and is considered to be a natural consequence of technological development. The critical question from a development perspective is: How will small-scale producers be able to cope with this? Smallholder livestock production can be competitive, as has been proved in the dairy sector of India (Operation Flood) and in the poultry sector of Bangladesh.

The expected high demand for meat, eggs, and dairy products will provide opportunities as well as challenges for development of smallholder and rural-based systems in the coming decades. The big issue in a development perspective is how to stimulate and support the livestock sector, in such a way that the growing demand for animal products will also benefit small-scale producers and lead to more equity and poverty reduction (Henriksen 1998).

With this goal in mind, governments should target the following objectives:

- Develop policies, infrastructure, and vertical integration that will promote private investment and interventions in the livestock sector
- Impose rules and regulations related to environmental impact of industrialized livestock production based on the "polluter pays" principle
- Develop veterinary rules and regulations required for protection of public health
- Promote development of modern smallholder livestock production systems capable of satisfying consumers' requirements for quantity and quality
- Empower producer organizations, so as to enable farmers to influence agricultural policies and strategies and to make them important players in the livestock industry
- Make use of the increased urban demand for livestock products as an opportunity for rural growth and poverty alleviation
- Generate improvement in environmental quality by providing the resources necessary for environmental investments, raising societal standards and pressures for improved environmental behavior, and pursuing institutional and policy change.

It has been very well established that smallholder livestock production (in the developing country context) acts as a tool for poverty reduction; hence it is recommended that smallholder production, processing, and marketing should be promoted as an integrated part of an agriculture sector support program, together with the industrialized production systems with measures to mitigate risks to the environment and human health.

References

Ahuja, V., M. P. G. Kurup, N. R. Bhasin, and A. K. Joseph. 2008. Assessment and reflections on livestock service delivery systems in Andhra Pradesh. CALPI Programme Series 4. Cited 29

January 2009. Available at http://www.fao.org/Ag/AGAInfo/programmes/en/pplpi/docarc/rep-ap_synthesis.pdf

Birthal, P. S. and Parthasarathy Rao. 2002. Technology options for sustainable livestock production in India. In *Proceedings of the Workshop on Documentation, Adoption and Impact of Livestock Technologies in India,* 18–19 January 2001. Patencheru, India: ICRISAT.

Birthal, P. S., P. R. Deoghare, S. Kumar, Riazuddin, J. Jayasankar, and A. Kumar. 2003. *Development of Small Ruminant Sector in India*. New Delhi: Ad hoc project submitted to the Indian Council of Research.

Birthal, P. S., A. K. Jha, and A. K. Joseph. 2006. *Livestock Production and the Poor*. New Delhi: CALPI.

Birthal, P. S. and V. K. Taneja. 2006. Livestock sector in India: opportunities and challenges for small holders. In *Workshop on Small holder Livestock Production in India: Opportunities and Challenges*, 31 January–1 February 2006, New Delhi, ed. P. S Birthal, V. K Taneja, and W. Thorpe, 5–64. Nairobi: International Livestock Research Institute.

CALPI-IWMI. 2005. *Livestock Environment Interactions in Semi arid Watersheds: A Study in Semi-arid India.* Hyderabad: International Water Management Institute.

CLRI. 1996. *Project Report for Setting up Leather Complex at Unnao*. Chennai: Central Leather Research Institute.

Dixit, P. K. 1995. *Influence of Government Policy and Analysis of Leather Industry: A Case of Kanpur.* M. Kanpur: Tech. thesis in Department of Industrial and Management Engineering, Indian Institute of Technology.

FAI. 1982. *Fertilizer Statistics*. New Delhi: Fertilizer Association of India.

FAI. 2002. *Fertilizer Statistics*. New Delhi: Fertilizer Association of India.

FAO. 2006. *Responding to the Livestock Revolution*. Livestock Policy Brief No 1. Rome: FAO.

FAOSTAT. 2006. Agriculture data. Rome: FAO.

Government of India. 1999. *National Sample Survey Organisation*. New Delhi: Ministry of Statistics and Programme Implementation, Government of India.

Government of India. 2000. *National Agricultural Policy.* New Delhi: Ministry of Agriculture, Department of Agriculture and Cooperation.

Government of India. 2001. *Census of India*. New Delhi: Ministry of Home Affairs.

Government of India. 2004. *Basic Animal Husbandry Statistics*. Department of Animal Husbandry, Dairying and Fisheries. New Delhi: Ministry of Agriculture

Government of India. 2005. *17th Indian Livestock Census 2003*. Department of Animal Husbandry and Dairying and Fisheries. New Delhi: Ministry of Agriculture.

Government of India. 2006. *Basic Animal Husbandry Statistics 2006*. Department of Animal Husbandry, Dairying and Fisheries. New Delhi: Ministry of Agriculture.

Henriksen, J. 1998. Small scale dairying: opportunities and constraints. In *Integrated Livestock/Crop Production Systems in the Smallholder Farming System in Zimbabwe*, ed. N. J. Kusina, 8–18. Proceedings of a workshop at the University of Zimbabwe, Harare, Zimbabwe, 13–16 January 1998.

IFPRI-FAO. 2002. *Livestock Industrialization, Trade and Social Health–Environment Issues for the Indian Poultry Sector.* Rome: FAO.

Indian Dairyman. 2006. *Challenges of Indian dairy sector.* 58(12). New Delhi: IDA.

Jodha, N. S. 1992. Common property resources and dynamics of rural poverty in India's dry regions. *Unasylva* 46: 180. Cited 16 March 2009. Available at: http://www.fao.org/docrep/v3960e/v3960e05.htm

Jones Lang LaSalle. 2005. India the next IT off shoring locations. *Tier III Cities Business Wire,* 27 October 2005. New Delhi.

Khan, A. and F. S. Bidabadi. 2004. Livestock revolution in India: its impact and policy response. *South Asia Research* 24(2): 99–122.

Kumar, P. and P. S. Birthal. 2004. Changes in consumption and demand for livestock and poultry products in India. *Indian Journal of Agricultural Marketing* 18(3): 110–123.

Mehta, R., R. G. Nambiar, S. K. Singh, S. Subrahmanyam, and C. Ravi. 2002. *Livestock Industrialisation, Trade and Social-Health, Environment Impact in Developing Countries: A Case Study of Indian Poultry Sector.* Phase I project report submitted to International Food Policy Research Institute, Washington, DC: IFPRI.

NATCOM. 2004. *India's Initial National Communication to United Nations Framework Convention on Climate Change.* Cited 16 March 2009. Available at: http://unfccc.int/resource/docs/natc/indnc1.pdf

Onyekakeyah, L. 2006. *Water in Livestock Production.* Nairobi: International Livestock Research Institute.

Parthasarathy Rao, O. 2003. *Addressing Resource Conservation Issues in Rice Wheat Systems of South Asia: A Resource Book.* New Delhi: Rice Wheat Consortium for Indo-Gangetic Plains-CIMMYT.

Parthasarathy Rao, O., P. S. Birthal, D. Kar, S. H. G. Wickramaratne, and H. R. Shrestha. 2004. *Increasing Livestock Productivity in Mixed Crop Livestock Systems in South Asia.* Patenchery, India: ICRISAT.

Schjolden, A. 2000. Leather tanning in India: environmental regulations and firms compliance. F-I-L Working Papers No. 21. Oslo: Forurensende Industri–Lokalisering.

Shankar, V. and J. N. Gupta. 1992. Restoration of degraded rangelands. In *Restoration of Degraded Range Lands: Concepts and Strategies*, ed. J. S. Singh, 115–155. Meerut, India: Rastogi Publications.

Sharma, V. P, S. Staal, C. Delgado, and R. V. Singh. 2003. *Livestock Industrialisation Project Phase II: Policy, Technical and Environmental Determinants and Implications of the Scaling up of Milk Production in India,* Annex III. Research Report of IFPRI-FAO. Washington, DC: IFPRI.

Singh, O. P., A. Sharma, R. Singh, and T. Shaw. 2004. Virtual water trade in dairy economy. *Economic and Political Weekly,* 31 July 2004. Mumbai, India.

UN. 2008. *World Population Prospects: The 2006 Revision.* New York: United Nations Population Division.

UNEP. 1991. *Tanneries and the Environment: A Technical Guide to Reducing the Environmental Impact of Tannery Operations.* Technical Report Series No. 4. Nairobi: United Nations Environment Programme.

Yadav, S. S. 1998. *Integrated Environmental Management Plan for Leather Tanneries: A case of Jajmau, Kanpur.* M. Tech. thesis. Ahmedabad: Centre for Environmental Planning and Technology, School of Planning.

5

Brazil and Costa Rica

Deforestation and Livestock Expansion in the Brazilian Legal Amazon and Costa Rica: Drivers, Environmental Degradation, and Policies for Sustainable Land Management

Muhammad Ibrahim, Roberto Porro, and Rogerio Martins Mauricio

Abstract

In Latin America a large percentage of virgin forest has been converted for cattle ranching and to some extent for industrial agricultural production. The conversion of forest to cattle ranching based on unsustainable management of grass monoculture pastures is associated with land degradation, leading to loss of farm productivity and environmental degradation.

The linkages between cattle ranching and deforestation and environmental degradation have been a subject of debate by conservationists and demand a critical analysis of the drivers that induced deforestation, and the impacts on the environment and natural resources, and what polices were instituted to restrain cattle-linked deforestation. This chapter analyzes the linkages between deforestation and cattle ranching using Costa Rica and the Brazilian Legal Amazon (the states of Acre, Amazonas, Roraima, Amapá, Pará, Rondônia, Mato Grosso, Tocantins, and Maranhão) as contrasting case studies of different policy approaches and different outcomes.

Policy similarities in the two countries included subsidized credits to establish pastures on deforested lands, provision of titles for land cleared and managed with cattle, and road construction. Various changing market forces were also responsible for deforestation, including international demand and good prices for beef, health and food safety issues in the European Union, and high demand for Brazilian soybean, soy meal, and soybean oil.

There are also policy differences. To curb deforestation, the Costa Rican government introduced environmental regulations and policies that led to progressive recovery of forest cover, including establishment of national parks and protected areas representing more than 35% of the total forest cover in 2005. Costa Rica has pioneered a payment for environmental services system that has successfully protected remaining forests as well as woodlands within agricultural areas. The Brazilian government has implemented similar policies with less success due to lack of resources (equipment, staff, etc.) to monitor forest cover and weak enforcement of the laws against illegal deforestation. Brazil has also been innovating in payment for environmental services, and there are huge potentials for saving remnant forest in the Amazon through the Reducing Emissions from Deforestation in Developing Countries program. Certification of cattle and soy products for environmental compliance represents another avenue for providing incentives for sustainable crop and livestock development while conserving forest resources.

Introduction

Tropical deforestation in Latin America progresses at a high rate with serious consequences for the environment, climate, and livelihoods of rural poor. The annual rate of forest loss in the region from 2000 to 2005 was 0.51%, compared to 0.46% during the 1990s. Since 1990, Latin America and the Caribbean lost more than 64 million ha of forest (FAOSTAT 2007). In Central America alone more than 10 million ha of primary forest were deforested and a large percentage of deforested land was converted to grass monoculture pastures, and similar trends were observed in Latin America (Kaimowitz 1996, Szott et al., 2000, Kaimowitz et al., 2004, Nepstad et al., 2006a,b). It is projected that the demand for beef, dairy, and other (pork and poultry) livestock products will increase significantly over the next four decades. This will exert more pressure on the use of the natural resources to meet this demand, and hence could have increased adverse effects on climate change, loss of biodiversity, and water resources.

The linkages between cattle ranching and deforestation have been a subject of discussion by many scientists and policy makers. Although cattle ranching has been widely blamed for deforestation in Latin America, there is considerable evidence that poor government policies and inadequate incentive schemes have been the fundamental driving forces behind the expansion and effects of cattle ranching (Kaimowitz and Angelsen 1998, Mertens and Lambin 1999, Laurance et al., 2002, Kaimowitz et al., 2004). It follows that a thorough analysis of the driving forces implicated in ranching expansion and unsustainable pasture management is an effective way to understand one of the major environmental issues in Latin America. This analysis will help in the design and implementation of policies and incentive mechanisms for the sustainable management of livestock and conservation of natural resources.

This chapter compares the geographical shifts and the driving forces related to cattle ranching and deforestation using Costa Rica and the Brazilian Legal Amazon as case studies. The time scales of deforestation in these two areas are different. In Costa Rica, 13.5% of the territory was deforested by 1900. Between 1950 and 1980 rapid deforestation occurred, and 58.3% of the area was deforested by 1980 (Fournier 1985). In the case of the Brazilian Amazon, deforestation rates were relatively low prior to the 1960s but then increased over time, reaching especially high levels in 1994–1996 and 2000–2004 (Nepstad et al., 2006b, Wertz-Kanounnikoff et al., 2008). Deforestation continues in the Amazon, although it tended to decrease in the years 2004 to 2008 (Schneider et al., 2002, INPE 2008).

Despite the differing time scales, there are similarities between the two countries as to how government policies have driven deforestation. For example, in both countries the major drivers of deforestation included government measures such as subsidized credit for the livestock sector, land tenure polices that provided titles for clearing the land, and construction of roads (Kaimowitz 1996, Schneider et al., 2002, Nepstad et al., 2006a). Market forces were also important drivers of deforestation, though there were differences between the two countries. In Costa Rica, the demand and high prices for beef (Trejos 1992) and to some extent increased local demand for milk and dairy products were the main driving forces (Kaimowitz 1996). By contrast, in the Brazilian Amazon, increased international demand for beef and more recently for soybean, soy meal, and soy oil, has been the driving force for deforestation, though there has been a declining trend over the last years (Schneider et al., 2002, Nepstad et al., 2006b). Meat exports from the Brazilian Amazon increased only after some zones were declared free of foot and mouth disease (FMD), which was not a problem for Costa Rica because it was free of this disease and did not have sanitary restrictions for exportation of beef (Kaimowitz 1996, Kaimowitz et al., 2004, Nepstad et al., 2006b). In addition, in Costa Rica there is little evidence of the impact of industrial crop production on deforestation, but production of subsistence crops (beans and maize) has had some impact.

By 1990 Costa Rica had lost a large percentage of its forest cover. Over the last two decades the government has implemented policies (laws and regulations) that have contributed to conservation of remnant forest and to an increase in the forest cover within agricultural areas (Sánchez-Azofeifa et al., 2001). In addition to environmental policies, the private agricultural sector implemented programs for sustainable livestock production and conservation of natural resources. For example, the Costa Rican dairy sector has been modernized and made more viable by the organization of farmers into cooperatives, together with technological innovations and marketing strategies.

In the Brazilian Legal Amazon, on the other hand, despite significant lowering of deforestation rates in the 2004–2008 period (INPE 2008), the expansion of soybean production and crops for biofuel tends to push ranching deeper into forested areas. Combined with expanded urban markets for timber and beef, this contributes to continued deforestation, unsustainable agroecosystems, and natural resource degradation, with profound impacts on local livelihood opportunities. The Brazilian federal and state governments have been implementing policies to curb deforestation in the Amazon region but with relatively little success compared to Costa Rica. The relatively high profitability of soybeans and the increasing demand for beef continue to be the main market forces driving deforestation in Brazil, whereas local governments lack resources and tools for implementing policies that will restrain deforestation (Volpi 2007).

Livestock Sector Trends in Costa Rica and Brazil

Compared to Brazil, Costa Rica is a small country (51,000 square kilometers and 4.5 million people) located in Central America. The economy is diversified with tourism and exportation of bananas and coffee being major components. Agriculture and livestock account for 11% of the gross domestic product (GDP) of the country, but when the value of agricultural processing industries is added, they account for 30% of the GDP. Outside the Central Valley (the main urban area), the agricultural sector accounts on average for 60% of provincial economies (INFOAGRO 2007). In Brazil's economy, the agriculture/livestock sector is also important. In 2007 it accounted for 25.1% of the GDP, of which the livestock sector accounted for 7.3%—a share that has been more or less stable since 2000, except in 2003 and 2004 when it was around 8.3% (ANUALPEC 2007).

The livestock sector in both countries has experienced structural transformations in the last 10 years. Forces that induced changes include increasing land values, expansion of industrial crop production, price variations, increased rural wages, fast growth of supermarkets, trade liberalization, and increased demand for livestock products (Smeraldi and May 2008). The trends in each sector varied, as we will summarize with special emphasis on the cattle sector, which had stronger linkages with deforestation in the two countries.

Poultry

In Costa Rica the poultry sector has seen the most rapid growth due to high integration in the industry, which allows grain imports (corn and soybeans), to be free of import duties. It includes three major integrated operations for processing, which has contracts with about two hundred farms. Egg production is a bit less concentrated than meat production, but even here just six companies are the major distributors to supermarkets and small food shops (INFOAGRO 2007). In Brazil, poultry meat production actually doubled between 1997 and 2007, and egg production also showed an increasing trend. Annual consumption of poultry meat trended upward from 1997 (28.6 kg/capita) to 2007 (38.1 kg/capita). This trend may be related to relatively cheaper prices for poultry meat than for beef (ANUALPEC 2007). In the case of Costa Rica, poultry meat consumption increased from 20.2 kg/capita in 2002 to 22 kg/capita in 2005, though it dropped in 2003 to 17.6 kg/capita (INFOAGRO 2007).

Pigs

Pig production has grown slowly in Costa Rica. However, the related processing industries grew more, due to imports of fat and other ingredients and high quality ham and sausage. Currently there are nearly a hundred small pig farms, but these are generally being pushed out of business because of low profitability, and there has been a trend toward consolidation of production in 12 large farms ((INFOAGRO 2007). In Brazil the population of pigs increased over time, especially between 1998 and 2005, and thereafter it has been stable. The tendency in Brazil is vertical integration of large farms (with over one thousand pigs) into markets (e.g., supermarkets) with the aim of increasing production and marketing efficiency. The data show that the swine population in the industrial sector has increased 4 to 5% per year (Miele and Girotto 2007). The location of large farms around urban and periurban areas has resulted in environmental pollution in both countries.

Beef and Dairy

In Costa Rica, beef and dual purpose operations have tended toward intensification and a clearer distinction between cow–calf operations and fattening operations. Farms oriented to beef production are generally located in the lowlands on the Atlantic and Pacific coasts and are generally larger than dual purpose and specialized dairy farms. On the Pacific coast the average sizes for large, medium, and small beef farms are 351.1 ha, 274.4 ha, 89.2 ha, respectively (Villanueva et al., 2003). The dual-purpose cattle farms are located closer to urban and periurban areas where market access and infrastructure are good. Milk production is concentrated in the Central Valley and in the San Carlos region (Cámara Nacional de Productores de Leche 2004). Specialized milk operations have grown in number, as well as the number of cows per farm and production of milk per cow. Important factors of change include improvements in genetics, reproduction and animal health, improved pastures, smaller plots with electric fences, silage, and forage banks. The specialized sector is also an important consumer of feed concentrate; however, the use of concentrates declined in 2008 because of high fuel prices (INFOAGRO 2007).

In Brazil the cattle herd grew over time, from 1975 to 2005, and recent data show that it decreased in 2007 (Table 5.1). In 2005, the national herd totaled 208 million and had increased 1.3% in comparison to 2004, but the data in 2007 indicate a reduction by 4.1% when compared to 2005. In the Center and South, increases were smaller, reflecting the displacement of cattle production by other agricultural activities, mainly sugar cane and soybean (IBGE 2005). In contrast, the north and northeast regions had the fastest herd growth rates. Between 1975 and 2005, the cattle herd in the North region (part of the Legal Amazon) grew by over 900%, from 4 million to more than 40 million, currently representing 20% of the Brazilian national herd. The fastest expansion occurred in the last decade, when cattle herds in the North region increased 141%, compared with 122% between 1975 and 1985 and 92% between 1985 and 1996. Whereas the Legal Amazon's total herd grew 110% in the 1996–2005 period (Table 5.1), the growth in other regions of Brazil (not shown in Table 5.1) was only 14%. However, recent data show that the cattle herd in the legal Amazon decreased from 75 million to 70 million head of cattle with the states of Mato Grosso and Pará showing a reduction of 3.6 and 15%, respectively (Table 5.1). This decrease may be associated with a recent outbreak of FMD and a decrease in the price of beef (Nepstad et al., 2006b).

The analysis of farm structure in the North of Brazil (Acre, Amapá, Amazonas, Pará, Rondônia, Roraima and Tocantins), which includes most of the states of the Legal Amazon (except for the states of Mato Grosso and Maranhão) and is the main focus of this chapter, indicate that 45% of registered farms are small farms, which account for only 4.1% of the total area. By contrast large farms, which represent only 4.9% of the number of registered farms, occupy 77.3% of the total area (Cardim et al., 2003).

In Costa Rica cattle production for beef and dual purpose operations is practiced in 35,000 farms, whereas specialized milk is produced in about four thousand farms. The 1300 largest and most efficient milk producers are grouped in the largest cooperative of the country, Dos Pinos, which has 1300 members and collects one million liters of milk per day (Cooperativa Dos Pinos 2009). At the local level there are other cooperatives and

Table 5.1. Cattle herd expansion in the Brazilian Amazon (1975–2007)

	1975	1985	1996	2005	2007
Brazil Total	101,673,413	128,041,757	153,058,275	208,330,325	199,752,014
Legal Amazon					
Maranhão	1,784,284	3,247,206	3,902,609	6,448,948	6,609,438
Mato Grosso	3,110,119	6,545,956	14,438,135	26,651,500	25,683,031
North region total	4,038,853	8,965,609	17,223,042	41,489,002	37,762,602
Rondônia	*155,392*	*770,531*	*3,883,712*	*11,349,452*	*11,007,613*
Acre	*120,143*	*334,336*	*847,208*	*2,313,185*	*2,315,798*
Amazonas	*203,437*	*425,053*	*733,910*	*1,197,171*	*1,208,652*
Roraima	*246,126*	*306,015*	*399,939*	*507,000*	*481,100*
Pará	*1,441,851*	*3,478,875*	*6,080,431*	*18,063,669*	*15,353,989*
Amapá	*62,660*	*46,986*	*59,700*	*96,599*	*103,17*
Tocantins	*1,909,244*	*3,603,813*	*5,218,142*	*7,961,926*	*7,395,450*
Legal Amazon total	8,933,256	18,758,771	35,563,786	74,589,450	70,055,071

Notes: Maranhão's data computed in this table are for the entire state, and not only for the share of its Legal Amazon territory. The 1975 data for Tocantins was estimated from municipalities located in the state of Goiás that in the 1980s were dismembered to form Tocantins.

Sources: IBGE 1975, 1985, and 1995–96 agricultural censuses; IBGE 2005–2007 municipal livestock production.

associations involved in milk processing and cheese production such as the Santa Cruz dairy association, which has more than 100 dairy farmers. Beef slaughter plants are 23 in total, but three of them (authorized for export) account for 70% of slaughtered cattle. In the dairy industry there are approximately one hundred micro rural processors, six midsize dairy plants, and one large plant. Improvements in the industry contributed to an increase in dairy exports. However, imports of dairy products also increased, including condensed and evaporated milk, specialized powdered milk, and high-quality cheese. The dairy industry is protected with a tariff of 55%, but there is a commitment to gradual reduction under the Central America–Dominican Republic–United States Free Trade Agreement (CAFTA-DR) (Cámara Naçional de Productores de Leche 2006, Cooperativa Dos Pinos 2009).

In Brazil, 24.6 billion liters of milk were produced in 2005, an increase of 4.7% over 2004. Production is concentrated in the south of the country, and Minas Gerais State is the main producer (28%), followed by Goiás (10.8%), Paraná (10.3%), and Rio Grande do Sul (10%) (IBGE 2005). Good prospects for milk are attracting investors from other sectors of the economy, such as the Opportunity Group that has accumulated some 450,000 head of cattle on nearly 510,000 ha in the south of Pará state. In parallel to a greater market integration of Amazon beef producers, a well-structured value chain has developed based on massive private investments in expanding beef and dairy industry capacity (ANUALPEC 2007). Milk is marketed through cooperatives and local and transnational companies, and medium- and large-size farmers are generally contracted by transnational companies. The milk farmers sell the milk to cooperatives and from there to big companies (Nestlé and Parmalat), whereas medium or large producers may sell directly to Nestlé. In Minas Gerais, Itambe Company, which is one of the largest, collects milk from 27 cooperatives representing 9067 dairy farmers and supplying 1,090,000,000 liters of milk per month. Nestlé collects milk from 5800 farms (1,800,000,000 liters per year) and Parmalat from 4457 farms (725,021,000 liters/yr, EMBRAPA 2008).

In summary, in both countries there have been changes in the livestock composition that may be due to several factors, including changing patterns of consumption of meat products, along with variations in prices of poultry, pork, beef, and dairy products and in prices of inputs.

Deforestation and Cattle Expansion in Costa Rica and the Brazilian Amazon

Deforestation in Costa Rica started from the 1940s and proceeded at a rapid pace, especially between 1960 and 1990 (Kaimowitz 1996). From an area of 26,000 km^2, covering more than half of the country in 1970, primary forests were reduced to only 16,000 km^2 in 1987, covering 31% of Costa Rica (Tschinkel 1988). It is estimated that at the end of 1989 only about 2700 km^2 of forest outside of national parks and forest reserves remained. Estimated annual deforestation rates were 60,000 ha between 1976 and 1980 and in 1982, and decreased significantly, reaching low values of 3033 ha in 2000 and 4737 ha in 2005 (Table 5.2).

Deforestation occurred around the rich volcanic soils in the Central Valley and shifted to the coastal Pacific regions and to the Atlantic region with the construction of roads. A large percentage of deforested lands were converted to pastures, which increased from 0.6 million ha in 1950 to 2.2 million ha in 1983 (Rodriguez and Vargas 1988, Van der Kamp 1990), while in the same period the cattle population increased from 0.6 million head in 1950 to 2 million head in 1980 (FAO 1980, Leonard 1987). Because of the negative effects of deforestation and pasture expansion on the environment, the government implemented policies to protect the remaining reserves of forests through the establishment of parks and protected areas (Sánchez 2009) and the development of incentive schemes to promote reforestation. In addition it invested in modernizing the cattle sector to improve production efficiency and reduce the environmental impacts of cattle husbandry.

By contrast, Brazil's leading share of global forest loss increased from 30% of the global total in the 1990s to 42% in the 2000–2005 period (FAOSTAT 2007), with 92% of Brazil's loss occurring in the Amazon region (INPE 2006). Prior to the early 1960s the Brazilian Amazon was protected from threats due to its isolation, poor access, and lack of development. However, the development of infrastructure (airports, hydroelectricity, roads, etc.) and government incentive programs resulted in deforestation from the 1960s onward, with varying annual deforestation rates. By 2005, forest cleared exceeded 690,000 km^2 or about 17% of the forested portion of the Legal Amazon (Fearnside 2005). One of the worst periods of destruction occurred from 1978 to 1988, when about 21,000 km^2 forest was destroyed each year. From 1988 to 1991 the rate of deforestation decreased because of Brazil's unraveling economy (Figure 5.1) (bank accounts were frozen); in 1992 13,786 km^2 was cleared. In 1993 and 1994, rainforest deforestation increased to an average of 14,896 km^2 per year and in 1995 it reached a record level of almost 30,000 km^2 as a result of the economic recovery. Deforestation in the Amazon decreased in 1996 and 1997 because agricultural land prices declined by over 50%. In 1998 forest loss was accelerated through the downward spiral of land use and forest fires because of the El Niño episode (40,000 km^2 of dried-out forest burned). In the first few years of this millennium deforestation climbed substantially to 27,400 km^2 in 2004 because of the increasing international demand for beef, soybeans, and soy products (Nepstad et al., 2006b). However, prices for beef and soy products decreased in 2005 and 2006, and the Brazilian currency (Real) gained strength against the US dollar. Together with better implementation of government polices, these factors led to a decline of deforestation rates over the last four years (INPE 2006). Increasing prices for fuel in 2008 and the use of corn for biofuel production have led to increasing prices of staple food, and this may be associated with increased deforestation rates.

Table 5.2. Estimates of annual deforestation rates in Costa Rica 1975–2006

Year	Area Deforested per Year, Hectares	Source
1976–80	60,000	Grainger 1993
1982	60,000	Nations and Komer 1982
1981–1990	50,000	FAO 1993
1991	40,000	Merlet et al., 1992
1996	18,000	Castro et al., 1997, CCAD 1998, FAO 2002
2000	3033	Castro et al., 1997, CCAD 1998, FAO 2002
2005	4737	Castro et al., 1997, CCAD 1998, FAO 2002

A large percentage of deforestation occurs in what is referred to as the Arc of Deforestation including the following areas: Southeast Maranhão, the South of Tocantins, the South of Pará, the North of Mato Grosso, Rondônia, the south of Amazonas, and the southeast of Acre (Volpi 2007). In absolute terms, Mato Grosso and Pará have had the majority of deforestation in the Brazilian Amazon, with most of this occurring between 2000 and 2005. However, the contribution of Mato Grosso to total deforestation rose from 35% in 2003 to 48% in 2004, whereas that of Pará decreased from 36% to 15% in the corresponding years (Alencar et al., 2004). In 2005, Mato Grosso was responsible for 40% of deforestation in the whole Brazilian Amazon (Volpi 2007).

In sum, deforestation decreased significantly in Costa Rica due to the implementation of government policies; however, in the Brazilian Legal Amazon, it continues at an alarming rate, though there have been some decreasing trends over the last two years.

Drivers of Deforestation and Pasture Expansion

Deforestation in Costa Rica and in the Brazilian Legal Amazon was linked to government policies and market forces (Kaimowitz 1996, Nepstad et al., 2006a,b). There are many similarities between the two countries in the way that government policies triggered deforestation in the two countries, though the time periods where deforestation peaked in Costa Rica and the Brazilian Legal Amazon are different. Market forces that drove deforestation in both countries include a demand for

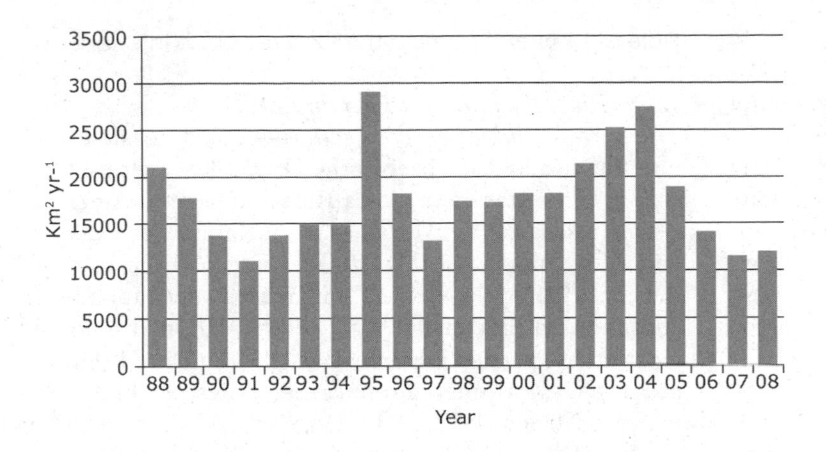

Figure 5.1. Rate of deforestation in the Brazilian Legal Amazon, 1988–2008.

beef and to some extent in Costa Rica increased local demand for milk. In the Brazilian Amazon the demand for soybean and soy products (soy meal and soy oil) on the international market has been the most recent driver of deforestation (Morton et al., 2006). This section will analyze how government polices and market forces influenced deforestation in Costa Rica and the Brazilian Legal Amazon.

Government Policies

A comprehensive analysis of how government polices influenced deforestation and the linkages with cattle ranching in Costa Rica was presented by Kaimowitz (1996). For many years the causes of deforestation in the Brazilian Amazon could be traced to federal policies designed to integrate the region with the Brazilian national economy and to defend it from international intervention (Hecht and Cockburn 1989, Nepstad et al., 2006b). Below is an analysis as to how the main federal polices affected deforestation.

Credit Policies

In Costa Rica, with increasing international demand for beef between 1960 and the 1970s (Myers 1981), the government increased the amount of credit for livestock so as to promote cattle expansion—by 1970 livestock credit was 37% of government agricultural loans (Williams 1986). The credit was subsidized such that real interest rates between 1970 and 1983 were negative, at times below 10% (Kaimowitz 1996). Subsidized credit in Costa Rica helped promote deforestation in several ways (Ledec 1992). It helped farmers to overcome capital constraints for pasture expansion and to invest in purchasing of animals. It also provided incentives to landlords to establish pastures on previously deforested lands so they could qualify for credit. However, in time the amount of livestock credit decreased because of the negative impacts of cattle ranching on deforestation, and cattle production was not so attractive because of relatively low prices for beef on the international market in 1975–1977 and in the 1980s (Trejos 1992).

Similarly, empirical studies of the Transamazon region of Brazil demonstrated that a large percentage of credit received by individual farmers was used to clear forest for cattle ranching (Moran 1981). Officially, tax incentives to finance extensive ranching in the forested areas of the Amazon ceased with a 1991 decree, and government-subsidized credits for investment in Amazon livestock projects had already been substantially reduced by 1988.

Road Construction

Highway paving stimulates deforestation by improving access to unclaimed or loosely titled land (Nepstad et al., 2000 and 2001, Soares-Filho et al., 2004). Among government measures, road construction was one of the most important factors associated with deforestation in both Costa Rica and the Legal Amazon. In Costa Rica, Sader and Joyce (1988) found that in 1977 the mean distance from the nearest road or railroad to nonforest locations was only 5.5 kilometers compared to a mean distance from forest locations of 14.2 kilometers. The annual expansion of roads in Costa Rica, for example, increased from 6.5% between 1974 and 1980 to 10.4% between 1981 and 1990; this includes the construction of the road between San José and Guapiles. With the development of road infrastructure, deforestation shifted to Alajuela and the Atlantic region (Kaimowitz 1996).

In the Brazilian Amazon, recently improved port facilities on the Amazon River have created even larger incentives to pave highways used to transport soybeans from Mato Grosso state, and promoted soybean cultivation in the Tapajós region, near Santarém (Nepstad et al., 2002; Nepstad et al., 2006a, Nepstad et al., 2006b). Kirby et al., (2006) found that paved roads were the best predictor of deforestation, with sites closer to paved roads more likely to be deforested. Paved roads explained 38% more variation in deforestation intensity than unpaved roads. Logging roads precede and accompany highways, opening up frontiers for investing timber profits in soy plantations and cattle ranches. Also, the occurrence of deforestation is increasingly associated with proximity to the emerging networks of dairy industries and secondary roads (Fearnside 2005). Laurance et al. (2001) noted that the immediate realization of all road projects proposed under Avança Brasil—a large-scale Amazon infrastructure development program of the Brazilian government to be implemented during 1999–2020—would result in an estimated 28 to 42% deforestation of the Amazon forest by 2020 (Laurance et al., 2001).

Land Tenure and Markets

Land tenure policies in both countries have been associated with deforestation and cattle expansion. For example, the Costa Rica Law 11 of 1941 permitted farmers to obtain title of larger areas of land if it was for pastures than if it was for crops (León et al., 1982). However, in 1990 the laws for land title were modified in a way that discouraged land clearance for pasture expansion (Utting 1993). In Brazil, the law confers ownership rights to those who demonstrate actual use of lands. The establishment of pasture has been the main strategy to obtain land tenure, so for many years cattle ranchers cleared forest to guarantee their tenure (Fearnside 2001), and as the most effective way to increase land value at lowest cost and for the long term (Schmink and Wood 1992, Veiga et al., 2001).

Despite the low profitability of cattle production and the reduction of government fiscal incentives, ranching has continued to expand, in part because of its utility to investors and land speculators in helping them claim title to land (Arima et al., 2005). Land titling in Brazil depends on demonstration of productive use, and one of the cheapest ways of achieving this is through the creation of pastures (Schmink and Wood 1992).

In Costa Rica, urbanization in the periurban areas of the Central Valley resulted in increased land prices, and this promoted the search for new areas for milk production, such as the northern portion of Alajuela (Camacho 1989). Furthermore, large ranchers in the Atlantic zone manage the land with maize cultivation and subsequently pasture and cattle production to speculate in increased land prices.

Government colonization schemes have also been major forces behind expansion of pasture area. For example, between 1974 and 1984, the Costa Rican government resettled 1801 families on 36,815 ha in Northern Alajuela, and in the following three years they resettled an additional 4604 families on 45,460 ha. A large percentage of the land disturbed was forestland (Girot 1989, Cruz et al., 1992). The most cited examples of policy leading to deforestation and land speculation are the Brazilian government subsidies and tax incentives for investment in cattle ranching in the Amazon (Mahar 1989, Schmink and Wood 1992, Hecht 1993, Moran 1993).

Role of Market Forces in the Cattle–Forest Interface

Costa Rica: Demand for Beef and Milk

The term *hamburger connection*—coined to describe how the demand for beef in the US market led to widespread deforestation in Costa Rica and Central America—was discussed by Myers (1981), Kaimowitz (1996), and many others. In the 1960s, rising real incomes in the United States led to a 20% increase in per capita beef consumption (Edelman 1985), and to high international meat prices, particularly between 1965 and 1974 (Howard 1987). Real international meat prices per kilogram increased from US$2.82 in 1965 to US$3.59 in 1970 (Trejos 1992). As a result beef exports in Costa Rica increased from US$5.6 million in 1966 to US$41.7 million in 1976 (Williams 1986). This coincided with the conversion of large areas of forested lands for pastures and cattle ranching—forest cover decreased from 63.4% of land area in 1960 and to 41.7% in 1977 (Fournier 1985), whereas the area of pastures rose from 1.3 million ha in 1960 to 1.7 million ha in 1978 (Rodriguez and Vargas 1988).

Between 1970 and 1990, international beef prices showed a decreasing trend except for a brief recovery in 1980. The combination of low prices, weak demand, and rising cost reduced the profitability of beef production (León et al., 1982). Beef production in Costa Rica oscillated around 90,000 tonnes between 1985 and 1995 and declined to 65,000 tonnes in 2005 (a 32% decrease since 1996), and this is reflected in a decreasing trend of beef exports between 1996 and 2003 (Figure 5.2a). The number of slaughtered beef cattle also followed a similar trend, declining from 467,000 head in 1996 to 275,000 in 2005 (FAOSTAT 2006). Inside Costa Rica per capita beef consumption rose, compensating to some extent for the decline in exports: per capita consumption rose from 20.4 kg in 1976/1983 to 22.9 kg in 1984. In more recent years consumption of beef has showed a decreasing trend, which may be associated with high beef prices compared with those of pork and poultry (FAOSTAT 2006).

Many farmers switched to milk and dual-purpose herds because of strong consumer demand (Table 5.3) and also because protectionist policies made milk production more profitable than beef production (Van der Kamp 1990), and Costa Rica switched from a net importer to exporter of milk (Figure 5.2b). Production and consumption of milk per capita showed an increasing trend from 152 liters per year in 1990 to 181 liters in 2000. There was a slight decline in 2005 to 172.2 liters/capita/yr (Table 5.3) (FAOSTAT 2006). On the other hand, milk exports showed an increasing trend between 1996 and 2005. This may be associated with intensification of dairy production through the use of improved technologies, leading to increased productivity—as evident from an increase in milk productivity of cows over the last years (Cámara Naçional de Productores de Leche

Figure 5.2a. Rate of beef exports and imports from Costa Rica, 1990–2004.

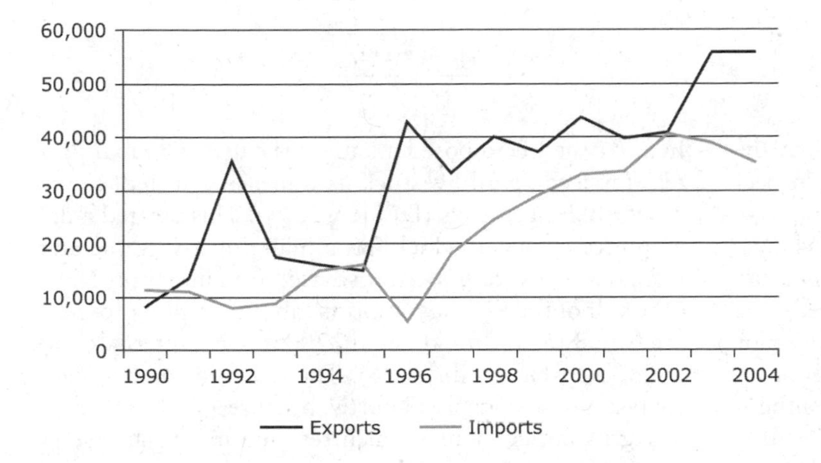

Figure 5.2b. Rate of milk exports and imports from Costa Rica, 1990–2004.

2006, Cooperativa Dos Pinos 2009). Over the last decade or so the area of pastures decreased, and the population of cattle showed a decreasing trend. On the other hand the percentage of tree cover showed an increasing trend, related to measures taken by the government to detain deforestation and to implement policies for reforestation (Sánchez 2009).

Brazil: Demand for Beef, Soybean, and Soy Products

The linkages between market forces and deforestation in the Brazilian Amazon have been analyzed by Nepstad et al. (2006b; Figure 5.3). These authors noted that exportation of beef and expansion of soybean cultivation were the main driving forces of deforestation, as discussed in the following section.

DEMAND FOR BEEF IN BRAZIL

Domestic beef consumption was one of the main drivers for the expansion of Brazil's cattle ranching activities between 1970 and the 1990s, but after this, several factors began to change the main drivers of deforestation from the domestic economy and policies to the international market. These factors include an increase in the demand for beef in Europe because of bovine spongiform encephalopathy

Table 5.3. Production and per capita consumption of fresh whole milk, Costa Rica, 1990–2004

	Year				Annual Growth Rate (%)		
	1990	1995	2000	2005	1990–1995	1995–2000	2000–2004
Production (1000 liters)	463.8	583.4	721.9	752.3	4.7	4.4	0.8
Consumption (liters/capita/year)	151.7	168.3	181.4	172.2	2.1	1.5	-1.0

Source: FAOSTAT 2006.

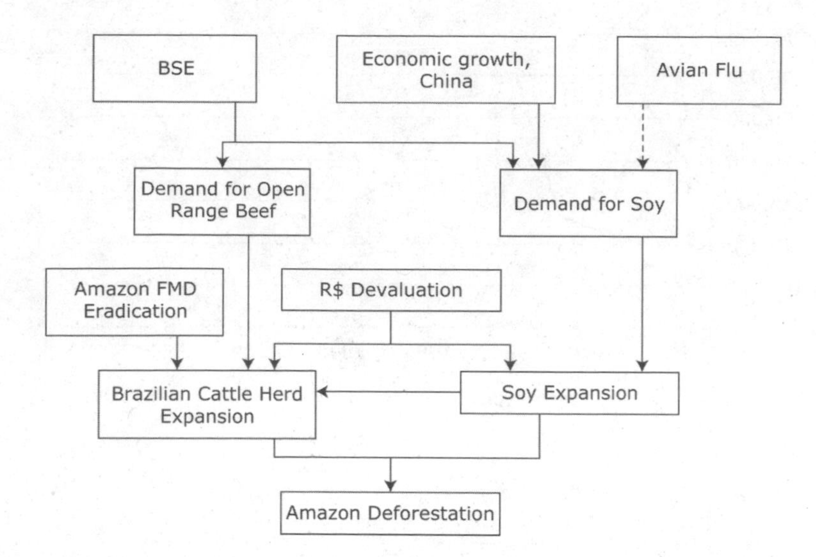

Figure 5.3. Linkages between market forces and deforestation in the Brazilian Legal Amazon.

(BSE), progress in eradicating FMD, devaluation of the Brazilian currency (the Real), and improvements in beef production systems. Among these, some authors have identified FMD eradication as the most important change that contributed to the expansion of the Amazon cattle industry. The FMD-free status was indeed conferred on a large area (close to 1.5 million km² [Figure 5.4 in the color well]) in nine forest regions in the Southern Amazon, including the states of Mato Grosso, Acre, and the southern half of the state of Pará (Kaimowitz et al., 2004, ABIEC 2005, Arima et al., 2005, Nepstad et al., 2006b) that has allowed the export of beef outside the Amazon.

Beef exports increased linearly from the mid-1990s till 2003, and thereafter there was a decline (Figure 5.5a) because after the discovery of FMD in the central Brazilian state of Mato Grosso do Sul in September 2005, 52 countries suspended the import of Brazilian beef (MAPA 2005). An important program to eradicate FMD (PAHO) may, however, contribute to a reduced incidence of FMD in the long term (Nepstad et al., 2006b). With respect to the dairy sector, Brazil imports a large quantity of milk, and there has been a strongly decreasing trend in the amount of milk imported (Figure 5.5b).

A large percentage of beef exports are produced in the Amazon. This is reflected in an increase in the cattle herd in the Legal Amazon, which expanded by 8.6% annually from 35.6 million head in 1996 to its 2005 level of 74.6 million head, followed by a decrease to 70.1 million head in 2007 (Table 5.1). The rapid increase in the cattle herd between 1996 and 2005 was correlated with a surge of deforestation between 2002 and 2004 in the Amazon (INPE 2006), which was also related to the expansion of soybean cultivation (Figure 5.1).

Demand for Soybeans and Soy Products

Over the last decade the international demand for Brazilian soybeans and soy products (soy oil and soy meal) has increased, triggering more deforestation in the Amazon. In 2001 the EU imposed a ban on the use of animal protein for feeding all livestock as a measure to reduce the risk of BSE outbreaks (DEFRA 2005). This created a demand for soymeal, which has a high nutritive value and is a good substitute for feeds based on animal proteins. One half of the EU's soy imports (about 6 million tonnes) are from Brazil (Brookes et al., 2005; LMC International 2003, Nepstad et al., 2006). The demand for Brazilian soybeans also increased partly because of China's rapidly growing economy, which resulted in an increase in the per capita consumption of soy-fed pork and poultry (Naylor et al., 2005). Soybean imports in China for 2003 were 21 million tonnes, an 83% increase over 2002. Some 29% of this amount was imported from Brazil (ASA 2003). Exportation of Brazilian products between 1997 and 2003 was enhanced by the devaluation of the Brazilian real (BACEN 2006).

Soy expansion into the Amazon began in the late 1990s, as new varieties were developed that tolerated the moist, hot Amazon climate (Fearnside 2001). The growing demand for soybeans, combined with low land prices and improved transportation infrastructure of Southeastern Amazonia, promoted major soy companies to invest in storage facilities in the region (Nepstad et al., 2006a). As a result the production of soybeans in the closed-canopy forest region of the Amazon (i.e., excluding savanna regions) increased 15% per year from 1999 to 2004 (IBGE-PAM 2005).

The expansion of the Brazilian soybean industry into the Amazon may have driven expansion of the Amazon cattle herd indirectly, through the effect on land prices, which have increased 5- to 10-fold in many areas of Mato Grosso (Nepstad et al., 2006b). Apparently many cattle ranchers who own properties suitable for soy production have sold off their holdings with enormous capital gains, enabling them to expand their herds and purchase land farther north in Amazonia where prices are lower (Naughton-Treves 2004).

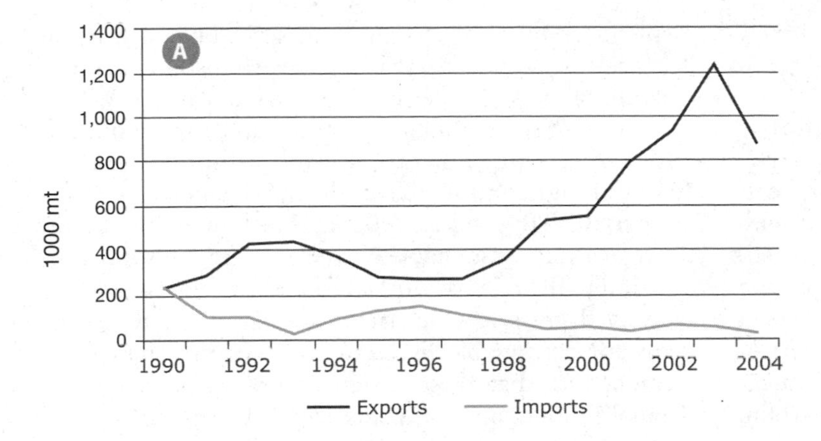

Figure 5.5a. Rate of beef exports and imports from the Brazilian Legal Amazon, 1990–2004.

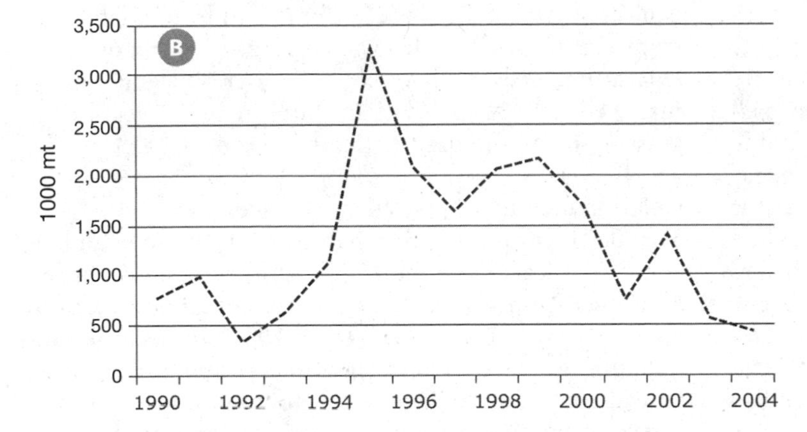

Figure 5.5b. Rate of milk exports and imports from the Brazilian Legal Amazon, 1990–2004.

This has led to recent expansion of large-scale mechanized agriculture at the forest margins; for example, in nine states of the Brazilian Legal Amazon, mechanized agriculture increased by 36,000 km², and deforestation was 93,700 km² during 2001–2004 (Nepstad et al., 2006b). In the state of Mato Grosso between 2001 and 2004, there was an 87% increase in cropland area and 40% of new deforestations in the Legal Amazon occurred (Morton et al., 2006). Favorable prices for soybean and its products in international markets, and higher profitability of soybean cultivation compared to cattle ranching, had impacts in the dynamics of forest clearing in the Legal Amazon. This is evident from the figures of deforestation and transition of deforested lands between 2002 and 2003 in Mato Grosso, where the share of deforested land converted to cattle pasture decreased from 78 to 66%, whereas the share converted to cropland increased from 13.5 to 23% (Morton et al., 2006). Deforestation for cropland cultivation in Mato Grosso during 2001–2004 was concentrated within the Xingu River basin and close to existing centers of crop production (Sinop, Sorriso, Lucas do Rio Verde, and Nova Mutum) along the Cuiaba-Santarem highway in central Mato Grosso state. Deforestation for cattle pasture predominated in the northern and western portions of the state (Morton et al., 2006).

The data show a decreasing trend in deforestation from 2004 and 2008 (Figure 5.1), which may be associated with low prices for beef and a decrease in the demand for soybeans. For example the spread of avian flu has reduced consumption of poultry and thus has decreased the demand for poultry rations and the soy meal they contain (Rocha and Boucas 2006). This together with a 25% increase in the strength of the Real and low prices for soybeans between 2003 and 2005 were important factors that decelerated both the expansion of the agroindustry and cattle ranching in the region as reflected in deforestation estimates for 2005 in Figure 5.3 (Nepstad et al., 2006b).

In summary, market forces have triggered deforestation in both countries, though there are differences in the major drivers. For example, international demand for beef in the United States was the driving force in Costa Rica, but with a decrease in price and demand for beef, and with discontinuation of some of government policies in the livestock sector, deforestation rates decreased significantly over the last two decades. On the other hand, in the Legal Amazon the international demand for beef and soybean and soy products are the main market forces driving deforestation. A decrease in demand for soy products and the outbreak of FMD in some areas in the Legal Amazon has been associated with a slight decrease in deforestation although it remained relatively high.

Environmental Consequences of Deforestation and Pasture Expansion

Large-scale deforestation for pasture expansion and cattle ranching in Latin America has been associated with negative impacts on the environment, climate, and livelihoods of farmers and the rural poor, and this has been a subject of major concern to policy makers and donors in the region. This section analyzes the negative impacts of deforestation and pasture expansion on the environment with regard to land degradation, biodiversity, carbon, and water resources.

Land Degradation

Throughout Latin America cattle production models are generally based on the use of grass monoculture pastures with little management. These pastures degrade over time, resulting in loss of productivity and negative environmental impacts. In Costa Rica and Brazil it is estimated that 40 to 50% of improved or established pastures are degraded, resulting in reduced carrying capacities. Normally 1.5 to 2 heads per hectare might be sustainably managed, but carrying capacities of degraded pastures are usually only 0.4 to 0.6 head of cattle per hectare at best (Szott et al., 2000). Recent studies in the subhumid tropics of Costa Rica have shown that pasture degradation resulted in a significant reduction in live weight gains and net income per hectare when compared to well-managed pastures (Figure 5.6, Lemus 2008).

In many areas, small farmers have abandoned degraded pasture farms and have migrated to the agricultural frontiers to practice slash-and-burn agriculture, which is usually followed by further cattle ranching (Kaimowitz 1996, Schelhas 1996). Some scientists have associated pasture land degradation with nutrient depletion, decreases in total porosity, water infiltration, and soil structure (Teixeira et al., 1996). However, studies conducted in the Brazilian Amazon showed that there was little linkage of pasture degradation to nutrient depletion, but rather to an increase in bulk density of surface soil layer, reflected in decreasing soil cover and pasture biomass (Miller et al., 2004). Because degradation of pastures is a major economic and environmental problem in Latin America, several institutions like the Tropical Agricultural Research and Higher Education Center (CATIE) and the Foundation Center for the Investigation in Sustainable Systems of Agricultural Production (CIPAV) have worked to develop and implement improved grass legume pastures and silvopastoral systems for sustainable livestock production. The results demonstrate that these systems sustain relatively high animal production, and under same conditions grass legume pastures had higher production than grass monoculture pastures (Table 5.4). For example, in the humid tropics of Costa Rica, liveweight gains of cattle on *Brachiaria brizantha* and *Arachis pintoi* grass legume mixture were 29.4% higher than those of the *B. brizantha* grass monoculture pasture (Hernández et al., 1995).

Deforestation and Impacts on Biodiversity

The destruction of forest habitats has been a global concern at least since the 1970s, although deforestation and other forms of land and environmental degradation have occurred at high rates since 1950. In Central and in South America, as in other tropical regions, the widespread conversion of forests to agricultural land poses a threat to biodiversity conservation. Deforestation leads to loss of native plant communities, loss of habitat and resources for wildlife, and disruption of ecological processes such as seed dispersal, pollination, and animal dispersal. As a consequence, deforestation is usually accompanied by biodiversity loss at the genetic, species, and ecosystem level (Harvey et al., 2005). Wassenaar et al. (2007) noted that the Brazilian *Cerrado* is a deforestation hotspot in Myers's biodiversity hotspots, having already lost some 80% of its primary vegetation (Myers et al., 2000).

For certain tree species in the still largely forested areas the projected distribution of hotspots may represent a serious threat. Looking at the example of large-leaved mahogany in the global tree conservation atlas under

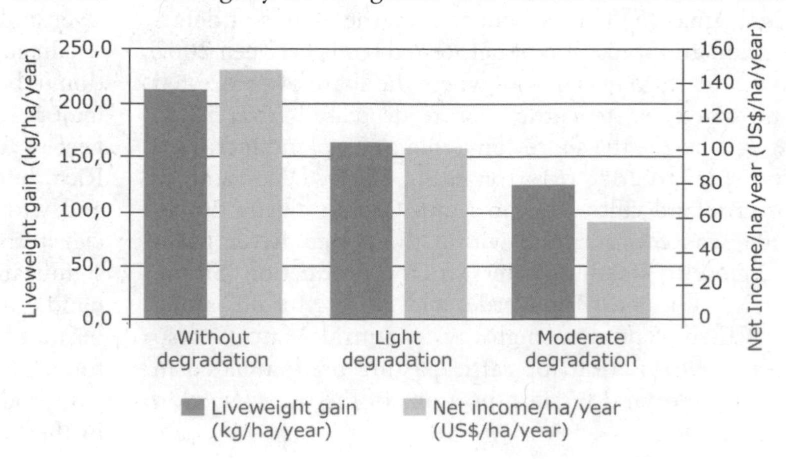

Figure 5.6. Live weight and net income per hectare in relation to pasture degradation in the subhumid tropics of Costa Rica.

Table 5.4. Liveweight gain (LWG) of cattle in different pasture and silvopastoral systems in Costa Rica and the Amazon of Brazil

System	Ecosystem	Stocking Rate (UA/ha)	LWG (kg/ha)	References
Costa Rica				
Brachiaria brizantha in monoculture	Humid tropical forest	6	714	
Brachiaria brizantha + *Arachis pintoi*	Humid tropical forest	6	924	Hernández et al., 1995
Brachiaria brizantha + *Leucaena leucocephala*	Subhumid tropical forest	1.8	404	Jiménez 2007
Brazil				
Brachiaria decumbens in monoculture	Humid tropical forest	0.7–1.6	241.5	Paciullo et al., 2007
B. decumbens + *Calopogonium mucunoides*	Humid tropical forest	4	385.0	CNPGC 1988
*Silvopastoral systems***	Humid tropical forest	0.7–1.6	283.6	Paciullo et al., 2007

** *Brachiaria decumbens* grass in alleys with woody species as *Acacia mangium*, *A. angustíssima*, *Mimosa artemisiana*, *Leucaena leucocephala*, and *Eucaliptus grandis*.

construction (Newton et al., 2003), it becomes clear that if deforestation continued in the Amazonian lowland hotspots of Ecuador, Peru, Bolivia, and Brazil, this would eliminate a substantial portion of the mahogany's natural distribution area (Wassenaar et al., 2007).

The conversion of forest to pasture systems results in a significant reduction in biodiversity, as seen for the Brazilian Amazon in Figure 5.7, which shows that the biodiversity index is lowest for improved grass pastures, compared to forest and agroforestry systems. Studies on landscapes dominated by cattle and pasture in different ecosystems and with different production systems indicate that there is a significant loss of biodiversity and a relatively small number of species of trees, birds, and butterflies, compared to the protected areas or biological stations as evident in Table 5.5, which shows that values for the different taxa mentioned were lower than those registered in the Santa Rosa national park and the La Selva Biological station of Costa Rica.

Over the past years, CATIE has been promoting the replication of complex silvopastoral systems (involving the integration of trees and shrubs in pastures) to improve farm productivity and for conservation of biodiversity. The results of various studies have shown that silvopastoral systems with high tree densities and multistrata live fences conserved a relatively large number of birds and butterfly species compared to degraded pastures, and values for bird species richness were comparable to those measured in secondary forest (Table 5.6). These studies show that silvopastoral systems can conserve endangered species, for example, the three-wattled bellbird (*Procnias tricarunculata*), the long-tailed manakin (*Chiroxiphia linearis*), the crested owl (*Lophostrix cristata*), and the white hawk (*Leucopternis albicollis*) (Tobar and Ibrahim 2008).

The conversion of forests into pastures has resulted in forest fragmentation (Harvey et al., 2005, Etter et al., 2006). Forest fragmentation generally results in the reduction of large habitat blocks into smaller areas, and the creation of edges between forest and nonforest habitat. These areas experience diverse physical and biotic changes associated with the abrupt margins of the forest fragment and the isolation of fragments from intact forest. All of these affect the habitat quality of forest fragments and their ability to maintain biodiversity (Saunders et al., 1991, Bierregaard et al., 2001, Kattan 2002, Harvey et al., 2005). In Costa Rica, Sánchez-Azofeifa and colleagues (2001) found that the loss of 2250 km^2 of forest from 1986 to 1991 was accompanied by a sharp increase in the number of forest fragments. The degree of deforestation and fragmentation was particularly severe in the tropical moist forest and premontane moist forest, where little forest cover remains and the mean forest patch is small (0.3 to 0.5 km^2). Though the conversion of forest into pasture and other agricultural land use has been occurring for decades, there has been no long-term study to quantify the impacts of the resulting forest fragmentation on the conservation of biodiversity. A long-term study of fragmentation in the Brazilian Amazon suggests that certain animal groups are more vulnerable to fragmentation than others, including understory birds, primates, shade-loving butterflies, and solitary wasps (Bierregaard et al., 2001; Harvey et al., 2005). Recently, cattle farms have been adopting silvopastoral systems, including the establishment of live fences and trees in pastures, and this has resulted in increased functional

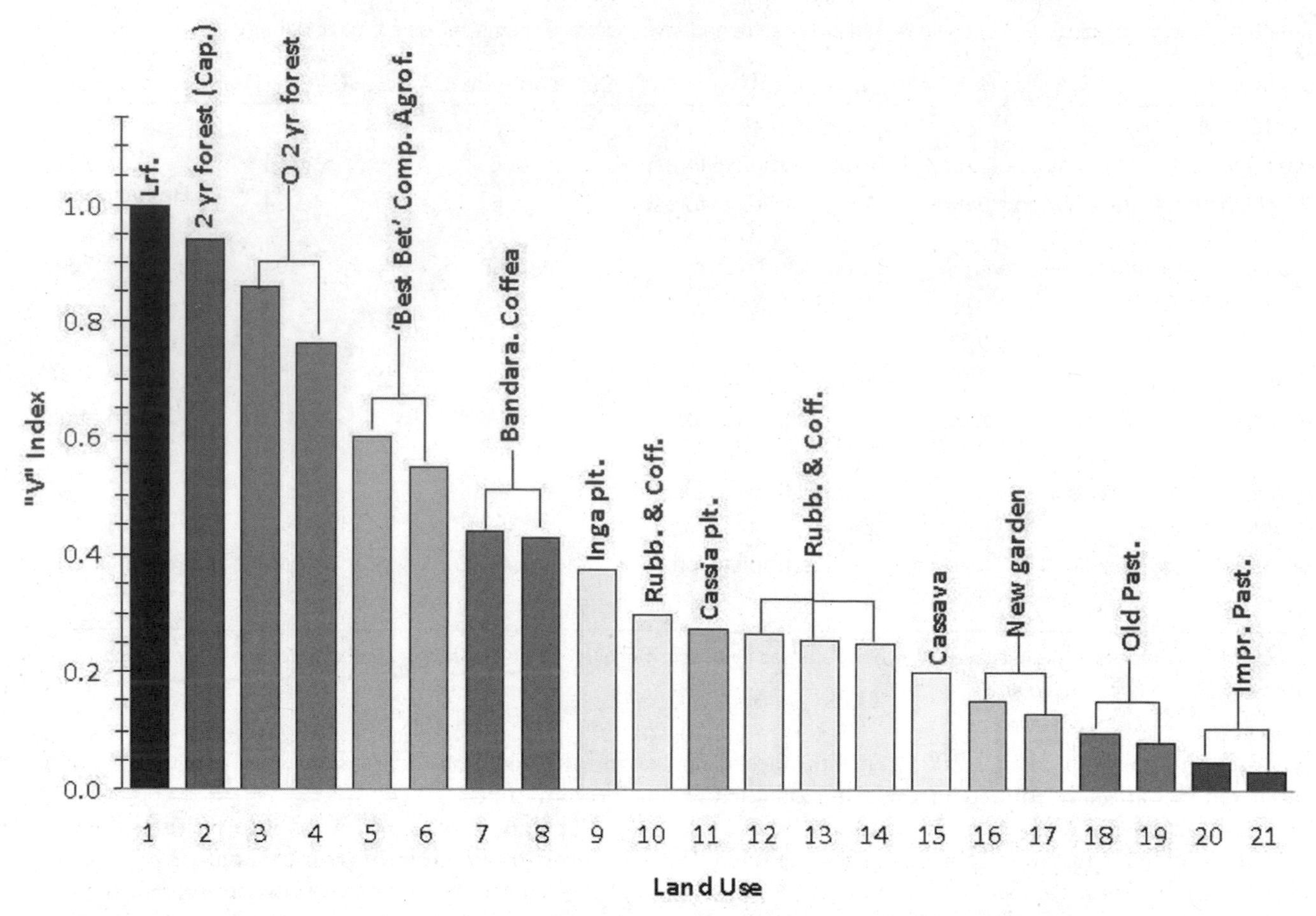

Figure 5.7. Conversion of forest to pasture systems and a reduction in biodiversity in the Brazilian Legal Amazon.

and structural connectivity in agricultural landscapes (Tobar and Ibrahim 2008).

Carbon Sequestration and Greenhouse Gas Effects

The quality of management of tropical pastures is critical to deciding whether soils under this land use represent a source or a sink of atmospheric carbon. Studies in Costa Rica and Brazil have shown that conversion of forest for the establishment of pastures results in a net loss of carbon. Veldkamp (1994) found a net loss of 2 to 18% of carbon stocks in the top 50 cm of forest equivalent soil after 25 years under pasture in lowland Costa Rica. However, many studies have shown that soil carbon stocks with well managed improved pastures are similar to those found in forest ecosystems (Bayer et al., 2002, Manfrinato et al., 2001), but, as mentioned earlier, grass monoculture pastures without fertilizer and proper grazing management generally degrade in time, and the loss of productivity of these pastures is associated with a reduction in carbon stocks. Total carbon stocks measured in different pasture and forest systems

Table 5.5. Status of biodiversity in agricultural landscapes dominated with cattle and pastures in the subhumid tropics of Esparza and Cañas and of the humid tropics of Río Frío, and the Santa Rosa National Park and "La Selva" Biological Station in Costa Rica

Number of Species by taxa	Esparza	Cañas	Santa Rosa NP*	Río Frío	La Selva Biological Station**
Tree species	68	83	243	85	323
Birds	158	122	236	214	375
Bats	NA	41	46	45	66
Butterflies	139	60	205	68	369
Dung beetles	NA	37	50	37	60

* Site close to Cañas.

** Site close to Río Frío.

Sources: Harber and Stevenson 2004, Harvey et al., 2006, Tobar et al., 2006, Saenz et al., 2007.

Table 5.6. Mean richness values of birds and butterflies monitoring in different land uses in Esparza, Costa Rica, 2003–2008*

Habitat	Birds	Butterflies
Degraded pastures	3.1 ± 0.41a	8.8 ± 1.2a
Pastures with low density trees	29.4 ± 2.71bc	25.6 ± 1.29b
Pastures with high density trees	39.8 ± 2.67c	31.8 ± 2.35cd
Simple live fences	22.2 ± 4.77b	27.6 ± 1.89bc
Multistrata life fences	26.8 ± 6.14b	36.8 ± 0.97d
Riparian forest	26.4 ± 2.87b	50.8 ± 2.62e
Secondary forest	33.2 ± 3.93bc	51.4 ± 2.87e

* Means in the same column and with the same letter are not significantly different ($P < 0.05$) using LSD Fisher test.

Sources: Saenz et al., 2007, Tobar et al., 2006, Tobar and Ibrahim 2008.

in the humid tropics of Costa Rica were significantly lower in degraded pastures compared to native forest. By contrast, improved grass–legume pastures and silvopastoral systems had relatively large amounts of carbon compared to degraded pastures and native forest. These results indicate that there is potential to derive income from the carbon sequestered in productive systems through carbon markets (Table 5.7).

Cattle production is also associated with emissions of methane and nitrous oxide. In Brazil it is estimated that methane emissions per head of cattle are 0.043 tonnes/yr (MCT Brazil 2004), and the increase in bovine population between 1997 and 2006 accounts for 9 to 12 billion CO_2 emissions (Smeraldi and May 2008). However, the integration of high-quality leguminous pastures and the use of woody fodder banks (e.g., *Leucaena leucocephala*) in ruminant production systems can help reduce the use of nitrogenous fertilizers for managing the pastures, and hence cut N_2O emissions and also reduce emissions of enteric methane.

Water Resources

The conversion of tropical forest for cattle ranching using grass monoculture pastures has also had negative effects on water storage and quality. Nepstad et al. (1994) found that in Pará, Brazil, during a severe dry season, plant-available soil water at 2 to 8 m depth declined from 380 mm in the forest to 310 mm in a degraded pasture. Average daily rainfall was 0.6 mm. Evapotranspiration was 3.6 mm for forest and 3.0 mm for degraded pasture. Less depletion of plant-available water in the degraded pasture indicates that this ecosystem can store less rainfall than forest and may therefore produce more seepage to the groundwater aquifer or subsurface runoff to streams in the wet season. Additionally, recent studies conducted by a global environment facility–funded silvopastoral project showed that water runoff was very significantly higher in degraded pastures (42%) compared to fodder banks with woody species (3%), pastures with high tree densities (12%) and young secondary forest (6%). These results support the assumption that land use changes with higher tree cover are beneficial for water harvesting, depending on the water requirements of tree species and ground cover (Andrade 2007). Pasture is a major land use in many watersheds in Costa Rica, and overgrazing results in sedimentation of rivers and reservoirs used for hydroelectricity production, which increases the cost of removing sediments.

Table 5.7. Mean total C stocks in soil (t/ha/1m-equivalent) and in tree biomass (t/ha) in the humid tropics of Pocora, Costa Rica*

System	Minimum age of use (yrs)	C in soil (t/ha)	C in tree aerial biomass (t/ha)	Total (t/ha)
Degraded pasture	> 30	107.9 d		107.9
I. ciliare	> 30	254.4 f		254.4
B. brizantha	> 19	153.0 e		153.0
Silvopastoral system with				
Acacia manginum + B. brizantha	15	160.9 e	12.8	173.5
B. brizantha + A. pintoi	16	186.8 e	174.2	186.8
Native forest		141.3 de		315.5

* Means in the column with the same letter are not significantly different ($p < 0.5$).

Source: Amezquita et al., 2008.

Policies and Incentive Mechanisms to Reduce Deforestation and Recover Forest Cover in Costa Rica and the Brazilian Legal Amazon

The analysis of the drivers of deforestation showed that government policies were critical as an underlying cause of deforestation in the early phases of deforestation in Costa Rica and in the Brazilian Amazon (for example tax allowances, credit subsidies, land tenure policies, construction of roads, etc.). However, at a later stage the main causes of deforestation changed from public policies to market forces, for example, good prices for beef, and in the case of Brazil, an increase in the demand for soybeans and more recently renewable energy (Kaimowitz 1996, Nepstad et al., 2006a). Measures required to overcome deforestation and land degradation pressures include policy dialogue and enhanced market linkages for environmental goods and services. The strengthening of local organizations is also crucial to more effectively develop and adapt technology and management innovations and to participate in policy dialogues, and innovative information and knowledge management (Nepstad et al., 2006a,b). In both Costa Rica and Brazil several measures have been and are being implemented to reduce deforestation, including command and control and incentive policy mechanisms. This section analyses the impacts of government policies to reduce deforestation and restore forest, and identifies barriers for policy implementation in both countries. There is also an emerging market for certified livestock products, which offers a good opportunity for developing cattle production systems in harmony with forest conservation. The potentials for certified livestock products are discussed later in the chapter.

Government Command and Control Policies

Government command and control policies have been very effective in reducing deforestation and in the progressive recovery of the forest cover in Costa Rica, and in the Brazilian Legal Amazon there is evidence that these policies have had some success in curtailing deforestation.

An inventory of policies implemented by the Costa Rican government to protect and recover forest resources was presented by Sánchez (2009). Policies on agrarian reform together with the creation of biological reserves have discouraged deforestation, especially in Costa Rica, in which the government approved policies for the creation of national parks, protected areas, and biological corridors (Sánchez 2009). Costa Rica first established its national park system in 1974, and it has expanded over time. The data in Costa Rica show that there has been a progressive recovery of forest cover over the last 10 to 15 years, reaching 48% in 2005 (excluding mangroves, forest plantations, and wetlands) of the national territory (FONAFIFO 2007). Of the area of forest cover, 45% is in the category of legal protection by the state, and 34% is contracted under payments for environmental services (PES). The Ministry of Environment implemented an executive decree (33106) to create national biological corridors for the conservation of forest resources and biodiversity—in 2008 there were 35 established biological corridors (Figure 5.8 in the color well, SINAC 2008).

Overall the Ministry of Environment and Energy of Costa Rica has been very successful in protecting the forested areas, in large measure because it has an adequately financed protection service and has vigorously enforced environmental law, including taking legal action over breaches (FONAFIFO 2007).

The Brazilian government has also tried to slow deforestation through a series of measures, including the banning of land titling in eight million ha of land along the BR 163 highway, and the designation of five million ha of new parks and reserves in the Eastern Amazon (Nepstad et al., 2006a,b). In addition, the Brazilian government has designated some two thousand military troops to protect the Amazon reserves and has taken measures to bring to justice environmental personnel found guilty of corruption (Soares-Filho et al., 2006).

In terms of forest conservation there are specific laws that govern the use of forests in Brazil. Any action to deal with deforestation has to be based on the Forest Law 4771 (September 1965, altered in August 2001 by Law 2166-67). In Brazil, conservation units (CUs) are protected areas, especially dedicated to conserve the original biodiversity and the natural or cultural resources. The CUs are managed by the National System for Nature Conservation Units (SNUC 2004).

Over the next ten years, the Brazilian federal government and the Ministry for the Environment are planning to create CUs covering 50 million ha of land, and to promote the sustainable development of these regions with inputs from partner organizations such as GEF, FUNBIO, KFW and WWF and GTZ and other mechanisms, to invest US$400 million in the Amazon (MMA 2007).

Ambitious government conservation policies in Brazil are recent, and there is a need for in-depth analysis to determine how these polices are curbing deforestation. Some policies are difficult to enforce, and some can have unwanted side effects because they act as a disincentive to sustainable forms of cattle and soybean production. For example, Brazil's environmental legislation requires that 80% of the forests and all riparian zones on private lands in the Amazon be maintained as reserves (Lima et al., 2005). This requirement can significantly reduce the profitability of cattle ranching and soybeans, which in turn can be a disincentive against developing good land stewardship for cattle and soybean production. The state government of Mato Grosso, where more than 40% of deforestation takes place, is using satellite-based monitoring of private forests. This system may become popular if the demand for environmentally friendly beef and soybeans increases (Nepstad et al., 2006b). However, lack of resources and infrastructure to implement government policies in the Legal Amazon is one of the main difficulties for enforcing environmental laws and regulations.

Payment for Environmental Services

Costa Rica is often considered a champion in the design and implementation of countrywide PES incentive mechanisms to conserve and restore forest resources, and the lessons learned from the Costa Rica experience are being used in the rest of Latin America. The PES program, which began in 1997, has been partly credited for helping the country—once known for the world's highest deforestation rates—to achieve negative net deforestation by the early 2000s (Castro et al., 1997, Chomitz et al., 1999). Forest Law 7575 of 1996 explicitly recognized four environmental services provided by forest ecosystems:

1. Mitigation of greenhouse gas emissions
2. Hydrological services, including provision of water for human consumption, irrigation, and energy production
3. Biodiversity conservation
4. Provision of scenic beauty for recreation and ecotourism.

The PES program is managed by the Fondo Nacional de Financiamiento Forestal (FONAFIFO), a semiautonomous agency with independent legal status. To date the bulk of PES program financing has been provided by allocating to FONAFIFO 3.5% of revenues from a fossil fuel sales tax (about US$10 million a year). From 2001 to 2006, the PES program was supported by a loan from the World Bank and a grant from the Global Environment Facility (GEF) through the Ecomarkets project (Pagiola 2006). In 2005, Costa Rica expanded the use of water payments by revising its water tariff (which previously charged water users near-zero nominal fees) and introduced a conservation fee earmarked for water conservation. Once fully implemented, this fee will generate an estimated US$19 million annually, of which 25% would be channeled through the PSA program (Fallas 2006). Additionally, FONAFIFO has established a marketing unit for environmental services, which has been very successful in negotiating funds from the private sector.

Forest cover of Costa Rica increased (Figure 5.8 in the color well) due to several factors, including the implementation of a forest legislation that banned the clearing of forestland. Changes in the profitability of livestock production have also reduced pressure to convert forests to pasture, particularly in marginal areas (White et al., 2001, Arroyo-Mora et al., 2005). But PES has also played a significant role. Studies have generally found that PES recipients have higher forest cover than nonrecipients. Zbinden and Lee (2005) found that PSA recipients in Northern Costa Rica had 61% of their farm under forest, compared to only 21% for nonrecipients. Likewise, Sierra and Russman (2006) found that PES recipients in the Osa Peninsula had over 92% of their farm under forest or bush compared to 72% for nonrecipients.

Brazil has not yet developed a national system like Costa Rica's, but there are some examples of PES for forest conservation by building public–private partnerships. An example is the Forest Stewardship Program (Bolsa Floresta) in the Amazonas state (Government of the State of Amazonas 2007). This is a public program of the State of Amazonas, implemented since September 2007 by the Amazonas sustainable foundation and two other Brazilian institutions—the Public Secretariat for the Environment and Sustainable Development and Bradesco, the largest private bank in Brazil. The key objectives are improved forest conservation (avoided deforestation) and livelihood improvements for traditional and indigenous communities in state-protected areas and sustainable-use reserves. Specifically the program rewards indigenous communities and long-term settlers for their commitment to avoiding deforestation. Payments are made according to categories—a family can receive about R$50 (US$30) per month and a community association can receive R$4,000 (US$2500) per year. A penalty is applied when participants deforest beyond a maximum limit or use unsustainable land-use practices. Presently, the program covers six reserves/protected areas and 2102 families in six conservation units of the Amazonas state.

Reducing emissions from deforestation and degradation (REDD) is currently being discussed as an additional strategy to mitigate climate change. This represents an enormous opportunity to save the Amazon forest, which is enormously important in the fight against global warming. Brazil contains 63% of the Amazon biome, and large-scale deforestation in the Amazon region contributes to carbon emissions and global warming. Nepstad et al. (2007) estimate that if current trends continue, 55% of the forests of the Brazilian Amazon will be cleared, logged, or damaged by drought by the year 2030, releasing 20 (± 5) billion tonnes of carbon to the atmosphere.

Many donors are interested and have started to fund the REDD program. However, REDD schemes will depend for their implementation on functioning and noncorrupt government institutions and effective law enforcement, and forest governance in the Amazon is weak. About 80% of deforestation in the Brazilian Amazon is estimated to be illegal (Wertz-Kanounnikoff et al., 2008). There is a need to look for new policies to put an end to open frontiers by enforcing existing land-use policies and by using environmental taxation or PES to establish a price for the use of traditionally abundant natural resources. PES and REDD programs should provide incentives to conserve forested resources and reduce deforestation, especially if long-term payments are programmed. However, this will depend on whether the level of compensation matches the opportunity cost, and whether cattle farmers are satisfied with economic and environmental benefits.

Certification

Globally, there is mounting pressure from donors and marketing companies, as well as from consumer and producer organizations, to reduce the negative ecological and social impacts of production systems (Clay 2004). In addition, there is a growing demand for improved animal welfare and food safety practices in animal production.

For example, there is some concern in importing countries—especially in the EU—that the production of soybeans and beef in Amazon is linked with deforestation, use of slave labor, and increased risk of diseases (Monbiot 2005). In this respect, the United Kingdom's National Beef Association called for a boycott of Brazilian beef because of its association with Amazon deforestation and its contribution to global warming (IcWales 2003). The pressure on Amazon beef and soy producers is also coming from within Brazil, as consumers demand beef produced according to good environmental and social standards. Already, a growing number of beef retailers in Southern Brazil (e.g., the supermarket chains Carrefour and Pão de Açúcar) and meat processors (e.g., Friboi, Bertim) are looking to the Amazon for reliable sources of high-quality beef produced on ranches that obey environmental legislation and use good land-management techniques (Nepstad et al., 2006b).

The growing demand for certified animal products (natural, organic, environmentally friendly, etc.) offers incentives for the development of sustainable cattle production systems in harmony with the environment. The high cost of environmental compliance has been a barrier for the adoption of good farming practices (e.g., silvopastoral systems, protection of forest, etc.) in cattle farms. Because certified products can command higher prices, a system to certify environmentally friendly and sustainable cattle farms will provide incentives for farmers to comply with environmental regulations. Currently, CATIE is collaborating with Rainforest Alliance to develop a certification system for sustainable cattle production. This system includes critical criteria for certification of farms for protecting the environment and ecosystem, animal welfare, food safety, and social working conditions (Sepulveda, personal communication). In the case of the Brazilian Amazon, certification could be provided to producers who demonstrate their compliance with forest-reserve legislation and who adopt best management practices such as those used by some soy and cattle producers (Nepstad et al., 2006b).

The demand for forage-based or grass-fed livestock products (beef and milk) has been increasing because of the health concerns associated with ration-fed systems (Nepstad et al., 2006b, Roosevelt 2006). Therefore, certification systems will have to include critical criteria, not just for farmers but along the whole value chain, to meet consumers' concerns about environmental, health, animal welfare, and social standards. This requires the development of good traceability systems to comply with international standards for exports. So as to target greater participation of Brazilian livestock products in European markets, the Ministry of Agriculture and Livestock in 2002 began the implementation of SISBOV, the Brazilian System of Identification and Certification of Origin for Bovine and Buffalo. Animals registered with SISBOV have an identification number containing the origin property, month of birth, gender, raising system, feeding, and sanitary data. In Costa Rica, the Ministry of Agriculture has introduced a system for traceability, as well as health measures to ensure good food safety (e.g., use of mandatory vaccines, testing for tuberculosis and brucellosis, suppression of the use of animal by-products for feeding animals, list of prohibited veterinary products, etc.). The Costa Rican government approved a law for organic agricultural and livestock production, but it has had little success—perhaps because of the demanding criteria for certification and lack of sufficient incentives for farmers making the shift from conventional to organic production systems.

Recent trends in the cattle value chain in the Amazon can be exemplified by a US$90 million loan from the World Bank's International Finance Corporation (IFC) to the Bertin Group, Brazil's second-largest meat processing company. The project was announced in March 2007 as setting new benchmarks for environmental and social standards in cattle ranching and meat processing in the Amazon. The loan will support Bertin's corporate investment program to expand and modernize its operations across the country and will help it develop a system, the first of its kind in Brazil, to ensure that cattle are sourced from ranchers that use sustainable practices and do not contribute to increased deforestation of the Amazon (IFC 2007). IFC reported that the loan evaluation process included a thorough assessment of the direct and indirect impacts that the expansion of Bertin's meat processing plants could have on deforestation, and assessed how the project could be used to address social issues such as forced labor and agrarian violence. In the immediate aftermath of this announcement, however, the Brazilian Forum of NGOs and Social Movements for the Environment and Development (FBOMS) contested the project, stating that the IFC board was misled because the project includes construction of six facilities in critical areas in the states of Rondônia, Mato Grosso, and Pará, although studies and public hearings were conducted in only one of the regions (Marabá, in Pará). In addition, they argued that a presentation to IFC management failed to mention that three of these facilities, including a meat packing plant, were to be located at the environmentally sensitive headwaters of the Xingu River (Amigos da Terra 2007).

Perhaps the most far-reaching driver of the reform of agroindustrial commodity producers is the Equator bank initiative, in which finance institutions representing more than 80% of the project finance worldwide, including four Brazilian banks, are developing environmental and social standards known as the Equator Principles, and beginning to apply these standards as conditions to loans extended to the private sector (BankTrack 2004).

Though certification of animal products can provide incentives to cattle farmers for compliance with environmental, health, social, and animal welfare standards, policies will have to be implemented to eliminate perverse effects of certification, such as, for example, the clearing of forestlands to produce certified products. As a measure to reduce perverse effects of certification of cattle products, zoning of the Amazon and other hot spots will be required so as to contain the production of certified products in areas where monitoring can be carried out on the forest cover, and so forth.

Conclusions

There are many similarities between Costa Rica and Brazil as to how government policies have triggered deforestation. Among these are subsidized agricultural credit, road construction, and inappropriate land tenure policies that gave land titles to recently cleared forest areas for cattle ranching and other agricultural activities. Some authors have noted that road construction, which opened access to forest areas, was the most important of these policies. However, the governments of both countries have taken steps to eliminate these policies because of their negative impacts on forest reserves and the environment.

Apart from government policies, market forces were also important drivers for deforestation, though there were differences between the two countries. Costa Rica, which is free of FMD, had access to US markets for beef. The resulting increase in demand for Costa Rican beef provided incentives for deforestation and cattle ranching. On the other hand, the outbreak of BSE in Europe and measures taken by the EU to eliminate the use of animal proteins for supplementing livestock created a large demand for Brazilian beef—and a large percentage of this was produced in the Amazon, which is free of FMD. However, recent outbreaks of FMD in some states of the Amazon have led to a decrease in the amount of beef exported from the Amazon. In contrast to Costa Rica, where most of the deforestation was linked to cattle ranching, recent demand for Brazilian soybeans and soy products in Europe and China has led to the expansion of soybean cultivation lands, often involving large-scale deforestation in the Legal Amazon.

Deforestation and unsustainable modes of cattle ranching have both been associated with environmental degradation, as evident in the percentage area, in watersheds where cattle is the dominant land use, of degraded pastures, forest fragmentation and loss of biodiversity, reduction of carbon stocks, emission of greenhouse gases, and reduction of water harvest.

In Costa Rica the implementation of government command and control policies and a system for payment for environmental services have been associated with a progressive recovery of the forest cover over the last years. A large percentage of Costa Rican forest is under national parks, protected areas, and biological corridors and the government has been successful in law enforcement for the protection of the forest. It has been innovative in the design and implementation of a PES system, which has had impacts in the recovery of the forest cover. By contrast, the Brazilian command and control policies have not been so successful in curbing deforestation in the Legal Amazon, which, despite a declining trend, has remained relatively high. This may be partly attributed to the differences in scales between the two countries: the Legal Amazon is an enormous area compared to Costa Rica, which makes it difficult to implement law enforcement to protect the forest reserves. There is also a high level of lawlessness of all kinds in the Amazon, partly due to corruption in local officials, whereas Costa Rica has had a more harmonious society, largely free of the dramatic inequalities in landholding found in the rest of Latin America.

References

ABIEC. 2005. Foot-and-mouth disease outputs, Brazil. Associação Brasileira de Indústrias de Carne, Sao Paulo, Brazil. Cited December 2005. Available at www.abiec.com.br.

Alencar, A., et al. 2004. Desmatamento na Amazônia: Indo Além da Emergência Crônica. Instituto de Pesquisa Ambiental da Amazônia, Belém, Brazil. Cited February 2006. Available at www.ipam.org.br.

Amezquita, M. C., E. Amezquita, F. Casasola, B. L. Ramirez, H. Giraldo, M. E. Gomez, T. Llanderal, J. Velásquez, and M. A. Ibrahim. 2008. C stocks and sequestration. In *Carbon Sequestration in Tropical Grassland Ecosystems*, ed. L. Mannetje, M. C. Amezquita, P. Buurman, and M. Ibrahim, 49–68. Wagenigen: Academic Publishers.

Amigos da Terra. 2007. IFC approves loan to the Bertin group after presentation different from information in the project. Available at www.amazonia.org.br.

Andrade, H. J. 2007. *Growth and inter-specific interactions in young silvopastoral systems with native timber trees in the dry tropics of Costa Rica.* Ph.D. Thesis. Bangor, Wales: CATIE-University of Bangor.

ANUALPEC. 2007. Anuário da pecuária brasileira. Available at www.fnp.com.br

Arima, E., P. Barreto, and M. Brito. 2005. Pecuária na Amazônia: Tendências e implicações para a conservação ambiental. Belém, Brazil: Instituto do Homem e Meio Ambiente da Amazônia. Available at www.imazon.org.br.

Arroyo-Mora, J. P., G. A. Sánchez-Azofeifa, B. Rivard, J. C. Calvo, and D. H. Janzen. 2005. Dynamics in landscape structure and composition for the Chorotega region, Costa Rica from 1960 to 2000. *Agriculture, Ecosystems & Environment* 106(1): 27–39.

ASA. 2003. China soybean imports running at record level. ASA Weekly Archives. St Louis, Missouri. American Soybean Association. Available at www.asa.org.

BACEN. 2006. Exchange rate. Brasilia: Banco Central do Brasil. Cited January 2006. Available at http://www.bacen.gov.br.

BankTrack. 2004. *Principles, Profits, or Just PR? Triple P Investments under the Equator Principles: An Anniversary Assessment.* Amsterdam: BankTrack.

Bayer, C., D. Pinheiro, G. Mafra, and K. Kinrad. 2002. Carbon stocks in organic matter fractions as affected by land use and soil management, with emphasis on no-tillage effect. *Ciência Rural, Santa Maria* 32(3): 401–406.

Bierregaard, J., W. Laurance, and J. Gascon. 2001. Principles of forest fragmentation and conservation in the Amazon. In R. O. Bierregaard, C. Gascon, E. O. Wilson, E. Salati, and R. Mesquita (eds.), *Lessons from Amazonia: The Ecology and Conservation of a Fragmented Forest.* New Haven: Yale University Press.

Brookes, G., N. Craddock, and B. Kniel. 2005. *The Global GM market: Implications for the European Food Chain.* Canterbury, United Kingdom: BrookesWest.

Camacho, A. 1989. *Factores que afectan la modernizacíon de la agricultura: el sector lechero en Costa Rica, 1967–1988.* San José: IICA.

Cámara Naçional de Productores de Leche. 2004. Información del Sector Lechero de Costa Rica. Cited 10 June 2004. Available at http://www.proleche.com/proleche/info_sector.htm.

Cámara Naçional de Productores de Leche. 2006. *Evolución del número total de hembras del hato de leche especializado y doble propósito (1984–1991).* San José, Costa Rica: CNPL.

Cardim, S. E., P. Vieira, and J. L. R. Viégas. 2003. *Análise da Estrutura Fundiária Brasileira.* Brasília–DF: Núcleo de Estudos Agrários e Desenvolvimento Rural do Ministério do Desenvolvimento Agrário.

Castro, R., F. Tattenbach, N. Olson, and L. Gamez. 1997. The Costa Rican experience with market instruments to mitigate climate change and conserve biodiversity. Paper presented at the Global Conference on Knowledge for Development in the Information Age, Toronto, Canada, 24 June 1997.

CCAD. 1998. *Incendios forestales en Centroamérica, balance, 1998.* San Salvador: Comisión Centroamericana de Ambiente y Desarrollo.

Chomitz, K. M., E. Brenes, and L. Constantino. 1999. Financing environmental services: The Costa Rican experience and its implications. *Science of the Total Environment* 240: 157–169.

Clay, J. 2004. *Agriculture and the Environment.* Washington, DC: World Wildlife Fund–U.S.

CNPGC. 1988. *Relatorio técnico annual do Centro Nacional de Pesquisa de Grado de Corte 1983–1985.* Campo Grande, MS, Brasil: Centro Nacional de Pesquisa de Grado de Corte.

Cooperativa Dos Pinos. 2009. Zonas de recolectores de leche. Consultado 15 ene. 2009. Cooperativa Dos Pinos. Available at http://www.dospinos.com.

Cruz, M. C., C. A. Meyer, R. Repetto, and R. Woodward. 1992. *Population Growth, Poverty, and Environmental Stress: Frontier Migration in the Philippines and Costa Rica.* Washington, DC: World Resources Institute.

DEFRA. 2005. BSE: Legislation. London: Department for Environment, Food, and Rural Affairs.Available at http://www.defra.gov.uk/animalh/bse/legislation/index.html.

Edelman, M. 1985. Extensive Land Use and the Logic of the Latifundio: A Case Study in Guanacaste Province, Costa Rica. *Human Ecology* 13: 153–185.

EMBRAPA. 2008. Relatório de Sustentabilidade. Available at http// www.cnpgl.embrapa.br.

Etter, A. C. McAlpine, K. Wilson, S. Phinn, and H. Possingham. 2006. Regional patterns of agricultural land use and deforestation in Colombia. *Agriculture, Ecosystems and Environment* 114: 369–386.

Fallas, J. 2006. *Identificación de zonas de importancia hídrica y estimación de ingresos por canon de aguas para cada zona.* San José: FONAFIFO.

FAO. 1980. *Production Yearbooks.* 1966, 1976, 1979, 1980, 1990, 1991, 1992. Rome: Food and Agriculture Organization of the United Nations.

FAO. 1993. *Forest Resources Assessment 1990, Tropical Countries.* FAO Forestry Paper 112. Rome: Food and Agriculture Organization of the United Nations.

FAO. 2002. *Global Forest Resources Assessment 2000, Main Report.* FAO Forestry Paper 140. Rome: Food and Agriculture Organization of the United Nations.

FAOSTAT. 2006. FAO database. Available at www.fao.org.

FAOSTAT. 2007. FAO database Available at www.fao.org.

Fearnside, P. M. 2001. Soybean cultivation as a threat to the environment in Brazil. *Environmental Conservation* 28(1): 23–28.

Fearnside, P. M. 2005. Deforestation in Brazilian Amazonia: History, Rates, and Consequences. *Conservation Biology* 19: 680–688.

FONAFIFO. 2007. Servicios Ambientales, Estadísticas PSA. Fondo Nacional de Financiamiento Forestal. Available at http://www.fonafifo.com/paginas_espanol/servicios_ambientales/sa_estadisticas.htm.

Fournier, L. A. 1985. El sector forestal de Costa Rica: antecedentes y perspectivas. *Agronomía Costarricense* 9: 253–260.

Girot, P. O. 1989. Formación y estructuración de una frontera viva: el case de la región norte de Costa Rica. *Goistmo* 3(2): 17–42.

Government of the State of Amazonas. 2007. *Amazonas Initiative on Climate Change, Forest Conservation and Sustainable Development. Secretariat for Environment and Sustainable Development.* Manaus, Brazil: Government of the State of Amazonas.

Grainger, A. 1993. Rates of Deforestation in the Humid Tropics: Estimates and Measurements. *Geographical Journal* 159: 33–44.

Harber, W. A. and R. D. Stevenson. 2004. Diversity, migration, and conservation in northem Costa Rica. In *Biodiversity Conservation in Costa Rica: Learning the Lessons in a Seasonal Dry Forest,* ed. G. W. Frankie, A. Mata, and S. Vinson, 99–114. Berkeley: University of California Press.

Harvey, C., A. Medina, D. Sánchez, S. Vilchez, B. Hernandez, J. Saenz, J. M. Maes, F. Casanoves, and F. L. Sinclair. 2006. Patterns of animal diversity in different forms of tree cover in agricultural landscapes. *Ecological Application* 16: 19–86.

Harvey, C., F. Alpizar, M. Chacón, and R. Madrigal. 2005. *Assessing Linkages between Agriculture and Biodiversity in Central America.* San José, Costa Rica: Mesoamerican & Caribbean Region Conservation Science Program, The Nature Conservancy.

Hecht, S. B. 1993. The logic of livestock and deforestation in Amazonia. *Bioscience* 43 (10): 687–695.

Hecht, S. and A. Cockburn. 1989. *The Fate of the Forest: Developers, Defenders, and Destroyers of the Amazon.* New York: HarperCollins.

Hernández, M, P. J. Argel, M. Ibrahim, and L. 't Mannetje. 1995. Pasture production, diet selection, and liveweight of cattle grazing *Brachiaria brizantha* with or without *Arachis pintoi* at two stocking rates in the Atlantic zone of Costa Rica. *Tropical Grasslands* 29(3): 134–141.

Howard, P. 1987. *From Banana Republic to Cattle Republic: Agrarian Roots of the Crisis in Honduras.* Ph.D. diss., University of Wisconsin, Madison.

IBGE. 2005. Produção da Pecuária Municipal 2005. Instituto Brasileiro de Geografia e Estatística. Available at http://www.sidra.ibge.gov.br/bda/pesquisas/ppm/default.asp.

IBGE-PAM. 2005. Indicadores da economia mundial. Instituto Brasileiro de Geografia e Estatística, Produção Agrícola Municipal. Cited December 2005. Available at http://www.sidra.ibge.gov.br/bda/agric.

IcWales. 2003. Call to boycott beef from Brazil. IcWales, Cardiff, Wales, United Kingdom. Cited January 2006. No longer available at http://icwales.icnetwork.co.uk/0100news/0200wales/contentobjectid=13402774method=fullsiteid=50082headline=-Call-to-boycott-beef-from-Brazil-namepage.html#storycontinue.

IFC. 2007. IFC TO FINANCE BERTIN: Project to set new benchmark for environmental and social standards in cattle ranching and meat processing in the Amazon. Press release, 8 March 2007. Available at: http://www.ifc.org/ifcext/media.nsf/content/SelectedPressRelease?OpenDocument&UNID=2A71421255949B1D85257298007216A9.

INFOAGRO. 2007. (Exportaciones e Importaciones de Cobertura Agropecuario) Available at www.infoagro.go.cr.

INPE. 2006. INPE divulga estimativa do desmatamento na Amazônia Legal para o período Agosto 2005–Agosto 2006. Instituto Nacional de Pesquisas Espaciáis. Available at: http://www.inpe.br/noticias/noticia.php?Cod_Noticia=856.

INPE. 2008. INPE divulga estimativa do desmatamento na Amazônia Legal para o período Agosto 2005–Agosto 2006. Instituto Nacional de Pesquisas Espaciáis. Available at: http://www.obt.inpe.br/prodes/prodes_1988_2008.htm.

Jimenez, J. A. 2007. *Diseño de sistemas de producción ganaderos sostenibles con base a los sistemas silvopastoriles para mejorar la productividad animal y lograr la sostenibilidad ambiental.* Thesis. Turrialba, Costa Rica: CATIE.

Kaimowitz D. 1996. *Livestock and Deforestation: Central America in the 1980s and 1990s: A Policy Perspective.* Jakarta, Indonesia: CIFOR.

Kaimowitz, D. and A. Angelsen. 1998. *Economic Models of Tropical Deforestation: A Review.* Bogor, Indonesia: Center for International Forestry Research.

Kaimowitz, D., B. Mertens, S. Wunder, and P. Pacheco. 2004. Hamburger connection fuels Amazon destruction: cattle ranching and deforestation in Brazil's Amazon. Bogor, Indonesia: Centro de Investigación Forestal Internacional. Available at http://www.cifor.cgiar.org/publications/pdf_files/media/Amazon.pdf.

Kattan, G. 2002. Fragmentación: patrones y mecanismos de extinción de especies. In *Ecología y Conservación de los Bosques Neotropicales,* ed. M. Guariguata and G. Kattan, 561–590. Libro Universitario Regional (EULAG-GTZ)

Kirby, K. R, W. F. Laurance, A. K Albernaz, G. Schroth, P. M. Fearnside, S. Bergen, E. Venticinque, and C. Da Costa. 2006. The future of deforestation in the Brazilian Amazon. *Futures* 38: 432–453.

Laurance, W. F., A. K. M. Albernaz, G. Schroth, P. M. Fearnside, S. Bergen, E. M. Venticinque, and C. Da Costa. 2002. Predictors of deforestation in the Brazilian Amazon. *Journal of Biogeography* 29:737–748.

Laurance, W. F., M. A. Cochrane, S. Bergen, et al. 2001. Environment: the future of the Brazilian Amazon. *Science* 291: 438–439.

Ledec, G. 1992. New directions for livestock policy: an environmental perspective. In *Development or Destruction: The Conversion of Tropical Forest to Pasture in Latin America,* ed. T. Downing, S. Hecht, H. Pearson, and C. G. Downing, 27–65. Boulder: Westview Press.

Lemus, G. 2008. *Análisis de productividad de pasturas en sistemas silvopastoriles en fincas ganaderas doble propósito en Esparza, Costa Rica.* M.Sc. Thesis. Turrialba, Costa Rica: CATIE.

León, J. C., et al. 1982. *Desarrollo tecnológico de la ganadería de carne.* San José: Consejo Nacional de Investigaciones Científicas y Tecnológicas.

Leonard, J. J. 1987. *Recursos naturales y desarrollo económico en América Central: un perfil ambiental.* Turrialba: Centro Agronómico Tropical de Investigacción y Enseñanza.

Lima, A., C. T. Irigaray, R. T. Silva, S. Guimarães, and S. Araujo. 2005. *Sistema de Licenciamento Ambiental em Propriedades Rurais do Estado de Mato Grosso: Análise de Lições na Sua Implementação (Relatório Final).* Brasília: Ministério do Meio Ambiente/Secretaria de Coordenação da Amazônia/Programa Piloto para a Proteção das Florestas Tropicais do Brasil/Projeto de Apoio ao Monitoramento e Análise.

LMC International. 2003. Supply chain impacts of further regulation of products consisting of, containing, or derived from, genetically modified organisms. Oxford: LMC International. Prepared for DEFRA and Food Standards Agency. Available at http://www.defra.gov.uk/environment/gm/research/pdf/epg 1–5-212.pdf.

Mahar, D. 1989. *Government Policies and Deforestation in Brazil's Amazon Region.* Washington, DC: World Bank.

Manfrinato, W., M. Piccolo, C. Cerri, M. Bernoux, and C. Pellegrino. 2001. Medición de la variabilidad espacial y temporal de Carbono del suelo con el uso de los isótopos estables, en una transición bosque-pradera en el estado de Paraná, Brasil. *Simposio Internacional Medición y Monitoreo de la Captura de Carbono en Ecosistemas Forestales.* 18–20 October 2001, Valdivia-Chile.

MAPA. 2005. Agroindustria. Ministério de Relações Exteriores, Available at http://www.mre.gov.br/cdbrasil/itamaraty/web/port/economia/agric/apresent/apresent.htm.

MCT (Ministério de Ciência e Technologia, Brasil). 2004. Indicadores sobre Ciência e Technologia no Brasil. www.mct.gov.br (investigación hecha en 08.10.2004).

Merlet, M., G. Farrell, J.-M. Laurent, and C. Borge. 1992. *Identificación de un programa regional de desarrollo sostenible en el trópico húmedo, informe de consultoría.* Paris: Groupe de Recherche et d'Echanges Technologiques.

Mertens, B. and E. Lambin. 1999. Modelling land cover dynamics: integration of fine-scale land cover data with landscape attributes. *International Journal of Applied Earth Observation and Geoinformation* 1(1): 48–52.

Miele, M., and A. F. Girotto. 2007. A suinocultura brasileira em 2007 e cenários para 2008. Embrapa Suinos e aves. Ministério do Meio Ambiente (Environmental Department–Brazil.

Available at http://www.mma.gov.br/index.php?ido=conteudo.monta&idEstrutura=48.

MMA. 2007. Ministério do Meio Ambiente. Available at http://www.mma.gov.br/index.php?ido=conteudo.monta&idEstrutura=48.

Monbiot, G. 2005. The price of cheap beef: disease, deforestation, slavery, and murder. *Guardian*, 18 October. Cited 28 June 2006. Available at http://www.guardian.co.uk/uk/2005/oct/18/bse.foodanddrink.

Moran, E. F. 1981. *Developing the Amazon*. Bloomington: Indiana University Press.

Moran, E. F. 1993. Deforestation and land use in the Brazilian Amazon. *Human Ecology* 21(1): 1–21.

Morton, D. C., R. S. Defries, Y. E. Shimabukuro, L. O. Anderson, E. Arai, F. del Bon Espirito-Santo, R. Freitas, and J. Morisette. 2006. Cropland expansion changes deforestation dynamics in the southern Brazilian Amazon. *Proceedings of the National Academies of Sciences* 103(39): 14637–14641.

Myers, N. 1981. The hamburger connection: how Central America's forests became North America's hamburgers. *Ambio* 10: 3–8.

Myers N., R. A. Mittermeier, G. G. Mittermeier, G. A. B. da Fonseca, and J. Kent. 2000. Biodiversity hotspots for conservation priorities. *Nature* 403: 853–858.

Nations, J. D. and D. Komer. 1982. Indians, immigrants and beef exports: deforestation in Central America. *Cultural Survival Quarterly* 6: 8–12.

Naughton-Treves, L. 2004. Deforestation and carbon emissions at tropical frontiers: a case study from the Peruvian Amazon. *World Development* 32(1): 173–190.

Naylor, R., H. Steinfeld, W. Falcon, J. Galloway, V. Smil, E. Bradford, J. Alder, and H. Mooney. 2005. Losing the links between livestock and land. *Science* 310: 1621–1622.

Nepstad, D., C. Carvalho, E. Davidson, P. Jipp, P. Lefebvre, N. Negreiros, E. da Silva, T. Stone, E. Susan, T. S. Vieira. 1994. The role of deep roots in the hydrological and carbon cycles of Amazonian forests and pastures. *Nature* 372: 666–669.

Nepstad, D., J. P. Capobianco, A. C. Barros, G. Carvalho, P. Moutinho, U. Lopes, P. Lefebvre, and M. Ernst. 2000. Avança Brasil: os custos ambientais para a Amazônia. Belém, Brazil: Instituto de Pesquisa Ambiental da Amazônia. Available at http://www.ipam.org.br.

Nepstad, D., G. Carvalho, A. C. Barros, A. Alencar, J. P. Capobianco, J. Bishop, P. Moutinho, P. Lefebvre, U. L. Silva Jr., and E. Prins. 2001. Road paving, fire regime feedbacks, and the future of Amazon forests. *Forest Ecology and Management* 154(3): 395–407.

Nepstad, D., D. McGrath, A. Alencar, A. C. Barros, M. Carvalho, M. Santilli, and C. Vera Diaz. 2002. Frontier Governance in Amazonia. *Science* 295(5555): 629–631.

Nepstad, D. C., C. M. Stickler, and O. T. Almeida. 2006a. Globalization of the Amazon soy and beef industries: opportunities for conservation. *Conservation Biology* 20(6): 1595–1603.

Nepstad, D. C., S. Schwartzman, B. Bamberger, M. Santilli, D. Ray, P. Schlesinger, P. Lefebvre, A. Alencar, E. Prinz, G. Fiske, and A. Rolla. 2006b. Inhibition of Amazon deforestation and fire by parks and indigenous reserves. *Conservation Biology* 20(1): 65–73.

Nepstad, D., B. Soares-Filho, F. Merry, P. Moutinho, H. Oliveira-Rodriguez, M. Bowman, S. Schwartzman, O. Almeida, and S. Rivero. 2007. *The Costs and Benefits of Reducing Carbon Emissions from Deforestation and Forest Degradation in the Brazilian Amazon*. Falmouth, MA: Woods Hole Research Center, WHRC-IPAM-UFMG.

Newton, A., S. Oldfield, G. Fragoso, P. Mathew, L. Miles, and M. Edwards. 2003. *Towards a Global Tree Conservation Atlas*. Cambridge: United Nations Environment Program-WCMC/FFI.

Paciullo, D., C. Branda de Carvalho, L. Aroeira, J. Morenz, F. Lopes, and R. Rossielo. 2007. Morfofisilogia e valor nutritivo do capim sob sombreamento natural e a sol pleno. *Pesq. Agrop. Bras.* 42(4). Available at http://www.scielo.br/scielo.php?pid=S0100-204X2007000400016&script=sci_arttext&tlng=pt.

Pagiola, S. 2006. Payments for environmental services in Costa Rica. Presented at ZEF-CIFOR workshop on Payments for Environmental Services: Methods and Design in Developing and Developed Countries, Titisee, Germany, 15–18 June 2005.

Rocha, A. do A., and C. Boucas. 2006. Propagação da gripe das aves já reduz preço da soja. *Valor* 1457: A1.

Rodriguez, S., and E. Vargas. 1988. *El recurso forestal en Costa Rica, Politicas públicas y sociedad*. Heredia: Editorial Universidad National.

Roosevelt, M. 2006. The grass-fed revolution. *Time* 11 June 2006. Available at http://www.time.com/time/magazine/article/0,9171,1200759,00.html.

Sader, S. A., and A. T. Joyce. 1988. Deforestation rates and trends in Costa Rica, 1940 to 1983. *Biotropica* 20: 1 l–19.

Saenz, J., F. Villatoro, M. Ibrahim, D. Fajardo, and M. Perez. 2007. Relación entre las comunidades de aves y la vegetación en agropaisajes dominados por la ganadería en Costa Rica, Colombia y Nicaragua. *Agroforesteria de las América* 45: 37–48.

Sánchez, O. 2009. El pago por servicios ambientales del fondo nacional de financiamiento forestal, un mecanismo para lograr la adaptación sobre el cambio climático en Costa Rica. In *Políticas y sistemas de Incentivos para el fomento y adopción de buenas prácticas agrícolas como una medida de adaptación al Cambio Climático en América Central*, ed. C. Sepúlveda and M. Ibrahim. Turrialba, Costa Rica: CATIE.

Sánchez-Azofeifa, G. A., R. C. Harris, and D. L. Skole, 2001. Deforestation in Costa Rica: a quantitative analysis using remote sensing imagery. *Biotropica* 33(3): 378–384.

Saunders, D., R. J. Hobbs, and C. R. Margules. 1991. Biological consequences of ecosystem fragmentation: a review. *Conservation Biology* 5: 18–32.

Schelhas, J. 1996. Land use choice and change: intensification and diversification in the lowland tropics of Costa Rica. *Human Organization* 55: 298–306.

Schmink, M., and C. Wood, eds. 1992. *Contested Frontiers*. Princeton, NJ: Princeton University Press.

Schneider, R., E. Arima, A. Veríssimo, C. Souza Jr., and P. Barreto. 2002. Sustainable Amazon: limitations and opportunities for rural development. World Bank Technical Paper No. 515. Environment Series. Washington, DC: World Bank. Cited 21 February 2006. Available at: http://www.amazonia.org.br/arquivos/13523.pdf.

Sierra, R., and E. Russman. 2006. On the efficiency of environmental service payments: a forest conservation assessment in the Osa Peninsula, Costa Rica. *Ecological Economics* 59: 131–141.

SINAC. 2008. *Guía práctica para el diseño, oficialización y consolidación de corredores biológicos en Costa Rica*. San José: Sistema Nacional de Áreas de Conservación. In press.

Smeraldi, R. and P. May. 2008. *O Reino do Fado: uma nova fase na pecuarização da Amazônia*. Sao Paulo: Amigos da Terra-Amazonia Brasilera.

SNUC. 2004. Lei nº 9.985 de 18 de julho de 2000; Decreto nº 4.340, de 22 de agosto de 2002. Sistema Nacional de Unidades de Conservação da Natureza. 5. ed. Aum. Brasília: MMA/SBF.

Soares-Filho, B., A. Alencar, D. Nepstad, G. Cerqueira, M. C. V. Diaz, S. Rivero, L. Solorzano, and E. Voll. 2004. Simulation of deforestation and forest regrowth along a major Amazon highway: the case of the Santarém-Cuiabá highway. *Global Change Biology* 10: 745–764.

Soares-Filho, B., D. Nepstad, L. Curran, G. Cerqueira, R. Garcia, C. Ramos, E. Voll, A. McDonald, P. Lefebvre, and P. Schlesinger. 2006. Modeling Amazon conservation. *Nature* 440: 520–523.

Szott, L., M. Ibrahim, J. Beer. 2000. The hamburger connection hangover: cattle pasture land degradation and alternative land use in Central America. Informe técnico/CATIE no. 313. Turrialba, Costa Rica: CATIE.

Teixeira, W. G., E. G. Pereira, L. A. Cruz, and N. Bueno. 1996. Influência do uso nas características físico-químicas de um Latossolo Amarelo, textura muito argilosa, Manaus, AM. In *Congresso Latino-Americano de Ciência do Solo, 12., 1996, Águas de Lindóia* (CD-ROM). Anais Campinas: Sociedade Brasileira de Ciência do Solo/Sociedade Latino-Americana de Ciência do Solo.

Tobar, D., M. Ibrahim, C. Villanueva, and F. Casasola. 2006. Diversidad de mariposas diurnas en un paisaje agropecuario en la región del Pacifico Central, Costa Rica. En *Memorias—IV congreso latinoamericano de agroforestería*. Varadero, Cuba, 24–28 October 2006.

Tobar, L. D., and M. Ibrahim. 2008. *Uso y valoración de la diversidad en paisajes agropecuarios*. Serie técnica. Informe técnico No. 350. Turrialba, Costa Rica: CATIE.

Trejos, R. 1992. El comercio agropecuario extraregional. In *La agricultura en el desarrollo económico de Centroamerica en los 90*, ed. C. Pomareda, 87–124. San José: IICA.

Tschinkel, H. 1988. *Forestry in Costa Rica: An Overview*. San José: USAID.

Utting, P. 1993. *Trees, People and Power: Social Dimensions of Deforestation and Forest Protection in Central America*. London: Earthscan.

Van der Kamp, E. J. 1990. *Aspectos económicos de la ganadería en pequeña escala y de la ganadería de la carne en la zona atlántica de Costa Rica*. CATIE/Wageningen/MAG. Field Report No 51.

Veiga, J. B., A. M. Alves, R. Poccard-Chapuis, M. C. Thales, P. A. da Costa, J. O. Grijalva, T. V. Chamba, R. M. Costa, M. G. Piketty, and J. F. Tourrand. 2001. *Cattle Ranching, Land Use and Deforestation in Brazil, Peru and Ecuador*. Gainesville, USA: Annual Report for the Inter-American Institute.

Veldkamp, E. 1994. Organic Carbon Turnover in Three Tropical Soils under Pasture after Deforestation. *Soil Science Society of America Journal* 58: 175–180.

Villanueva, C., M. Ibrahim, C. Harvey, and H. Esquivel, 2003. Tipología de fincas con ganadería bovina y cobertura arbórea en pasturas en el trópico seco de Costa Rica. *Agroforestería en las Américas* 10(39–40): 9–16.

Volpi, G. 2007. Climate mitigation, deforestation and human development in Brazil. In *Human Development Report 2007/2008*. New York: UNDP Human Development Report Office.

Wassenaar, T., P. Gerber, P. H. Verburg, M. Rosales, M. Ibrahim, and H. Steinfeld. 2007. Projecting land use changes in the Neotropics: the geography of pasture expansion into forest. *Global Environmental Change* 17: 86–104.

Wertz-Kanounnikoff, S., M. Kongphan-Apirak, S. Wunder. 2008. Reducing forest emissions in the Amazon Basin: a review of drivers of land-use change and how payment for environmental services (PES) schemes can affect them. CIFOR Working Paper No. 40. Bogor, Indonesia: Center for International Forestry Research.

White, D., F. Holmann, S. Fijusaka, K. Reategui, and C. Lascano. 2001. Will intensifying pasture management in Latin America protect forests—or is it the other way round? In*Agricultural Technologies and Tropical Deforestation* ed. A. Angelsen and D. Kaimowitz. Wallingford, UK: CABI Publishing.

Williams, R. 1986. *Export Agriculture and the Crisis in Central America*. Chapel Hill: University of North Carolina Press.

Zbinden, S., and D. R. Lee. 2005. Paying for environmental services: an analysis of participation in Costa Rica's PSA program. *World Development* 33(2): 255–272.

6

China

The East–West Dichotomy*

Ke Bingsheng

Abstract

Drivers for expansion of the livestock sector in China have been very strong over the past two decades. Factors contributing to an ever-rising demand for livestock commodities include overall population growth, the increasing urban share of population, fast income growth for urban and rural households, improved marketing infrastructure, and the expansion of international trade. On the input side, technical innovation, rapid development of the processed feed industry, and improvements in public services for animal disease control have all contributed to the improvement of livestock productivity.

These driving forces have resulted in two major developments: rapid production growth of all livestock commodities, and increased intensification of livestock operations for all species. Intensification has proceeded in two dimensions: the emergence of very large scale livestock farms operating in an industrial manner, and the spatial concentration of livestock animals leading to very high density in certain regions.

The positive impacts of the changing livestock sector lie mainly in the improvement of national food security, animal disease control, and a cleaner living environment for small farmers. The adverse impacts are mainly environmental, including air, surface water, and ground water pollution; nitrogen overload caused by large intensive farms in the eastern part of the country; and soil erosion and desertification caused by overgrazing of extensive pastoral systems in the western regions of the country. The development of China's livestock sector also has significant international implications, in particular as a strong stimulus for export and production of soybeans in the Americas, North and South.

A number of policy measures have already been taken to address the environmental problems. Although these have achieved some positive results, there is still much room for improvement, in particular in enforcement of regulations. Because all the driving forces discussed are still at work, further growth and intensification of livestock production in China are inevitable and irreversible. To mitigate the adverse impacts associated with this changing livestock landscape, joint efforts by all stakeholders are needed.

Introduction

As the largest producer of livestock commodities in the world, China has a stock of 502 million pigs, 4.51 billion chickens, 117 million cattle, and 369 million goats and sheep, representing, respectively, 51%, 26%, 9% and 19% of the world total in 2007 (FAO 2008). The total meat production of China in 2007 was 68 million tonnes, accounting for 25% of the world total.

The livestock sector in China has developed rapidly over the past two decades. Total meat production quadrupled over the period 1985–2007 (MOA 2007a). There have also been significant changes in livestock farm structure and in the geographical distribution and composition of livestock commodities. The intensification of livestock production in China started in the late 1980s and has advanced progressively ever since. This process has been fueled both by the rapidly rising demand for meat and other livestock products, and by the emergence and growth of the modern feed industry and other technical advances. However, there is uneven development among different regions within the country. In some of the most developed coastal regions, small backyard livestock raising systems have been completely replaced by large-scale industrial operations, whereas in

* Paper presented at the 2007 AAAS Annual Meeting Session ID 170: Livestock in a Changing Landscape: Drivers, Consequences and Responses, 19 February 2007, San Francisco, USA. Thanks to Hennning Steinfeld, Pierre Gerber, and Harold Mooney for their comments and suggestions. Views and opinions in this chapter are of the author's own responsibility and do not necessarily represent those of MOA of China.

the remote western inland provinces, farmers still depend heavily on traditional extensive systems. The coexistence of highly advanced and integrated livestock corporations with very traditional pastoral and backyard operations constitutes a strong geographical dichotomy in China's livestock sector.

The importance of studying livestock development in China is twofold. First, with its sheer scale of production, the development of China's livestock sector has significant global implications in itself. Second, the dichotomy of the country's livestock sector sheds valuable light on world livestock issues.

Drivers and Changes

Many factors have contributed to the changes in the livestock sector. In the following, the major factors or drivers are analyzed in two groups. In the first group are the "pull" forces that affect total demand for livestock commodities. In the second group are the "push" forces that impact livestock production from the input side.

Drivers of Livestock Demand

The major drivers on the demand side include the expansion of total population and the urban–rural composition, income growth of the consumers, improvement of infrastructure, per capita consumption of livestock commodities, and trade in livestock products.

Demographic Changes

China's total population increased from 987 million in 1980 to 1.32 billion in 2007. As a result of the national population control policy, the growth rate fell from 1.5 to 0.6% over this period. Population growth in terms of numbers peaked in the late 1980s, when there were 15 million newborns per year. The figure has declined by half in recent years but still runs at nearly 8 million annually. China's total population is expected to peak at 1.5 billion by 2030, according to authoritative projections (SPFP 2006).

The urban population (including rural migrant workers) has grown at a much faster pace, more than doubling over the past two decades. This growth trend will continue into the future—the urban share in the total population is projected to reach 60% by 2030 (Figure 6.1). This rapid urbanization implies more demand for livestock commodities because urban households consume more meat and other livestock commodities per person than the rural population in China, according to household survey statistics (discussed in detail in the next section).

Income Growth of Urban and Rural Households

Probably the most powerful driver has been the rapid rise in incomes for both urban and rural populations.

Incomes of urban residents have risen rapidly over the past three decades. Deflated in 1978 Yuan, the per capita income of urban households in 2007 is over six times higher than in 1978 (Figure 6.2) and has been doubling roughly every ten years. Because Chinese consumers still have a relatively low consumption of livestock products, the income effect on demand is large. According to research findings, the income elasticity of all livestock products is larger than zero. This implies that demand for livestock products will rise as incomes grow. According to various estimates, the income elasticity of demand for pork is about 0.12 to 0.5. For beef and mutton it is about 1.0 to 1.5, implying that demand for these products rises faster than overall income growth (Jiang 2002, Meng 2002, Lu and Mei 2008, Yu 2008). For milk the elasticity of demand is 0.5 (Sheng et al., 2004, Lu and Mei 2008).

The improvement of Chinese urban consumers' purchasing power can also be seen from the changes in the share of food in total household expenditure (Engel's coefficient). This coefficient has declined from 58% in 1978 to 36% in 2007, which implies that Chinese consumers now have more flexibility in consumption. If they wish, they are more able to increase expenditure on livestock products. With further improvement in income, they will consume more meat, milk, and other livestock commodities, especially among low-income sections of the population where current meat consumption level is

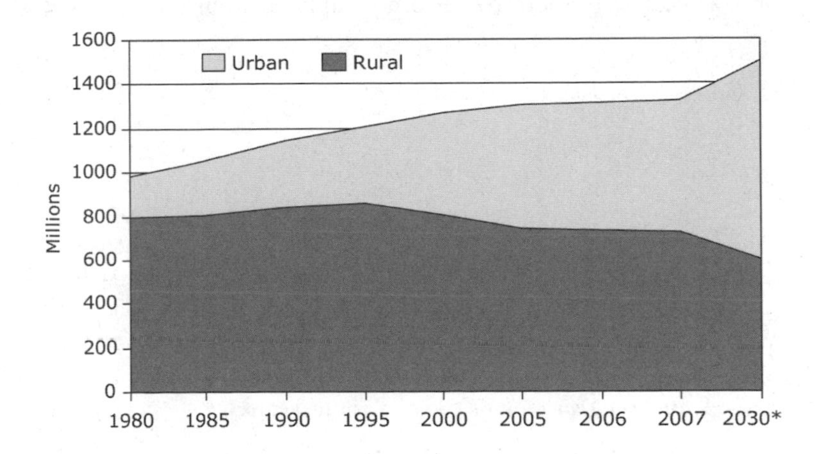

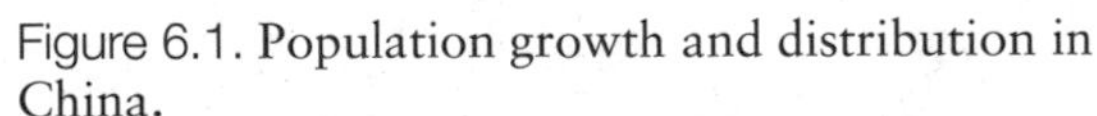

Figure 6.1. Population growth and distribution in China.

* 2030 Projections by the State Population and Family Planning Commission (SPFP).

Source: National Bureau of Statistics of China (NBS): *Statistical Yearbook of China,* various years.

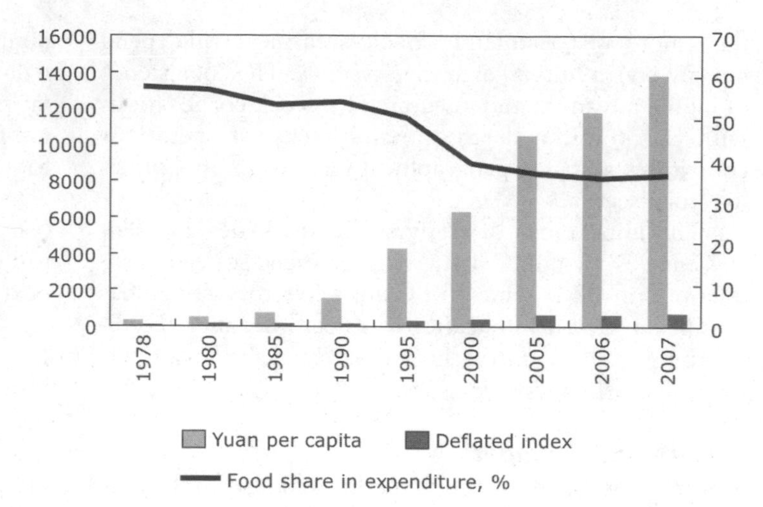

Figure 6.2. Changes in per capita income for urban households in China.
Source: National Bureau of Statistics of China (NBS): *Statistical Yearbook of China,* various years.

relatively low and the room for increase is large. A closer look at meat consumption for different income segments reveals a close correlation between income and meat consumption, as will be discussed in more detail.

For China's rural population, per capita incomes have also improved dramatically in the past three decades. The growth path and pace are very similar to those of urban households (Figure 6.3).

The impact of rural income growth on the livestock sector is more complex than that of urban residents. Just as it did in urban households, rapid income growth in rural households has stimulated the demand for more livestock products.

However, rural households are not just consumers of livestock products. A large proportion of them are also livestock producers. Thus increased rural income does much more than just drive demand for livestock products; it also directly affects the production structures of the livestock sector in several ways.

First, the overall importance of the livestock sector for farmers' incomes has declined. The income share earned from livestock in the total household income of farmers fell from 14% in 1990 to 8% in 2007 (Table 6.1). This is the average for the whole country. In the eastern regions, the share is even lower. One important reason for traditional farmers to raise a couple of animals in the yard in the past was that they could earn cash income by selling the animals—usually pigs, chickens, or eggs. This source of cash income has become much less significant as nonfarm salary has risen dramatically. The rural population has two ways to earn nonfarm salaries: either to work in small town and village enterprises or to go to the cities as migrant workers. According to various estimates, the total number of rural migrants working in cities is in the range of 120 to 140 million. Thus backyard livestock rearing has lost its traditional importance in providing cash income.

Secondly, rising rural incomes also imply higher agricultural labor costs. Because raising livestock is a relatively labor intensive activity in China, rising labor costs make small-scale livestock raising less attractive. It takes a lot of work to raise a couple of pigs, and the earnings from it have become less and less competitive with nonfarm activities. Most young people in the eastern part of the country have left agriculture and found jobs in nonfarm sectors.

Last but not least, with rising incomes, the rural population has become less tolerant of the environmental and quality of life problems of backyard livestock raising, especially dirt, odors, and flies. As a result, small backyard livestock raising has disappeared in many villages of the coastal provinces, such as Zhejiang, Jiangsu,

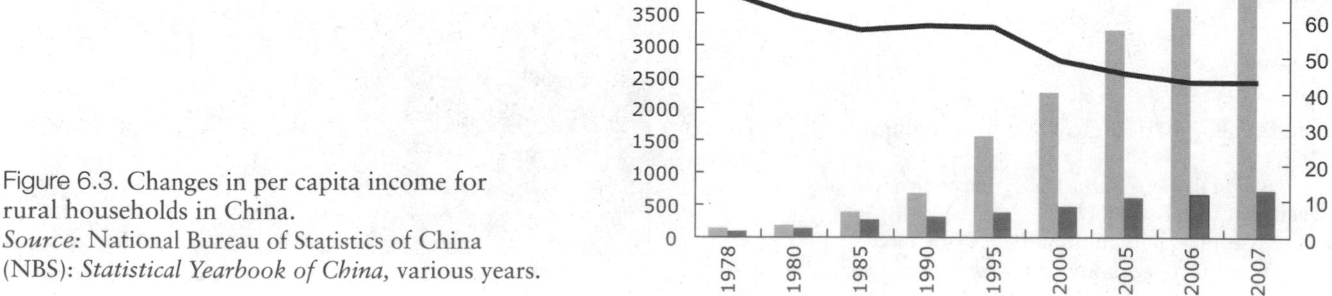

Figure 6.3. Changes in per capita income for rural households in China.
Source: National Bureau of Statistics of China (NBS): *Statistical Yearbook of China,* various years.

Table 6.1. Per capita net income of farmers' households in China

	Total Yuan	Nonfarm salary Yuan	%	Livestock Yuan	%
1985	398	72	18	52	13
1990	686	139	20	97	14
1995	1578	354	22	128	8
2000	2253	702	31	207	9
2005	3255	1175	36	284	9
2006	3587	1375	38	266	7
2007	4140	1596	39	335	8

Source: National Bureau of Statistics of China (NBS): Statistical Yearbook of China, various years.

Shandong, and Guangdong, where most farmers have significantly improved their living conditions. Livestock raising in these provinces is now more concentrated in large-scale intensive farms.

Infrastructure Development

The dramatic improvements in China's transport systems over the past two decades have had a considerable impact in making livestock products more widely accessible. The railway system is the major means for long distance transport such as interprovincial transportation of goods. The length of the railway network grew by one third between 1985 and 2007, and in addition the quality and efficiency of the rail system are much improved. Average speeds and the extent of double-track rails have grown, so that the capacity of the network has increased significantly. In particular, the number of refrigerated cars nearly doubled in the last two decades, from 3991 in 1985 to 7419 in 2005. This is especially important for transporting frozen livestock products and has facilitated interregional marketing.

The improvement in the highway transportation system and capacity has been even greater. The length of paved roads has doubled in the last two decades. More importantly, the construction of expressways in China has seen a spectacular increase. China opened its first expressway as recently as 1990. Now China boasts an expressway length of over 41,000 km, second only to the United States. The number of trucks has nearly quintupled, from 2.2 millions in 1985 to 10.5 million in 2007.

Retailing facilities in China have also improved significantly in the past two decades, especially in recent years. Supermarkets and chain stores have been booming across the country, first in large and medium-sized cities, and now in towns and even villages in the economically more developed regions. One major reason for many farmers to have backyard raising in the past was to meet the needs of self-consumption. This has also lost its importance over time because shops, weekly markets, and supermarkets and other marketing facilities selling livestock products are widely developed. Farmers now have easy access to a great variety of livestock commodities, from fresh meat to processed products.

Growth of Per Capita Consumption

The consumption of livestock products has increased continuously over the past two decades. Per capita consumption of livestock commodities is a reflection of the combined effects of income, price, preference, physical accessibility of the commodities, and other factors.

Annual sample household surveys are conducted separately for urban and rural areas in China. The results of these surveys for livestock products consumption are given in Tables 6.2 and 6.3. Some important observations can be drawn. Urban and rural areas have some factors in common: the consumption of livestock products has risen significantly in the last two decades. Pork was and still is predominant in meat consumption, but consumption of poultry and milk has seen the fastest growth.

However, there are urban–rural differences. A comparison of Figure 6.4 and Figure 6.5 reveals that the urban population consumes about 69% more meat, 119% more eggs, and over four times more milk than the rural population. This is due to a number of factors, including differences in dietary preference. However, the principal and most decisive reason is the income disparity between urban and rural populations. The average income level in urban areas is over three times that in rural areas (Figure 6.2 and 6.3). About 12% of the average urban household income is spent on consumption of livestock products (NBS 2008).

A closer look at consumption patterns among different income groups of urban households reveals a very clear correlation between disposable income and the per capita consumption level of all livestock products (Table 6.2). The survey results clearly show that, as income rises, grain consumption falls and livestock product consumption rises. This is particularly true for the processed meat products and processed poultry products. This same correlation can also be observed with provincial data. This very positive correlation of income and livestock consumption suggests that demand for livestock products will continue to rise, given further economic growth and income improvements.

Trade of Livestock Products

China's international livestock trade has developed in line with the general trend of agricultural trade in China (Table 6.3). There are several features to be noticed. First, both exports and imports of livestock products have increased markedly over the last decade. However, the value of imports has risen more rapidly than that of exports, making China a net importer of livestock products since 2000. Secondly, the share of livestock in

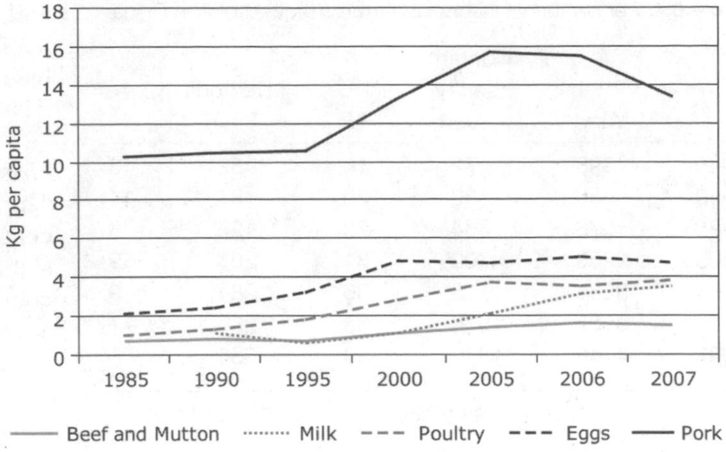

Figure 6.4. Consumption of livestock products in China–urban.
Source: National Bureau of Statistics of China (NBS): *Statistical Yearbook of China,* various years.

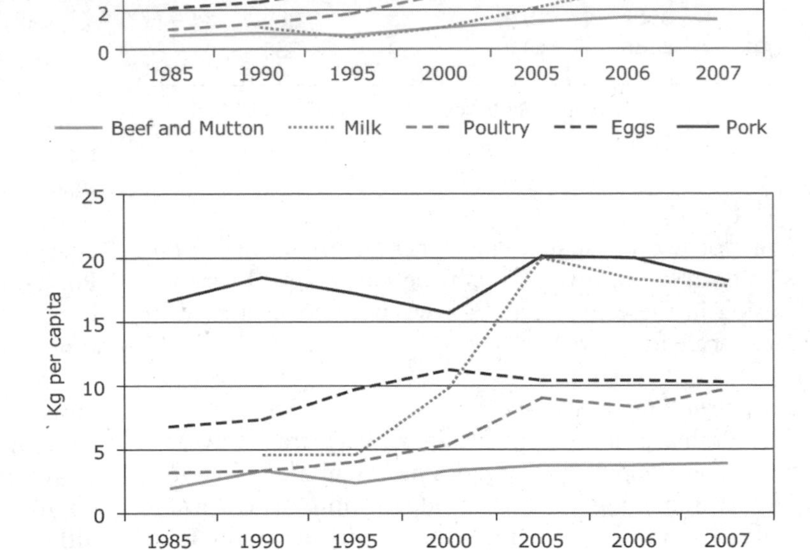

Figure 6.5. Consumption of livestock products in China–rural.
Source: National Bureau of Statistics of China (NBS): *Statistical Yearbook of China,* various years.

overall agricultural exports has fallen slightly, whereas the share in imports has risen remarkably. Thirdly, a closer look at the composition of China's livestock trade reveals that the predominant share of livestock products imported into the country are nonedible raw materials for industrial uses (Table 6.4). Wool and hides together constituted 60% of total imports of livestock products in 2005. Among imported edibles, dairy and poultry products are the most important. Large import volumes of those products can be partially attributed to the rising demand and the low import tariff. The actual import tariff rates are 1% for wool, 5% for hides, and 10% for most dairy products. On the export side, pork and poultry products account for over half of the total.

Table 6.2. Consumption of urban households by income group in China, 2006, kg/per capita

Income Group	Average	Lowest	Low	Lower Middle	Middle	Higher Middle	High	Highest
Income, RMB	12,719	3871	5946	8104	11,052	15,110	20,610	34,834
Grain	75.92	78.07	77.92	78.24	76.68	74.98	74.38	66.65
Meat total	27.59	20.67	24.45	26.56	28.6	30.27	31.69	30.82
Pork	20.00	16.30	18.54	19.43	20.55	21.35	22.31	21.50
Beef	2.41	1.59	2.00	2.34	2.65	2.72	2.74	2.63
Mutton	1.37	0.89	1.08	1.40	1.49	1.59	1.58	1.36
Other meat	0.24	0.15	0.22	0.25	0.29	0.27	0.23	0.21
Processed meat	3.56	1.74	2.61	3.14	3.61	4.34	4.84	5.12
Poultry	8.34	5.43	6.88	7.78	8.54	9.39	10.29	10.66
Processed Poultry	1.84	0.88	1.33	1.56	1.92	2.23	2.49	2.76
Fresh milk	18.32	8.80	12.91	16.26	19.16	22.29	24.52	25.91
Eggs	11.07	8.56	9.94	11.00	11.61	11.99	12.36	11.4
Fishery	12.95	7.50	9.46	11.09	12.91	15.30	17.66	19.36

RMB = Renminbi

Source: Price and urban household survey data yearbook 2007, China Statistics Press.

The impact of international trade on the domestic livestock sector varies significantly among different commodities and regions. On the import side, the impact of trade on domestic production is very significant for wool because imported wool (especially fine wool) accounts for half of domestic consumption. This has a huge direct impact on the domestic price of wool. Powdered milk imports also have important impacts on domestic dairy farmers, especially for those in the three northernmost provinces of Heilongjiang, Inner Mongolia, and Xinjiang. For all other food commodities in the livestock sector, imports do not have a significant effect because the import volume is very small compared with the huge quantity of domestic production.

On the export side, the impact of international trade is not sizable enough to significantly impact national supply chains. Only poultry products exported (mainly to Japan) and swine products exported (to Hong Kong) are of some importance, and for very few regions such as Shandong Province. However, total agricultural trade has kept on increasing since China's entry in the World Trade Organization in December 2001. The increase in total agricultural imports is due mainly to oil seeds and edible oil, whereas for agricultural exports it is due to the water products and vegetables.

Drivers on the Input Side

Drivers on the input side are those that have a "push" effect on livestock sector development and include advances in technology, the feed industry, and public services to agriculture.

Table 6.3. Trade of livestock commodities in China, billion US$

	Export billion US$	Import % of agricultural trade	Export	Import
1993	1.7	0.9	15	23
1994	2.2	1.1	16	16
1995	2.8	1.5	19	12
1996	2.9	1.4	20	13
1997	2.7	1.4	18	14
1998	2.5	1.3	18	16
1999	2.2	1.9	17	23
2000	2.6	2.7	17	24
2001	2.7	2.8	17	24
2002	2.6	2.9	14	23
2003	2.7	3.4	13	18
2004	3.2	4.0	14	14
2005	3.6	4.2	13	15
2006	3.7	4.6	12	14
2007	4.0	6.5	11	16

Source: Ministry of Agriculture of China (MOA), compiled data based on unpublished custom statistics.

Technology Innovation and Application

Technology is by far the most important factor affecting the input side of livestock production in China. New higher-productivity animal breeds, new feeding systems, new rearing facilities, and new methods of livestock production management have all improved livestock production efficiency, pushing the sector toward more intensification.

Mechanization has also significantly contributed to changes in livestock production systems and helped to promote the commercialization of the livestock sector in China. Agricultural machinery capacity has quadrupled in the past two decades. Draft animals in many parts of the country have been partially or completely replaced by tractors and harvesters. Many small farmers have shifted their animal stock from draft cattle into dairy cattle. Numbers of other draft animals (horses, donkeys, and mules) have also declined significantly (Figure 6.6).

Feed Industry and Feed Production

The emergence and development of the processed feed industry have played a decisive role in shaping the structure of the livestock sector in China. Starting in the late 1970s, the feed industry developed virtually from scratch over the past 20 years. Industrial feed production soared from a mere 2 million tonnes in 1980 to 123 million tonnes in 2007, including complete feed, concentrate feed, and premixed feed (Table 6.5). About 10% of total feed production is used in aquaculture. The quality and reliability of industrial feedstuffs have also been gradually improved. Many livestock producers, including traditional sectors, have lost their resistance toward processed feed and have become accustomed to it. The robust development of the industrial feed sector has been

Table 6.4. Composition of livestock trade in China, 2005

	Export million US$	Import % of livestock trade	Export	Import
Beef	181	96	5.0	2.3
Mutton	60	57	1.7	1.3
Poultry	915	355	25.4	8.4
Pork	946	179	26.3	4.2
Rabbit	21	0	0.6	0.0
Milk	82	459	2.3	10.9
Eggs	75	0	2.1	0.0
Wool	66	1212	1.8	28.7
Honey	87	1	2.4	0.0
Hides	2	1317	0.1	31.2
Others	1168	551	32.4	13.0
Total	3604	4227	100	100

Source: MOA, compiled data based on unpublished custom statistics.

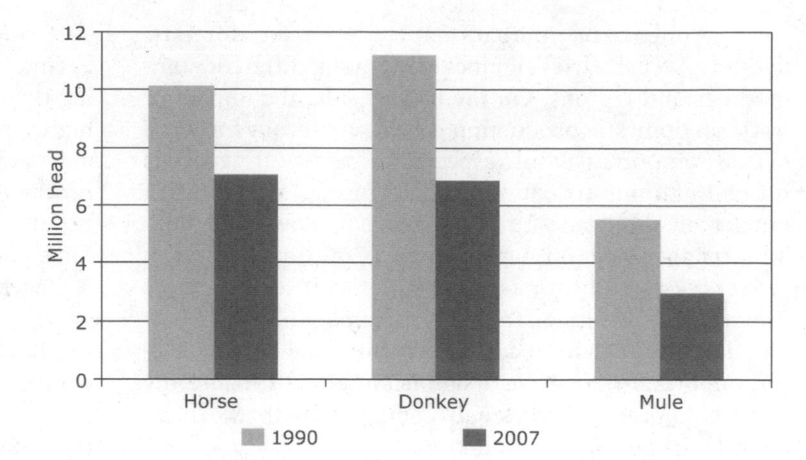

Figure 6.6. Decline of draft animals in China.
Source: National Bureau of Statistics of China (NBS): *Statistical Yearbook of China,* various years.

the decisive factor in the rising intensification of livestock systems, especially in the poultry and pig sectors.

Geographically, the feed industry is mostly concentrated in the eastern parts of the country, in a pattern that reflects the scale structure of the livestock sector in China. Large-scale pig and poultry farms are concentrated in the coastal provinces, as will be discussed. Of the total feed processed in 2006, 52% comes from the eastern zone, 30% from the central zone, and 18% from the western zone (MOA, 2007c). Figure 6.7 shows the regional structure.

Foreign investment and foreign companies have played an important role in the development of the Chinese feed industry. By the end of 2006, there were 486 overseas-funded feed companies in China, mostly located in the east coast of the country. Foreign companies have brought the concept of scientific animal nutrition, which was completely new in China until the early 1980s. These foreign companies have played key roles as pioneers and catalysts. Following their successful examples, domestic feed companies have been set up, including many private ones. By the end of 2006, there were 15,501 registered feed mills, of which only 838 were owned by the government or collectives (MOA 2007b). China's feed industry is still dominated by a great number of relatively small companies. In 2007, there were 157 feed companies whose production exceeded 100,000 tonnes, the top 10 Feed Groups/Cooperatives produced about 30% of the total feed production in the country (MOA 2008). However, consolidation of the sector is under way, and competition has become increasingly fierce in recent years.

Public Services

Improvements in public services have also played a very important role in China's changing livestock landscape. Among these, the most important are publicly originated technical innovation, technical extension, animal disease prevention and control, quality standards and information, marketing information, transportation and storage facilities, and so on.

Public financial inputs for these services have increased by a large margin (although precise figures are not available). The central government budget for agricultural support and service in general grew nearly sixfold in the past decade, from RMB 77 billion Yuan in 1995 to RMB 431 billion Yuan in 2007. This is not only because the central government has increased financial inputs, but also provincial, prefectural, and county governments make large contributions to the public service system for the livestock sector.

Before the economic reforms initiated at the end of the 1970s, the livestock sector in China was under strict direct government control. This control encompassed all phases of the economic process, through all the stages of production and distribution to consumption. This system has been completely changed, through the various reforms introduced since 1978. In the main, a market system has been established, and there are very few government interventions.

Subsidies to large pig farms persisted as a very popular policy in China until the mid-1990s. In that era, large livestock farms were normally state owned or collectively owned and located in suburbs of large cities. For example, in Beijing in the mid-1990s, state pig farms received

Table 6.5. Processed feed production in China, million tonnes

	Total	Complete	Concentrate	Premix
1980	2.0	2.0		
1985	15.0	15.0		
1990	31.9	31.2	0.5	0.2
1995	52.7	48.6	3.5	0.6
2000	74.1	59.1	12.5	2.5
2005	106.8	77.6	24.5	4.7
2006	110.7	81.2	24.6	4.9
2007	123.3	93.2	24.9	5.2

Source: MOA, Yearbook of Animal Husbandry in China 2008, China Agricultural Press.

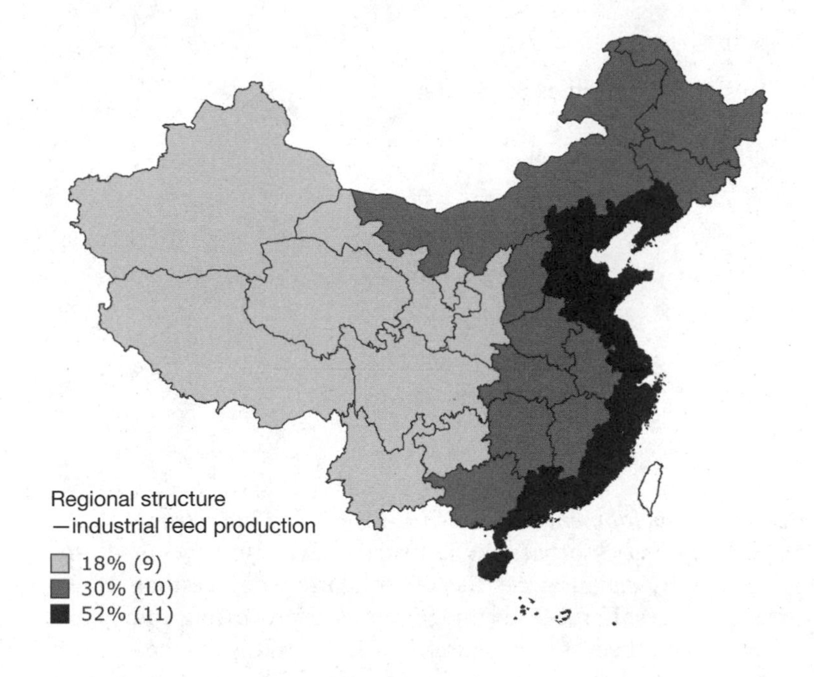

Figure 6.7. Regional structure of industrial production (%). Numbers in parentheses indicate number of provinces.
Source: MOA 2007c.

50 Yuan for every pig they sold to the state slaughtering houses. The main goal was to keep the system going so as to provide the urban population with sufficient meat, milk, and eggs. This goal was seen as crucial for social and political stability in a time of food shortage. Today, given the improved market supply brought about by the booming private economy, regular subsidies to large livestock farms have been abolished as part of the process of economic reform in the state sector.

Providing technical extension to promote production has long been a major policy area in the livestock sector. There are national programs managed by central government, and local programs initiated by local governments at provincial, prefecture, or county level. One of the priority areas for technical extension is the establishment and maintenance of a nationwide livestock breeding system, including introduction of high-quality species, breeding farms, and artificial insemination services.

Another important area of technical extension is the demonstration farm program. Under one type of program first launched in 1985, counties in major grain-producing areas are selected and encouraged to use ammonia-treated straw to raise beef cattle. The central government grants subsidies to improve the related facilities and investment conditions. There are now about 200 such model counties. A similar program was implemented for promoting hogs with leaner meat and less fat. By now, over 400 counties have been designated as "lean meat pig-raising base counties."

Training is also a key area for governmental action, especially for prefecture and county governments. This can take many forms, including distribution of information sheets, printed technical materials, books and manuals, and training courses. Training activities are often an integrated part of the demonstration program.

Changes in Livestock Production

The drivers analyzed above have led to rapid growth in China's livestock production, an increase in livestock's share of total agricultural production, intensification of operations, and changes in the geographical structure of the livestock industry.

Growth of Overall Production and by Species

The share of the livestock sector in the overall agricultural sector has steadily increased over the past two decades. Measured by output value, the livestock sector accounted for 22% of the agricultural total in 1985, rising to 33% in 2007 (Table 6.6).

The production of livestock commodities has risen continuously in the past two decades. Total meat production nearly quadrupled over the period 1985 to 2007 (Figure 6.8). Beef has seen the fastest growth, followed by poultry and mutton, whereas the growth rate of pork production is slowest. Growth of milk and egg production has also been rapid: between 1985 and 2007 milk production grew nearly 15-fold and egg production nearly fivefold. As a result of this uneven development, the share of pork in the meat total dropped from 85% in 1985 to 62% in 2007. In contrast, poultry's share increased from 8% to 23% over the same period. This change is a reflection of the more favorable feed conversion ratio for poultry than for pork in commercial production systems. This in turn leads to lower prices of poultry and increased demand.

These production statistics seem to disagree with the results of consumption data based on household

Table 6.6. Composition of agricultural output value in China, billion RMB (1978 = 100)

	Total	Cropping	Forestry	Fishery	Livestock	Livestock % of Total
1980	147	111	6	2	27	18
1985	217	150	11	8	48	22
1990	280	181	12	15	72	26
1995	385	225	13	32	115	30
2000	609	339	23	66	181	30
2005	815	406	30	83	275	34
2006	833	439	33	81	247	30
2007	842	425	32	77	278	33

Source: NBS, various years.

survey (described in previous sections). There are several possible explanations for this. First, the statistical definitions used in production and consumption are different. Secondly, the production data could have been overreported. The Chinese pyramid reporting system, which was quite reliable before 1980, has become less accurate since the dismantling of collectives and the introduction of the individual household-based production system. There has also been a tendency for officers at various levels to exaggerate the production figures because production growth is often regarded as an important indicator for the performance of local governments. The agricultural census of 1997 revealed that livestock production data were overreported by about 15% in 1996. The latest census was done in 2007, but the detailed data have not yet been published.

Thirdly, there might be bias in the sampling of the households. Putting all aforementioned factors together, it can be estimated that the actual disparity between meat production figures and consumption figures should be within the range of 15 to 20%. The real picture can be obtained when the 2007 agricultural census results are available.

Development of Livestock Processing Industries

Livestock processing industries have also developed rapidly, encouraging the consolidation of livestock enterprises. Large-scale integrated meat processing companies have been set up nationwide, especially in the eastern part of the country. The largest private meat processor in China is the Shuanghui Group Co. With the most modern facilities imported from all over the world, it has annual slaughtering capacity of 20 million hogs. In the company headquarters alone, over 30,000 hogs are needed every day to feed the processing lines. This quantity can be supplied only by large-scale pig farms. Thus the processing industry relies on intensive livestock production operations, and in turn stimulates the development of such large-scale operations.

Intensification of the Livestock Sector

The natural result of all the drivers discussed in the previous sections is that the intensification process of China's livestock sector has developed rapidly over the past two decades. The process has taken place at two levels: "complete" intensification—large industrial enterprises, including those that are highly vertically integrated; and

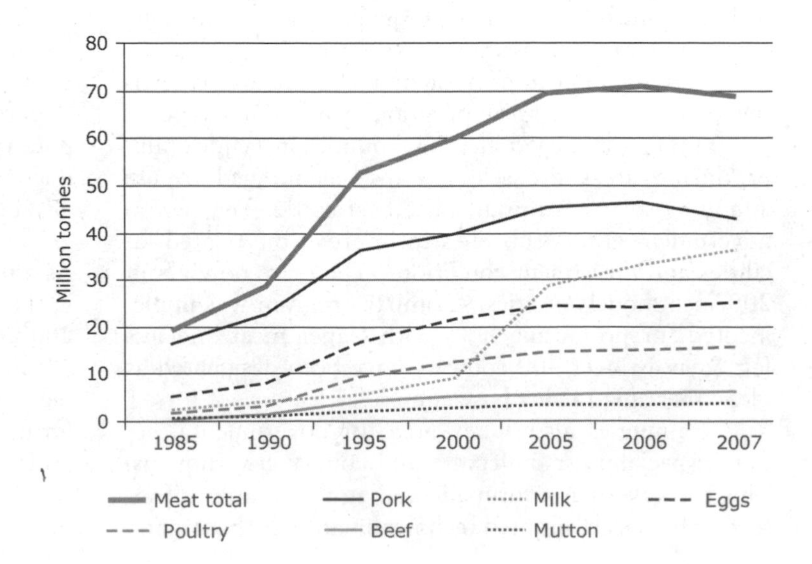

Figure 6.8. Production of livestock commodities 1985–2007.
Source: National Bureau of Statistics of China (NBS): *Statistical Yearbook of China,* various years.

partial intensification—small farmers producing with more purchased inputs (feeds) and for the market (more than for self-consumption).

Definitions of intensive and extensive systems in China vary among animal species and regions. Generally speaking, a pig farm with a yearly production of over 50 hogs is taken as a large pig farm. Smaller pig farms are categorized as backyard farms or extensive raising. Generally applied definitions for major livestock species are listed in Table 6.7.

Table 6.7. Criteria for intensive farms in China

Farm Type	Criterion for Intensive Farms
Pig	Annual production of 50 or more head of hogs
Beef cattle	Annual production of 50 or more head of cattle
Dairy cattle	With a stock of 20 or more dairy cattle
Broilers	Annual production of 2000 or more birds
Layers	With a stock of 2000 or more birds
Sheep and goat	Annual production of 30 or more sheep/goats

Source: MOA, internal reports.

The intensification process has not been the same for all animal species. Generally speaking, the poultry sector, including broilers and layers, has shown the fastest intensification process, followed by the pig sector. The cattle sector, including beef cattle and dairy cattle, has experienced a slower process of intensification.

As indicated in Table 6.8, 789 million broilers were produced by farms with an annual output of 10,000 birds or more in 1996, whereas the corresponding figure increased to 2794 million in 2006. It is estimated that about half of the broiler production is from large chicken farms with a capacity of over 10,000 birds a year.

Pig production has shown a similar trend (Table 6.9). In 1996, large farms produced only 48 million hogs, or less than 20% of the total, whereas by 2006 the figures rose to 316 million and 64%, respectively.

Intensification is taking place not just among large-scale producers but also among smallholders because they increasingly produce with purchased inputs (feed), and for the market rather than for self-consumption.

East–West Dichotomy and Spatial Concentration

The intensification process has been uneven among different regions in China. An apparent east–west dichotomy has been observed. In the economically more developed eastern parts of the country, livestock production

Table 6.8. Intensification of broiler production in China

1996	**Farm size, birds**	**50–200**	**200–1000**	**1000–10000**	**> 10,000**		**Total**
	Broiler output, million birds	220	283	968	789		2260
	% of total	9.7	12.5	42.8	34.9		
2006	**Farm size, birds**	**2000–10,000**	**10,000–50,000**	**50,000–100,000**	**10,000–500,000**	**500,000–1,000,000**	**Total**
	Broiler output, million birds	1733	1796	520	337	141	4527
	% of total	38.3	39.7	11.5	7.4	3.1	

Source: MOA Yearbook of Animal Husbandry in China 2007.

Table 6.9. Intensification of pig production in China

1996	**Farm size, head**	**50–200**	**200–1000**	**> 1000**			**Total**
	Output, million head	13.1	9.4	25.2			47.7
	% of total	27.5	19.7	52.8			
2006	**Farm size, head**	**50–100**	**100–500**	**500–3000**	**3000–10,000**	**10,000–50,000**	**Total**
	Output, million head	105.6	103.8	60.7	27.9	20.5	315.8
	% of total	33.2	32.6	19.1	8.8	6.4	

Source: MOA Yearbook of Animal Husbandry in China 2007.

is concentrated in a small number of large operations, and traditional smallholders have almost disappeared in many places. By contrast, in the vast western regions, traditional extensive systems, including small backyard raising and extensive pastoralism, are still predominant.

Figure 6.9 in the color well shows the geographical distribution of pig production in China. It shows clearly that most pigs are produced in the eastern half of the country. The only exception is Sichuan Province, still the largest pig producer in China. Formally Sichuan belongs to the western region, but it is geographically located between the west and the east. However, when we look at the farm size structure or the degree of intensification, Sichuan Province is among the lowest in the country, as shown in Figure 6.10.

In contrast, the southeastern provinces have the highest intensification of pig production, measured by the share of pigs produced by very large farms (production of 500 hogs or more per year). The share is 65% for Tianjin, 63 for Shanghai, 46% for Beijing, 40% for Zhejiang, 39% for Guangdong, and 38% for Fujian. All these are coastal provinces in the east. The share of very large farms in pig production is under 10% for all provinces in the west.

The geographical location of the feed processing industry matches this geographical pattern of intensification. Guangdong, Shandong, and Hebei are the three lead feed producers in China, all located in the eastern coast region.

Increased Production Efficiency

As part of the intensification process, the productivity of livestock production in China has improved continuously over the past two decades. As indicated in Table 6.10, hogs are now produced in a much shorter time span, and with more weight per head. When combined, these two factors imply a great rise in production efficiency, measured by the meat produced per head of pig stock. This parameter has risen from only 54 kg in 1985 to 98 kg in 2007 (Table 6.10).

There is a close correlation between farm size and technical indicators (Table 6.11). As revealed by a survey on pig farms conducted by the Research Centre for Rural Economy in 1999, both hog slaughtering weight and fattening days decline as the size of farms increases (RCRE 2000). Taking both of those factors into consideration, the weight gain per day is significantly higher in larger farms than in smaller farms. This is due to better technical production facilities in large farms—for example, the share of processed feed in total feed use also increases progressively with the size of the farm.

A similar result was found in the chicken sector in the same survey by RCRE, which shows clearly that the feed–meat conversion efficiency (kg of feed per kg of meat produced) increases with the rising number of birds on the farm, from 2.76 for farms with less than 100 birds to 1.76 for farms with over 5000 birds (Table 6.12).

Consequences

Based on the various factors discussed in previous sections, the intensification trend of livestock production in China is inevitable and irreversible. We will now consider the major consequences of this trend, including the positive and negative impacts, as well as the variations across different regions and among different stakeholders. The

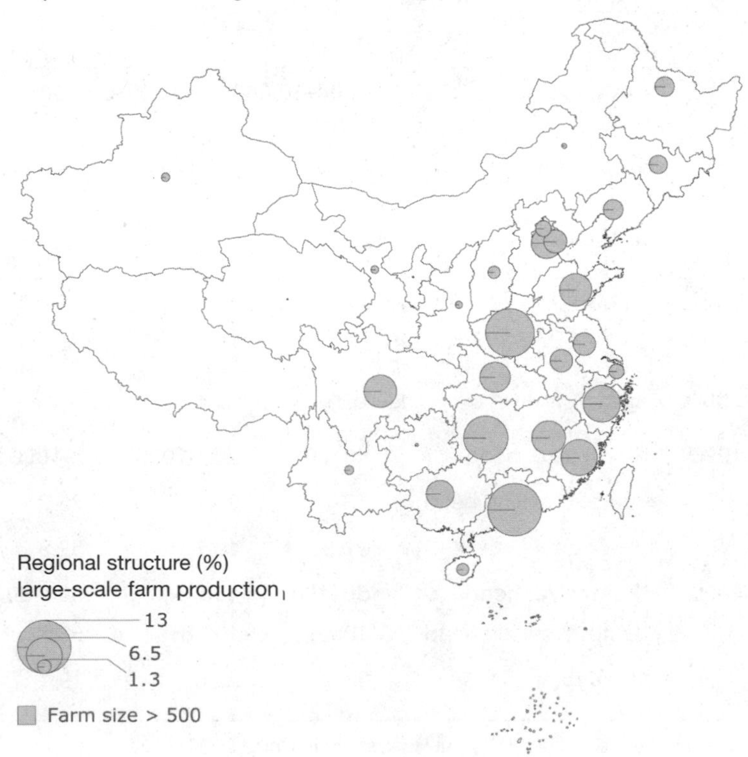

Figure 6.10. Hog farm size structure of China, 2006

Note: the values in this map are shares of hogs produced by very large farms (annual production > 500 hogs) in total hog production.

Table 6.10. Improvement in productivity of pig production in China

Year	Inventory at year beginning	Slaughtered	Meat output	Slaughtering rate	Carcass weight	Productivity per year
	Million heads	Million Heads	Million t.		Kg/head slaughtered	Kg/head of stock
1985	306.8	238.8	16.6	78	70	54
1990	352.8	309.9	22.8	88	74	65
1995	414.6	480.5	36.5	116	76	88
2000	416.4	518.6	40.3	125	78	97
2005	433.2	603.7	45.6	139	76	105
2006	418.5	612.1	46.5	146	76	111
2007	440.0	565.1	42.9	128	76	98

Note: Slaughtering rate = heads slaughtered as a percentage of inventory.
Carcass weight = meat output divided by heads slaughtered.
Productivity = meat output divided by inventory.

Source: National Bureau of Statistics of China (NBS): Statistical Yearbook of China, various years.

Table 6.11. Technical indicators for pig sector by farm size

	1–5 Heads	6–15 Heads	16–30 Heads	31–50 Heads	over 50 Heads
Slaughtering weight (kg)	109	86	89	86	84
Fattening days	234	158	132	130	122
Weight gain (kg/day)	0.47	0.54	0.67	0.66	0.69
Share of processed feed (%)	17	39	60	73	92

Source: RCRE 2000.

Table 6.12. Feed conversion ratio for chicken in China by farm size (lower = more efficient)

Birds/farm	≤100	100–500	500–1000	1000–5000	≥ 5000	Average
Feed ratio	2.76	2.51	2.40	1.77	1.76	2.25

Note: Feed Conversion Ratio = number of kilos of feed used to produce one kilo of meat.

Source: RCRE sample survey 1999.

changing landscape of China's livestock sector also has significant international implications in terms of its tendency to import huge and rising quantities of feed materials.

Beneficial Social and Economic Impacts

The positive impacts of the changing livestock sector in China are seen mainly in three areas: national food security, animal disease control, and cleaner living environments for small farmers who give up livestock production.

Food security is the top priority in China's agricultural policy making. China has had a long history of food shortage. The food supply situation substantially improved only after the rural reform policy was introduced in the late 1970s. In recent years, due to rapid industrialization and urbanization, farmland has been declining at an alarming speed. This, together with a shortage of irrigation water and other constraints, has led to the stagnation of grain production in the past 10 years while demand keeps rising. The increase in grain demand is mostly from the livestock sector because direct per capita consumption of food grains has been declining (see earlier discussion). However, the intensification of livestock production in China has improved feed use efficiency, which has partially offset the shortage of grain supplies. Without the improved efficiency of the livestock sector, the food security issue would be much more pressing.

Another positive effect is the improvement of living environments for farmers who formally raise in their backyard, especially in the more developed regions, generally in the suburbs of large cities and in the coastal areas of the eastern provinces. In these areas, nonfarm jobs have become the major income sources, and farmers no longer depend on backyard livestock raising for cash income. Given higher incomes and living standards, farmers have much lower tolerance for the bad smells, flies, and waste treatment problems associated with backyard livestock raising. A large share of animal production has been shifting from backyards to concentrated big farms. Therefore the negative impacts of animal pollution have been moved from the small farmers' backyards and have been concentrated in certain areas surrounding the large intensive operations.

A further, very important, benefit is that animal disease control measures are easier to implement with intensive farms than with extensive systems. In fact, the great number of small backyard operations raising chickens and ducks across the country is the biggest difficulty faced by the animal disease control authority. Large intensive operations are easier to contact, and the owners are more willing to implement vaccination, more alert about possible outbreaks of animal disease, and more serious about implementing eradication measures. This advantage of large intensive operations is best illustrated by the pattern of avian flu cases in recent years. According to monitoring results of the animal epidemic control authority, most of the 84 bird flu cases that occurred since early 2004 were found in the extensive system.

Adverse Environmental Impacts

Adverse impacts from the changes in China's livestock sector are mainly on the environment. The environmental problems also show an east–west dichotomy. In the eastern region, where the intensive system predominates, the most outstanding problems are water and soil pollution. In the vast western part of the country the major problems are soil erosion and desertification caused by overgrazing in the extensive pastoral system.

Surface and Ground Water Pollution from Intensive Systems

In the eastern region where intensive rearing predominates, the most apparent problem is pollution of surface and groundwater by nutrients from animal manure. This is the more serious because China already has a very high level of fertilizer inputs in agriculture, especially in the eastern provinces.

Based on some experimental parameters (Wang et al., 2006), the nitrogen content of livestock manure disposal can be calculated. As shown in Figure 6.11, the nitrogen load from livestock manure disposal is well above 100 kg per hectare of cropland in all the east provinces, and over 200 kg/ha in some of them. The total nitrogen load from livestock manure and chemical fertilizer is over 400 kg/ha for many provinces in the east, as shown in Figure 6.12.

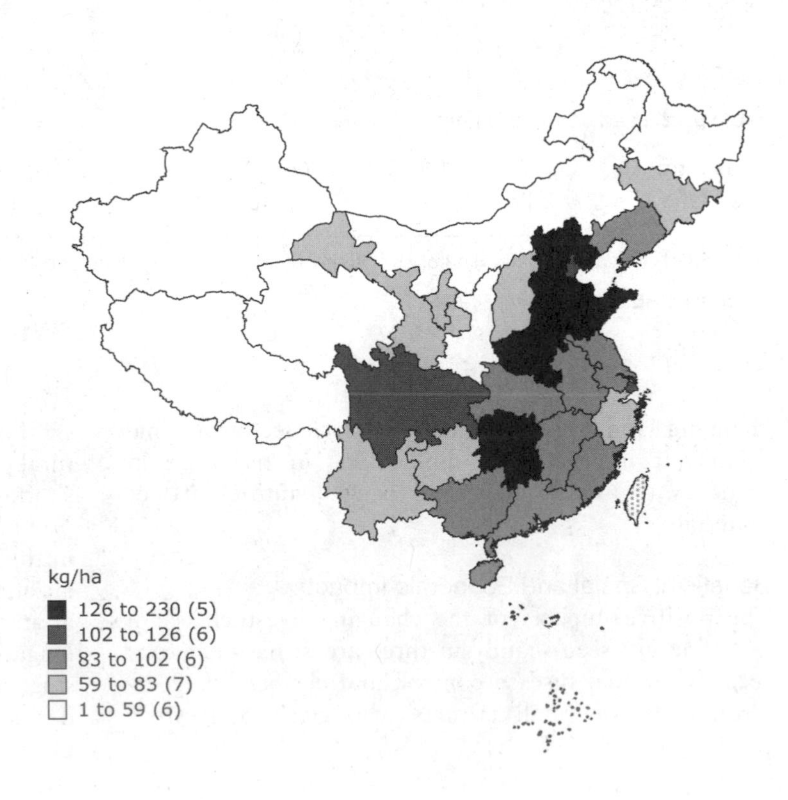

Figure 6.11. Nitrogen load from livestock manure in China, kg/ha cropland, 2005.

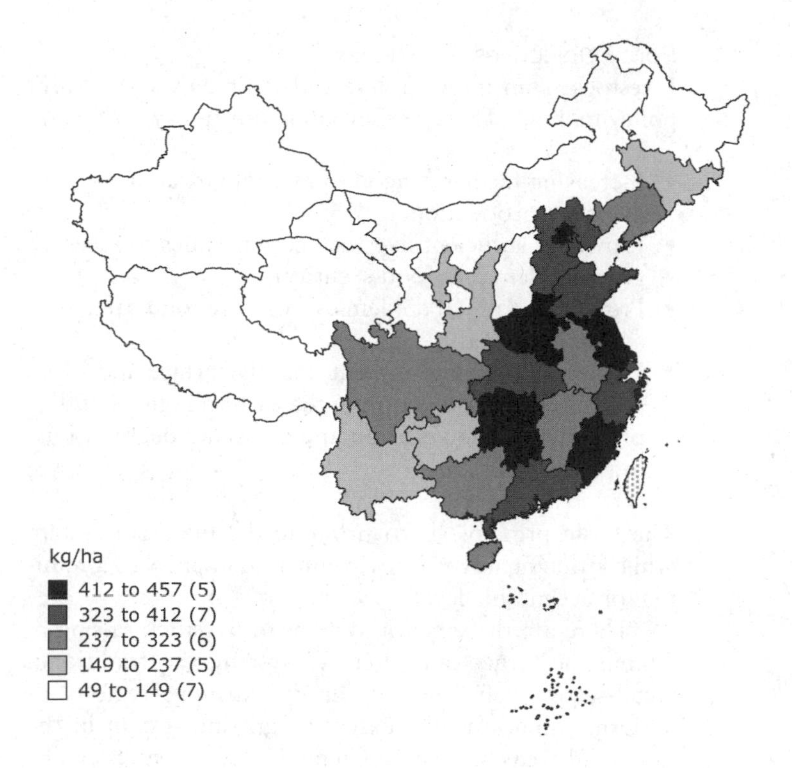

Figure 6.12. Nitrogen load from livestock manure and chemical fertilizer in China, kg/ha cropland, 2005.

The environmental consequences of deficient waste management are introduced by Menzi et al. (2009) in *Livestock in a Changing Landscape: Drivers, Consequences, and Responses.*

Land Degradation and Desertification in Pasture Areas

In western China, especially Inner Mongolia, land degradation and desertification are the leading environmental problems associated with livestock production. Overgrazing of grasslands is worsening as animal numbers keep rising, leading to desertification and increasing sandstorms. Although desertification directly affects only local people, sandstorms are felt across Northern China, especially in the Spring. Though the sandstorms were well controlled in recent years (Table 6.13), governments paid a high price.

Overgrazing is caused on the one hand by ever-increasing animal stocks, and on the other by extensive and unimproved use of grassland. In the Inner Mongolian region the number of goats and sheep has more than doubled in the last two decades, from about 25 million head in 1985 to over 56 million head in 2006. Many local herdsmen still use traditional ways of raising animals and continue to apply nomadic practices. There is little investment in improving grassland quality and productivity. Some efforts have been made in recent years, but the effects still remain very limited.

Overgrazing is also emerging as a problem in Tibet. The number of yaks, the predominant livestock animal in the region, has risen from 5.4 million to 6.3 million over the period of 1995 to 2005, placing increased pressure on the vulnerable plateau grassland system.

International Trade Consequences

Chinese grain production practically stagnated over the 1996–2007 period, whereas meat production rose by about 50%. The growth in meat production was made possible partly by improvements in feed use efficiency and partly by soaring imports of feed commodities, in particular soybeans. China's soybean imports soared from a negligible low level of 0.3 million tonnes in 1995 to 30.8 million tonnes in 2007 (Table 6.14). The import volume in 2007 was nearly twice that of domestic production. The share of China's imports of soybeans in world exports rose from 1% to 42% between 1995 and 2007, making China by far the largest importer of soybeans in the world. Of the 30.8 million tonnes of soybeans imported in 2007, 37% came from the United States, 34% from Brazil, and 28% from Argentina. China's soybean imports constitute about 39% of the soybean exports of

Table 6.13. The sandstorms in China in recent years

Year	2000	2001	2002	2003	2004	2005	2006	2007
Times per Year	15	18	12	10	19	19	18	10

Source: http://www.eedu.org.cn/Article/eehotspot/Desertification/200806/27162.html.

Table 6.14. Soybean imports of China

	Production (million tonnes)	Imports (million tonnes)	Imports as % of Chinese Production	Imports as % of World Exports
1995	13.5	0.3	2	1
1996	13.2	1.1	8	3
1997	14.7	2.9	20	7
1998	15.2	3.2	21	8
1999	14.3	4.3	30	11
2000	15.4	10.4	68	19
2001	15.4	13.9	90	26
2002	16.5	11.3	69	19
2003	15.4	20.7	135	37
2004	17.4	20.2	116	31
2005	16.4	26.6	162	41
2006	15.5	28.2	182	42
2007	15.6	30.8	197	42

Sources: MOA Agricultural Development Report of China 2008, FAOSTAT.

the United States, 45% of Brazil's, and about 70% of Argentina's.

The large and rising volume of imports by China has stimulated world soybean production, which has increased from 127 million tonnes in 1995 to 216 million tonnes in 2007. Most of the growth in world soybean production has taken place in the United States, Brazil, and Argentina, as indicated in Figure 6.13. This has had environmental consequences in the exporting countries, most notably in Brazil, where it has been associated with rapid deforestation and the attendant loss of biodiversity.

Responses

This section analyzes current responses to the changes in the livestock landscape, with a focus on government responses. Responses that are desirable but not yet effected will be presented in the conclusions and recommendations section.

Policy Objectives

Livestock is an important sector in China's agricultural policy making. There are several major policy goals:

- Increasing farmers' incomes by promotion of livestock production
- Providing sufficient livestock commodities to meet a growing demand (food security)
- Preventing animal epidemics to secure food safety and human health
- Protecting the environment (mainly surface and ground water protection in the eastern regions and prevention of soil erosion and grassland degradation in the western).

The basic principle is to promote the intensive system while reducing adverse environmental impacts to a minimal or acceptable level.

There are three major systems of livestock raising in China: the large-scale intensive system, the traditional small-scale raising system in the farming regions of eastern China, and the extensive grazing system in the grassland areas in western China. Policy measures can be divided into those promoting overall livestock production and those addressing problems in the three systems, respectively.

Policy Measures to Promote Overall Livestock Production

In the past, local governments often provided investment subsidies for the establishment of large livestock farms in order to secure supplies for nearby urban consumers. This practice was largely terminated at the beginning of this century, but there are still occasional investment subsidies to the livestock sector in some localities, usually to promote some special forms of operation. For example, in one county of the Beijing Municipality, the local government designated certain sites for livestock operations and granted investment subsidies to farmers who moved their livestock operation to those sites, either directly or

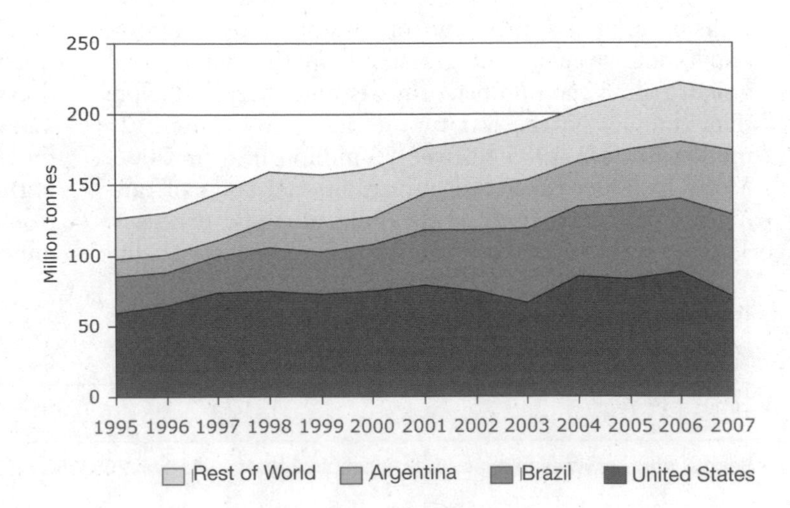

Figure 6.13. Soybean production of world major producers.
Source: FAO 2008.

through subsidized credit programs. This measure aimed to attract farmers away from backyard livestock rearing so as to reduce environmental problems in the villages. In other places, local government provided investment subsidies to promote the experimental integration of livestock production with greenhouse vegetable production. In yet other places—usually economically less developed regions—land is provided on favorable terms for the establishment of large livestock farms.

Agriculture is one of the top-priority areas for government support of high technology. There are 16 project areas under agriculture, 7 of which relate to the livestock sector, including animal genetics, environmentally sound facilities for intensive livestock farming, high-quality animal breeding, facilities for organic manure, pasture seed development, and straw treatment for feed and feed additives. These guidelines help various governmental agencies to determine the priority areas for various kinds of support, such as research funding, investment loan programs, and so forth.

Animal health, epidemic prevention, and quarantine have always been key areas of governmental action. Public awareness and concerns about the quality and safety of livestock products have become stronger than ever, in line with the rise in personal incomes and living standards. The issue of animal health has been brought into sharper focus since the outbreak of avian flu in 2004. On the other hand, animal disease control is not an easy task because the Chinese livestock sector is still characterized by a great number of very small-scale producers.

In the area of promoting feed production, the main policy instruments involve tariff and tax exemptions. Technical equipment for producing feed additives and for developing animal protein resources is exempted from import tariffs and value-added tax (normally 13% of the product's market value). Both imported and domestically produced feed and feed additives are also exempt from value-added tax. This policy has been crucial for the development of the feed industry in China.

Feed safety has become an increasing concern in China in recent years, boosted by the problem of dioxin contamination in poultry feed in Belgium. Heated disputes in North America and Europe on the use of growth stimulants, including hormones, have also motivated responses in China. In order to control feed safety issues more strictly, the State Council issued a special decree on production, import, and use of feed in 1999. Among other regulations, the decree prescribes that all new feed products and additives have to pass feed safety and environmental impact examinations before they are permitted to go into production. It also banned the use of hormones in feed.

After China's entry in the WTO in 2001, in order to meet the high quality standard requirement of export and ensure the food safety of exported food products, the Chinese government further strengthened the quality standards of agricultural products, including reducing the use of additives in feed, improving the production environment, and so on. Meanwhile, China's own food safety domestically is an important consideration. According to officials of MOA, one consideration is that livestock entrails are very popular in Chinese eating habits, and the entrails have a much higher content of additive residues and therefore higher risk levels. The ban on hormone use in feed may cause some trade disputes with the United States, especially, in the future when China would import pork or other meats.

Laws and Regulations to Address Environmental Problems Associated with Intensive Operations

Policies to protect the environment from the negative impacts of livestock production can be analyzed from four aspects: location of production facilities, waste discharge standards, economic incentives on waste management, and policy enforcement.

There are no national or local regulations addressing the location of livestock farms. Interview results from the pig farm questionnaires conducted by an FAO project team in Jiangxi Province show that half of the investigated farms are located within less than one kilometer from the nearest residential area. When questioned about the reasons for choosing their current location, almost all indicated access to good transportation, water sources, and so forth, as the main reasons. Very few mentioned the distance to residential or other sensitive areas. In a recent decree issued by the State Council, prompted by increasing worries about declining grain production and possible shortages, "basic farmland" is to be strictly protected and not allowed to be used to build livestock farms, fish ponds, or fruit orchards (State Council 2005). "Basic farmland" refers to fertile land suitable for field crop production; it is mainly located in plains areas.

In terms of waste discharge standards, the State Environment Protection Administration[1] and the State Administration for Quality Monitoring, Inspection and Quarantine jointly issued a decree in 2003 setting standards for waste emissions from livestock production. The decree set ceilings on daily emissions of waste water and gases, and on harmful contents of solid waste. Table 6.15 lists the daily water pollutant emission standards.

For gas emissions, the stench emission from intensive livestock farming is 70 (dimensionless). For livestock and poultry industry waste residues of sound environmental standard, there are two control items, the mortality rate of roundworm eggs should not be less than 95%, and the number of fecal coliforms should not be greater than 10,000 per kg. The detailed standards *Discharge Standard of Pollutants for Livestock and Poultry Breeding*

1. In March, 2008, the State Environment Protection Administration was reformed into the Ministry of Environmental Protection (MEP).

Table 6.15. Daily water pollutant emission levels for intensive livestock and poultry breeding

Items	Standard
Five-day biochemical oxygen demand (mg/L)	150
Chemical oxygen demand (mg/L)	400
Suspended solids (mg/L)	200
NH_3-N (mg/L)	80
Total phosphorus (P) (mg/L)	8
Fecal coliform (number/mL)	10000
Roundworm eggs (number/mL)	2

Source: Ministry of Environmental Protection of China 2008.

(GB 18596-2001) can be found at the Ministry of Environmental Protection of China website (http://www.zhb.gov.cn.htm).

Also in 2003, a decree on waste disposal fees was jointly issued by the State Development Planning Commission, the Ministry of Finance, the State Environment Administration, and the State Economic and Trade Commission. According to the decree, all enterprises should pay fees for the environmental pollutants they produce, including solids and wastes, air pollution, and sound pollution. Waste disposal from livestock production is included in the decree. The decree defines different waste units as a basis for fee charges. For livestock, the pig is taken as the unit, with 0.1 cattle and 30 poultry equivalent to one pig. Livestock farms with an inventory of more than 500 pigs, or 50 cattle, or 5000 poultry birds must pay the waste discharge fees. For waste disposal equivalent below national limits, 0.7 Yuan is charged for each head of pig per month. If the waste disposal equivalent exceeds the national standards, then the fee is doubled.

At the farm level there is a low degree of public awareness of livestock–environment issues, and little attention is paid to them. This situation was confirmed by a field visit of the author in Jiangxi and Guangdong provinces. Pig farm managers generally acknowledged that their farms were creating pollution problems in local water bodies, and that local governments were concerned with the situation. However, none of them were able to name any related laws or regulations, either national or local. The same was true for the heads of the three townships visited. More surprisingly, even the accompanying officials from the county environmental protection administrations could not list the names of the few related laws and regulations, let alone the exact content and standards prescribed in the regulations. In one county, the county environmental official did not know a single law or regulation on livestock pollution.

It is clear, given this low degree of awareness, that the enforcement of the national and provincial laws and regulations on livestock–environment issues must be weak or nonexistent in some places. First, the requirement for environmental assessments (EAs) is apparently not well enforced. Only one pig farm indicated that an EA had been conducted, but it could not provide the document. Secondly, waste discharge standards are not implemented. There is no systematic monitoring of the quality of the water bodies to which liquid wastes are discharged. Thirdly, no waste discharge fee has been collected, nor have any fees been charged for wastes exceeding the standards. Fourthly, though there are a large number of punishment articles in the laws and regulations, these are not implemented either.

The lack of incentives, sanctions, and pressures contributes to the low interest of livestock farms in environmentally sound waste treatment. Generally, a single pig farm with 3000 pigs uses about 200 tonnes of water to flush the waste. This results in a huge amount of wastewater discharged to the environment, but also in the overuse of groundwater resources because the flushing water is drawn from aquifers 20 meters deep underground.

There are several reasons why policy is weak or not enforced. The goal of economic development is usually given higher priority than environmental protection. There are institutional constraints: environmental officials are appointed by the local government and hence tend to yield to the economic growth priorities of the latter. There are loopholes in policy design. Policies include unrealistic requirements. Resources for implementation are inadequate, including human capacity, equipment, and funds for monitoring and testing at local environmental administration level.

Policy Measures to Address Environmental Problems for Smallholders

As mentioned in previous sections, environmental problems among smallholders have gained increased attention. Nationwide subsidy schemes for biodigesters are among the major policy measures aimed at treating animal manure and reducing water pollution and emissions of greenhouse gases. Simple biogas generating units are supported to treat animal and human excreta at the household level. They generally consist of a cement tank, a floating cover, a mechanism for removal of residues, and some pipes. Biodigesters serve the dual purpose of supplying renewable energy to rural households while partially mitigating pollution issues. The process reduces the organic matter content of waste; however, it does not greatly reduce its nutrient content, so the remaining effluents can be recycled on crops.

Biogas is generally used for cooking, and in some cases also for lighting. The use of biogas has a long history in China, but until recently with only limited spread due to technical constraints. Technical improvements in recent years have led to lower investment costs, more reliable functioning, and easy removal of residues. This has produced great enthusiasm among farmers to adopt the practice. The government has paid considerable attention to promoting the practice more widely and has significantly increased funding for the Rural Biogas Program. Shared between central and local governments, the program provides subsidies for about one third to two thirds of the total cost. It is currently the largest single program managed by the Ministry of Agriculture of China. In the past three years, about 3 million farmers have participated. The goal is to increase the number to 40 million by 2010 (State Council 2006).

Another measure to address the environmental problems associated with smallholders is the creation of "concentrated livestock raising areas." Farmers from the same village are encouraged to move animals from their own yards into a specific area on the outskirts of the village. Individual farmers have separate stalls in this area. The aim is that this spatial concentration and individual ownership of smallholders can allow easier animal disease control, technical extension, and environmental control, as well as removing animals and their smell, dirt, and flies from residential areas. Subsidies to establish these concentrated raising areas have been provided. However, the success of this practice so far is not as obvious as the biogas program. A number of factors complicate the situation, including the sharing of investment and operation costs, collective decision making, and coordination.

Policy Measures to Address Grassland Degradation Problems

Two regions are especially vulnerable to land degradation caused by extensive livestock production: the Loess Plateau and the grasslands of Inner Mongolia. A massive project was introduced into the Loess Plateau region to reverse the trends of environmental degradation in the late 1990s. It was a combination of penalties against extensive nomadic raising and incentives to promote intensive in-stall raising. This program has already achieved very positive initial success in reducing land degradation.

Overgrazing leading to desertification and sandy storms has been the major environmental problem created by the nomadic livestock raising system in Inner Mongolia. Similar measures to those implemented in the Loess Plateau have been taken in Inner Mongolia. The government has provided a huge amount of investment in building fenced grassland in an effort to change the nomadic and extensive way of raising livestock in the region. The grassland is still owned by collectives, but it is easier to control and manage. With fencing, the productivity of grassland can be increased and the pressures of excess extensive raising can be reduced. By the end of 2006, 44 million ha of grassland had been fenced, and nomadic grazing had been banned on 38 million ha of grassland (MOA 2007a). This ban could reduce the overgrazing and improve the sustainable development of grassland, but it could also have a negative impact on traditional nomadic culture. This raises some heated discussion on the development of nomadic culture.

However, in Inner Mongolia it is not possible to completely stop the nomadic way of raising livestock as it has been in the Loess Plateau. Increasing the grass yield involves not just building fences but also irrigation, and water availability is a major constraint in many parts of the region. As a result, the effect of intensification so far is limited.

Conclusions

In summary, major conclusions can be drawn as follows.

Due to drivers from both demand and input sides, Chinese livestock production has achieved rapid growth, and the intensification of livestock operations for all species has been improved. Many large-scale livestock farms operate in an industrial way, and the spatial concentration of livestock animals has reached very high densities in certain regions.

This changing landscape of the livestock sector has a series of consequences, both positive and negative. National food security, animal disease control, and farmers' backyard living environments are all improved. However, some issues are not yet improved, including air and water pollution, nitrogen overload caused by large intensive farms, and soil erosion and desertification caused by overgrazing by the extensive pastoral system. Apart from the domestic economic, social, and environmental impacts, the development of the livestock sector has also significant international implications, in particular as it has strongly stimulated the export and production of soybeans in North and South America.

In order to address these environmental problems, the Chinese government has taken a number of policy measures, including new laws and regulations. However, there is still considerable room for improvement, especially in the enforcement of regulations. Implementation of many regulations remains a big challenge because production and economic growth goals usually take priority over environmental considerations, especially from the long-term perspective.

Looking ahead, the trend of livestock development in the recent past will continue into the future because all driving forces discussed in this chapter are still in place and in effect. The further growth and intensification of livestock production in China are unavoidable and irreversible. To mitigate the adverse impacts associated with this changing livestock landscape, joint efforts by all stakeholders are needed.

From a public policy–making viewpoint, it is of immediate importance to enhance the awareness of all stakeholders and general public, to improve the formulation of laws and regulations, to strengthen the enforcement of existing regulations, and to provide more investment in providing public services such as research, extension, disease control, food safety, and information services. In the future, the following should be given higher priority:

1. Establishing a modern breeding and extension system for new varieties: In 2006 China had about 9709 farms with livestock and poultry breeding stock, but due to either the low quality of breeding stock or a lack of market information, most of them faced difficulties in sales that hampered continuous production.
2. Strengthening the disease prevention and control system: Each animal has its own scientific prevention and control "standard" and "approach," and their implementation requires a series of systems and technical measures. Due to difficulties in the implementation of disease prevention and control in the traditional small and scattered household-raising style, the mortality levels of livestock and poultry in China are high. The mortality rates for pigs, poultry, and sheep are 10%, 20%, and 8%—these levels are about twice those in the controlled animal disease-free zone (MOA 2006).
3. Reforming and improving the feed crop production and processing system: Feed crop production should be based on natural resources; about 17% of food grain production is used as feed, which is not economical. Feed crops should be given the same priority as grain and cash crops. China's feed industry has an annual production of 700 million tons of crop straw, which after ammoniation yields about 280 million tons of feed. This efficiency should be strengthened in the future.
4. Establishing and promoting modernization of facilities and equipment: This is the core of the modern livestock and poultry industry. It includes feed processing equipment, feeding equipment, disinfection facilities, testing equipment, room cleaning equipment, waste collection and processing equipment, and so on.
5. Further improving modern environmental standards: This includes modern standards to protect the environment from pollution by livestock and poultry manure, waste, dead animals, sewage, and so on.
6. Establishing modern management systems: The government should subsidize farmer training on professional and technical management knowledge. Given the unemployment challenge in the current economic recession, this kind of subsidy is human capital investment, which can have long-lasting effects.

References

FAO. 2008. FAOSTAT statistics database. Rome: FAO. Available at http://www.fao.org/FAOSTAT.

Gerber P., P. Chilonda, G. Franceschini, H. Menzi. 2005. Geographical determinants and environmental implications of livestock production intensification in Asia. *Bioresource Technology* 96: 263–276.

Jiang, N., X. Xin, and J. Yin. 2002. Analysis of the factors influencing consumption of livestock products in urban households in China. *Chinese Rural Economy* No. 12.

Lu W. and Y. Mei. 2008. Empirical research on consumption structure of animal products of Chinese urban and rural households. *Technology Economics* 27: 2.

Meng, X. 2002. *Analysis of the Beef and Mutton Market and Outlook in 2003*. Beijing: Agricultural Publishing House.

Menzi, H., O. Oenema, C. Buton, O. Shipin, P. Gerber, T. Robinson, and G. Franceschini. 2009. *Livestock in a Changing Landscape: Drivers, Consequences and Responses.* Washington, DC: Island Press.

Ministry of Environmental Protection of China. 2008. Discharge standard of pollutants for livestock and poultry breeding. Available at: http://www.zhb.gov.cn/tech/hjbz/bzwb/shjbh/swrwpfbz/200301/t20030101_66550.htm.

MOA. *Yearbook of Animal Husbandry,* various years. Beijing: Ministry of Agriculture.

MOA. 2006. Internal records of the Veterinary Department. Beijing: Ministry of Agriculture.

MOA. 2007a. *China Agricultural Development Report 2007*. Beijing: Ministry of Agriculture.

MOA. 2007c. *Yearbook of Feed Processing Industry of China 2007*. Beijing: Ministry of Agriculture.

MOA. 2008. Chinese feed industry: the rise of the great powers. *Farmers' Daily* Dec. 12.

NBS. *Statistical Yearbook of China*, various years. Beijing: National Bureau of Statistics of China.

RCRE. 2000. *Survey of Livestock Holders*. Internal report. Research Centre for Rural Economy. Beijing: Ministry of Agriculture.

Sheng W., L. Xinqjia, and Y. Jing. 2004. *Consumption and Its Determinants of Dairy Products in Urban Households of Beijing*. Beijing: Agricultural Publishing House.

SPFP. 2006. Internal report. Beijing: State Population and Family Planning Commission.

State Council. 2005. *No. 1 Document of the State Council on Enhancing Agricultural Productivity.* Beijing: State Council.

State Council. 2006. *The 11th Five Year Plan for Economic and Social Development*. Beijing: State Council.

Wang, F., et al. 2006. Estimation of livestock manure amount and its environmental effects in China. *China Environmental Sciences* 26: 5.

Yu J. 2008. Pork Income Elasticity and Prediction of Pork Consumption Growth in China. *Modern Business* No. 6.

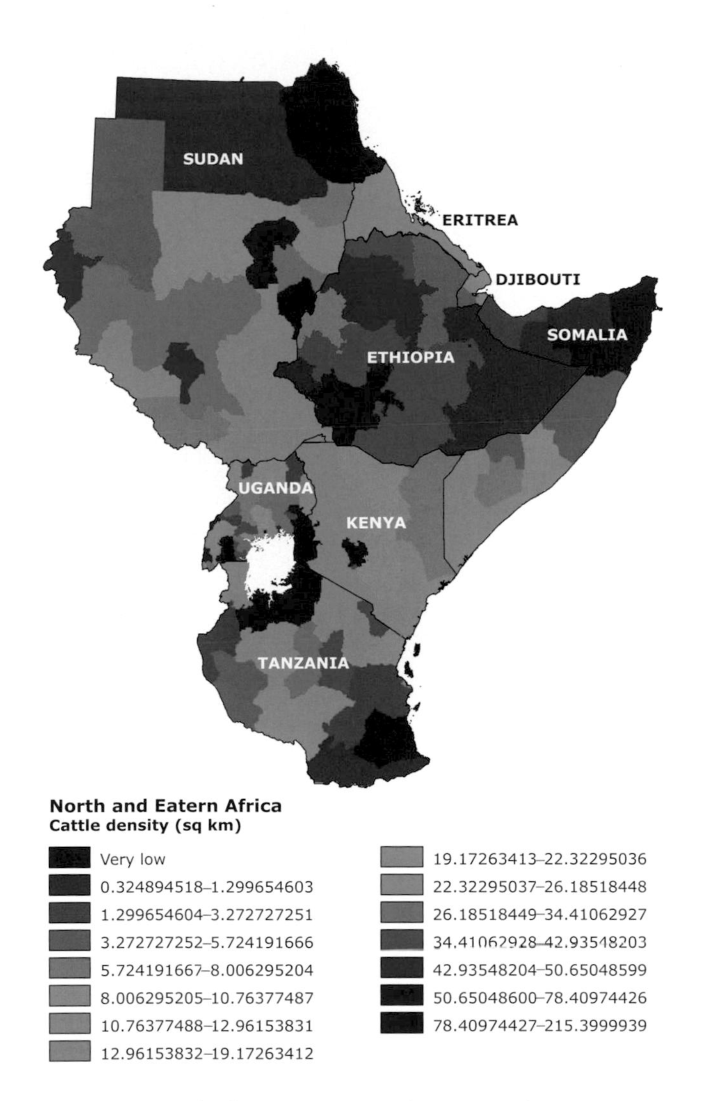

Figure 2.3. Cattle density Km^{-1} in the Horn of Africa.
Source: FAO 2008.

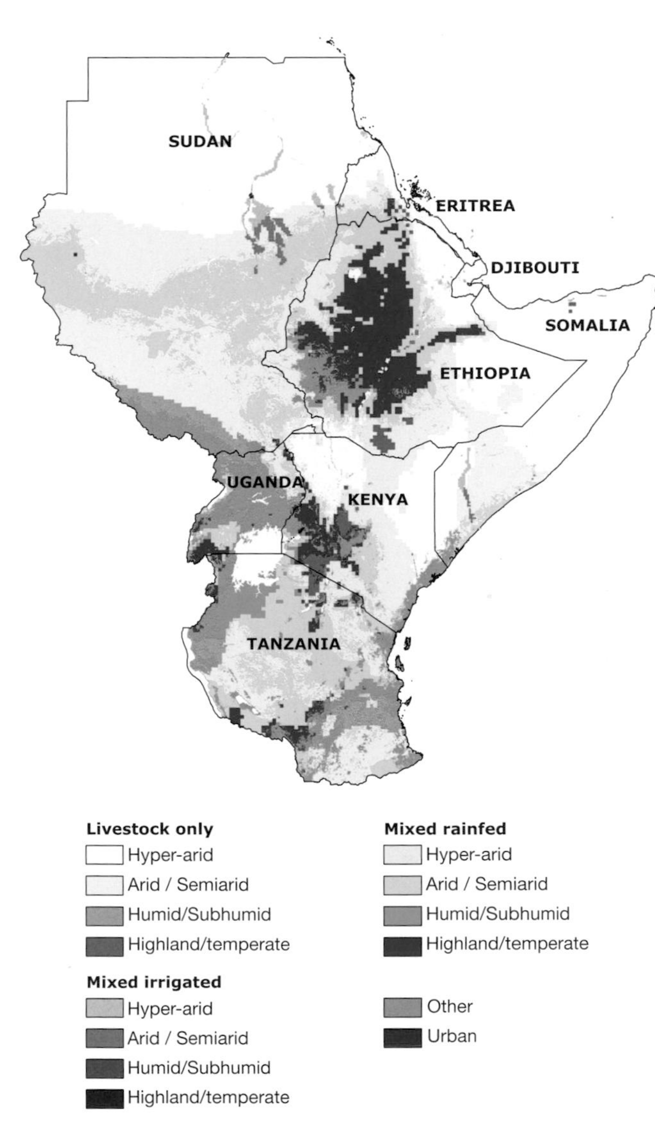

Figure 2.5. Distribution of livestock production systems in the Horn of Africa.
Source: ILRI GIS (www.ilri.org/gis).

Agroecological zone
Hyper-arid
Arid
Semiarid
Subhumid
Humid

MALI
NIGER
SENEGAL
GAMBIA
BURKINA FASO
GUINEA BISSAU
GUINEA
BENIN
COTE D'IVOIRE
GHANA
NIGERIA
SIERRA LEONE
LIBERIA
TOGO
CAMEROON

Figure 3.1. Agro-ecological zones and countries of West Africa.
Source: Adapted from FAO 2005.

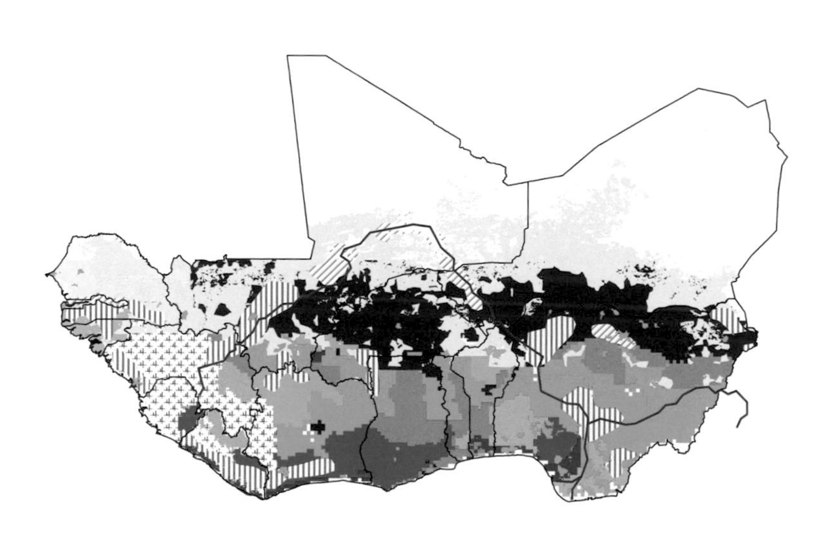

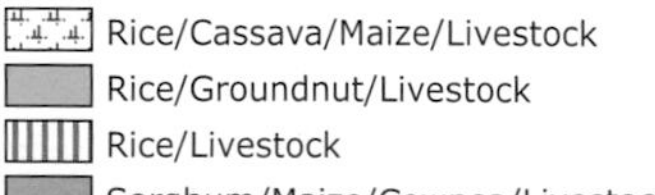

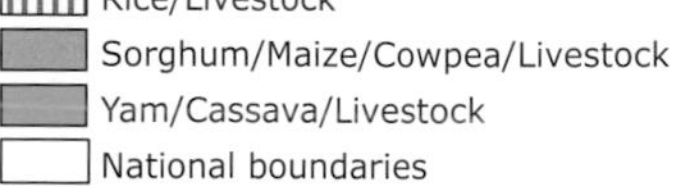

Figure 3.2. Distribution of major livestock production systems in West Africa.
Source: Fernandez-Rivera et al., 2004.

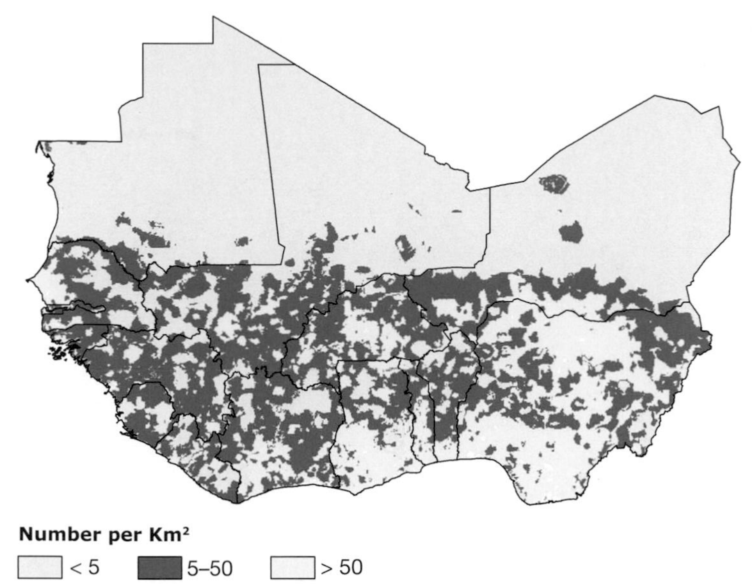

Figure 3.4. Human population density in West Africa, 2000.
Source: Thornton et al., 2002.

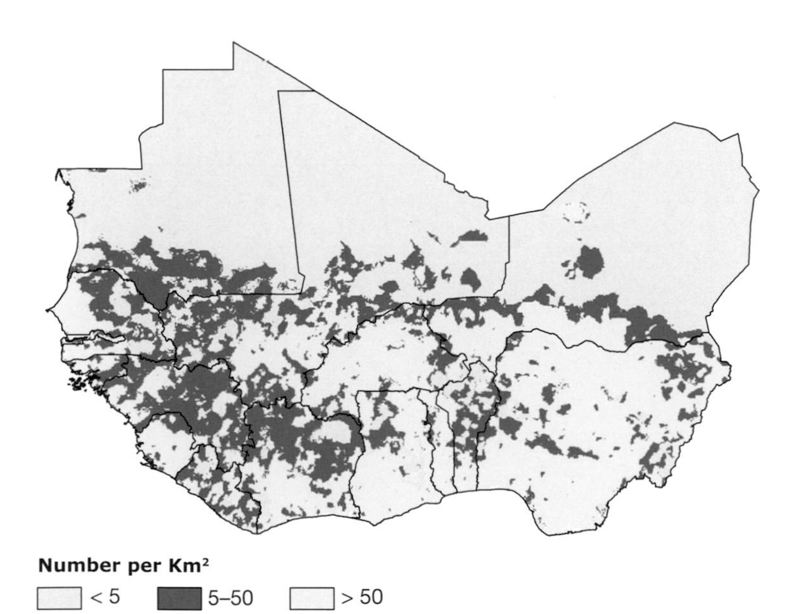

Figure 3.5. Projected human population density in West Africa, 2050.
Source: Thornton et al., 2002.

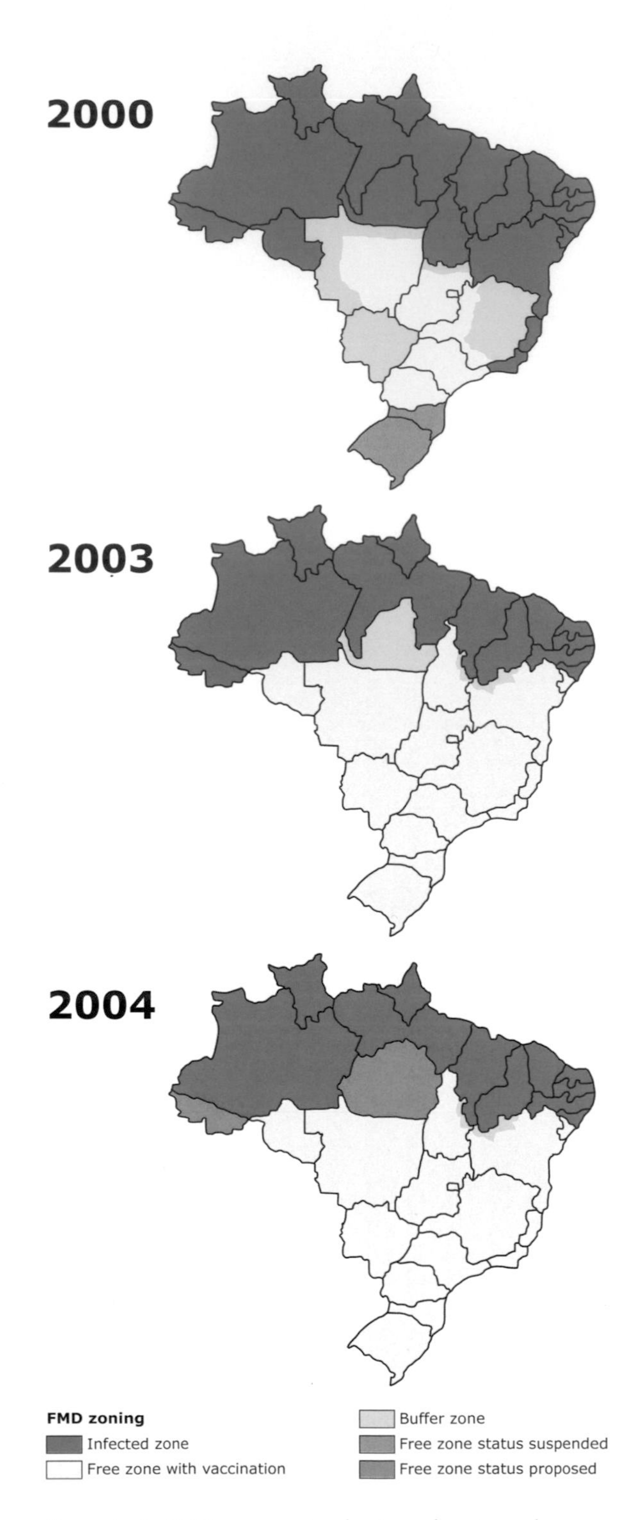

Figure 5.4. FMD zoning in the Brazilian Legal Amazon, 2000, 2003, 2004.

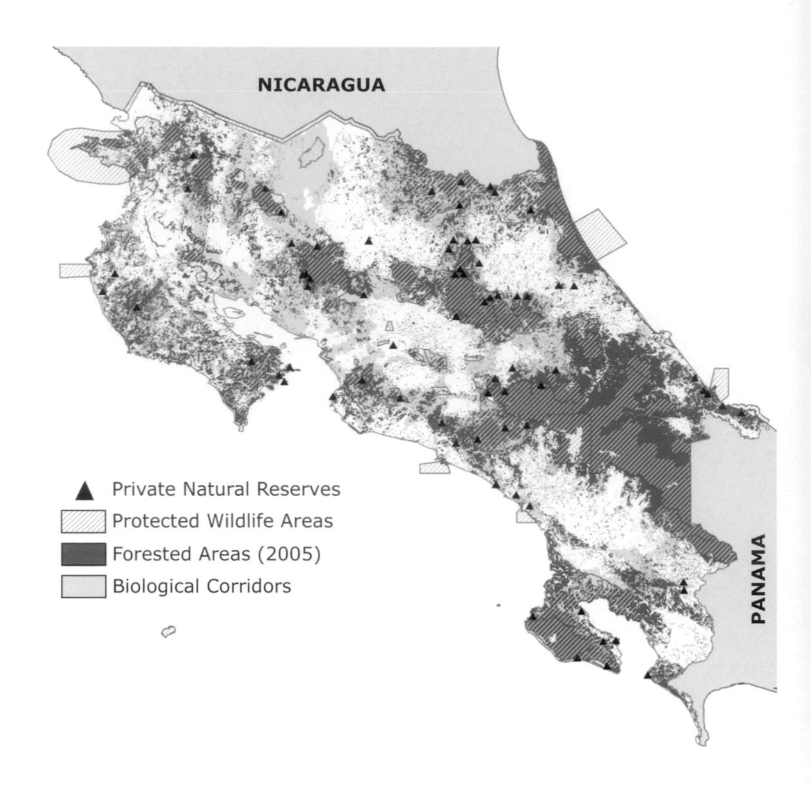

Figure 5.8. Biological corridors in Costa Rica for the conservation of forest resources and biodiversity.

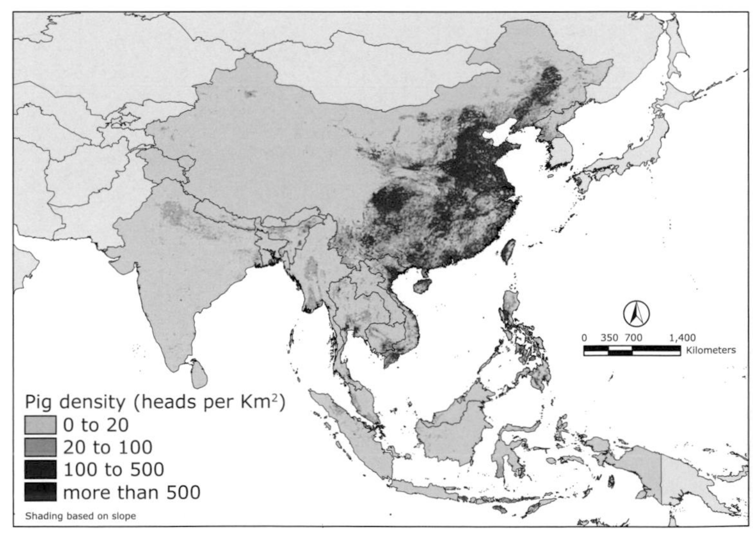

Figure 6.9. Geographical distribution of pig production in China.
Source: Gerber et al., 2005.

7

The United States

Trends in the Dairy Industry and Their Implications for Producers and the Environment

J. Mark Powell, Michael P. Russelle, and Neal P. Martin

Abstract
Animal agriculture in the United States continues to be transformed by changes in consumer demand, production economies of scale, enhanced animal genetics and nutrition, and the widespread use of historically inexpensive feeds, diet supplements, and fertilizers. The ongoing trend toward fewer and larger dairy farms has encompassed a greater use of imported feed and has led to production of quantities of manure nutrients that can exceed the recycling capacity of associated pastures and croplands.

The liberal use of relatively inexpensive fertilizers, in combination with manure and other agricultural nutrient sources, can result in numerous adverse environmental impacts, including damage to water quality through runoff and leaching, and gaseous emissions that can adversely affect human health, fertilize natural ecosystems, and contribute to global climate change. Federal and state regulations and local ordinances have been created to mitigate nutrient loss and environmental risks associated with animal production. On many farms, nutrient use can be reduced by matching livestock rations to their nutritional needs more closely, and by increasing the availability of nutrients in feed. For dairy, this would maximize feed conversion into milk and minimize nutrient concentrations in manure, without losses in productivity.

In some dairy systems, manure transport for land spreading can be made easier by reducing water use during manure collection and storage. Cost-effective methods of manure handling, treatment, and storage are available, although the level of adoption varies by farm size and the planning horizon of producers.

Current environmental policies focus on the largest livestock operations, but small- and medium-scale livestock farms often lack the resources to improve manure collection and management. Other important farm operational features and management should also be considered as targets for environmental regulations: for example, the balance between livestock numbers and pasture/cropland available for manure land spreading, optimal livestock feeding, and the abilities of farmers to collect and land-spread manure under the diverse biophysical and socioeconomic conditions they face.

Trends in the US Dairy Industry

Over the last half-century, global agriculture has been dramatically transformed by mechanization, inexpensive fuel, feed grain, fertilizers, and the use of other petroleum-based products. During this period, dairy farmers in the United States gradually specialized in milk production rather than raising multiple livestock types and selling various products. They increased their dairy herd sizes and shifted from grazing to feeding harvested forages and grain to cattle that rarely leave barns (Harper 2000).

The dairy sector is now a substantial element in the US livestock industry. During the last agricultural census in the United States (USDA 2004a), total sales of agricultural products equaled US$200.6 billion, of which approximately one half was derived from livestock, poultry, and their products. Dairy products accounted for $20.1 billion, or about 20% of all sales of livestock and poultry products (Table 7.1).

Great changes continue apace in the US dairy industry. There are shifts in the geographic regions where milk is produced, and increases in herd stocking densities (number of cows per unit area of cropland or pasture). There is a greater use of purchased feeds, manure storage, and contracting services for manure hauling and land spreading. On a national scale, efficiency has increased: more milk is produced today with fewer cows than in the past. The proportion of the nation's milk being produced on the largest farms continues to increase.

Table 7.1. Livestock and poultry inventories and sales in the United States, 2002

Livestock and Poultry Type	2002 Inventory (million head)	2002 Cash Receipts (billion US dollars)
Beef cattle and calves	33.4	45.1
Dairy cows (lactating)	9.1	20.1
Hogs and pigs	60.4	12.4
Poultry, layers ≥ 20 weeks old	334.4	24.0
Poultry, broilers	1389.3	

Source: USDA 2004a.

During 2008, the price of liquid milk was at a historic high, fueled by a great increase in global demand and escalating feed grain prices.

More Milk from Fewer Cows on Fewer Dairy Farms

Currently there are approximately 9.1 million dairy cows and 4.1 million replacement heifers in the United States. Dairy heifer replacements weighing less than 220 kg probably number from two to three million. During the past 20 years or so, the number of dairy cows in the United States has declined by about 25%, yet milk production continues to increase (Figure 7.1). Increases in the amount of milk produced on dairy farms are due to steady, consistent increases in milk production per cow. This has been attributed primarily to enhanced genetics, better nutrition and disease control, and reproductive management, along with other less important factors (CAST 1999). These trends of declining dairy cow numbers and increasing national and per cow milk production are expected to continue. In 2006 in the United States 9.1 million cows were producing an annual average of 9048 kg milk per head. Projections for 2016 are for 8.5 million cows producing 10,496 kg milk annually per head (USDA 2007).

There has been a steady trend of concentration in dairy farms. From 1969 to 1992 there was a 70% decline in the number of dairy farms in the United States (McBride 1997). This decline is ongoing and quite rapid. The number of dairy farms has fallen from about 181,270 in 1991 to 75,140 in 2006. It is projected that most future increases in milk production (Figure 7.1) will come from the largest dairy farms. Less than 10 years ago most milk was produced on dairy farms having fewer than 200 cows; today most milk is produced on farms having more than 500 cows (Figure 7.2). This trend led to the regulatory term *concentrated animal feeding operation* (CAFO), defined as animal operations having more than 1000 animal units (one AU = 454 kg), equivalent to approximately 700 adult dairy cows the size of Holsteins. CAFOs currently represent about 4.5% of the 450,000 animal feeding operations in the United States, yet they account for approximately 47% of the total manure generated on all US animal feeding operations (Aillery et al., 2006a,b).

A recent survey of dairy farmers strongly indicates that the trend toward fewer small farms and more large farms will continue (MacDonald et al., 2007). Seventy percent of the farmers milking fewer than 50 cows expected to be out of business within 10 years. At greater farm sizes, fewer expected to exit dairying: 48% among farms with 50 to 99 cows, and only 20% of farms milking at least 1000 cows. Because current returns to milk production on small dairy farms do not cover costs (Table 7.2), more small farms are leaving dairy farming than entering. The small farms that do continue to produce milk well into the future will have to be exceptionally well managed, and/or will have favorable input or product prices that provide them with above-average profits (MacDonald et al., 2007). Some farms will adopt

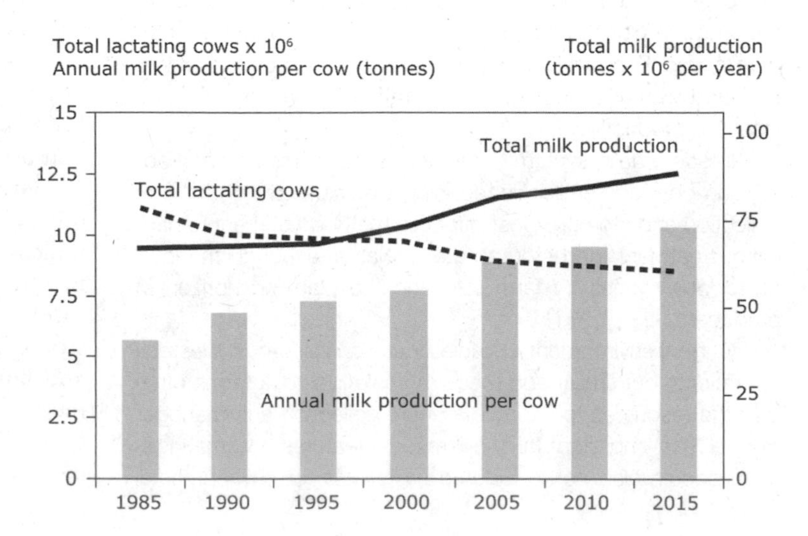

Figure 7.1. Cow population and milk production trends and forecasts in the US dairy industry. *Source:* USDA 2007.

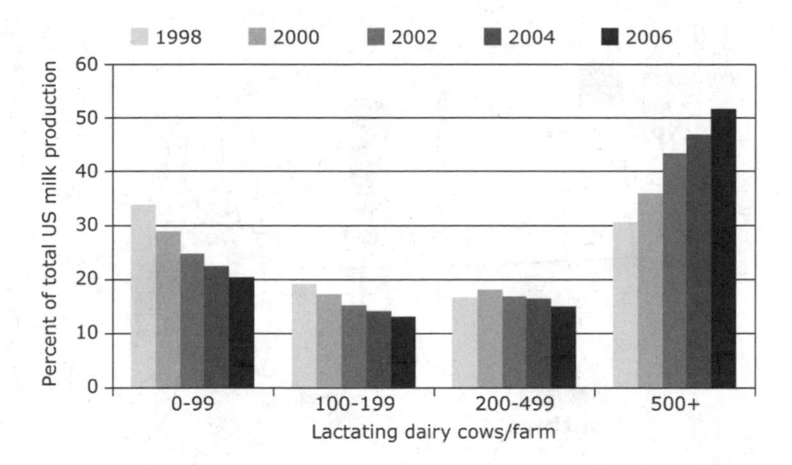

Figure 7.2. Percentages of national milk production in the United States produced on different farm sizes, 1998–2006.
Source: USDA 2007.

alternative production strategies, such as organic dairy production, to meet niche market demands.

Organic dairy farming is one of the fastest-growing animal agricultural sectors in the United States. Although organic milk makes up approximately 3% of total fluid milk sales, this share is growing rapidly (Greene 2007, Huffman 2008). Despite this rapid growth, organic milk cows currently account for only about 2% of the total dairy cow populations in California and Wisconsin, the two top dairy states for both organic and conventional milk production. The cost of production for organic dairy farms tends to be greater than for conventional dairy farms of similar size, in part due to higher organic feed costs and higher labor and capital costs per unit of milk produced. Higher costs are offset by the premium prices organic farmers receive for their products, enhancing overall financial returns. In 2005 about 37% of the organic operations with 50 to 99 cows covered all costs, except for capital recovery, compared to only 25% of conventional dairy farms of similar size (MacDonald et al., 2007). Given that organic standards require that cows have access to pasture, most expansion of organic dairy production will likely come from farms with small and medium herd sizes.

Geographic Redistribution of Dairy Farms

The US Midwest produces the highest percentage of the national milk supply, although this percentage is declining (Figure 7.3). Over the past 10 years or so, increases in milk production have occurred on dairy farms situated in the US Southwest. Whereas the Midwest has historically been the major milk production region, today the Southwest produces approximately the same percentage of milk as the Midwest. Dairy farm expansion in the western state of Idaho has been particularly strong—this state recently surpassed the traditional dairy states of Pennsylvania (northeastern region) and Minnesota (Midwest) in total milk production, and now ranks fourth nationally in milk production, behind California, Wisconsin, and New York (USDA 2007). New Mexico and Arizona have led dairy farm expansion in the Southwest, now ranking seventh and thirteenth of all states, respectively, in total milk production.

Table 7.2. Cost and profits of milk production on US dairy farms, by herd size

	Herd Size (Milk Cows)					
	1–49	50–99	100–199	200–499	500–999	1000+
Item	USD per 100 kg of fluid milk produced					
Gross value of production	39.40	38.72	37.93	38.03	36.51	36.47
Operating costs	27.12	28.53	25.38	24.94	24.41	21.48
Overhead costs	39.27	27.69	20.52	14.57	11.02	8.49
Unpaid labor	23.37	13.45	6.90	5.62	1.19	0.37
Capital recovery	11.60	10.05	8.58	5.62	4.48	3.66
Total costs	66.35	56.23	45.91	39.51	35.43	29.97
Net returns	−26.45	−17.51	−7.98	−1.48	1.08	6.50

Source: Adapted from MacDonald et al., 2007.

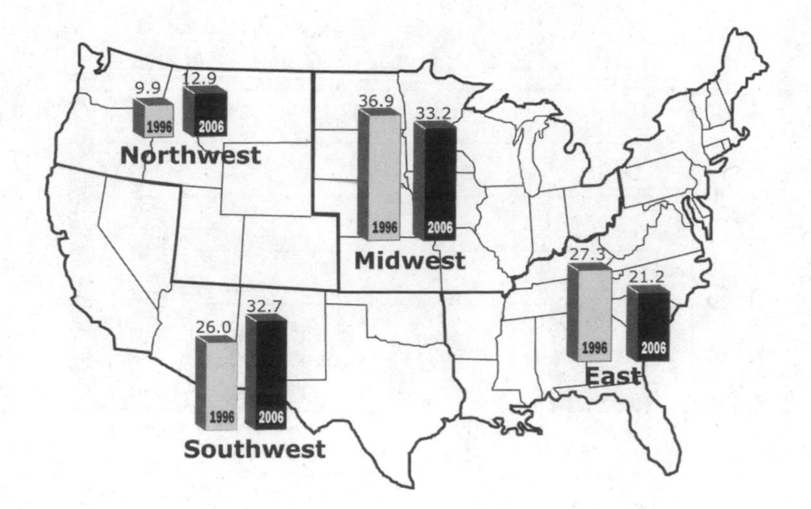

Figure 7.3. Regional percentages of national milk production in the United States, 1996 and 2006. *Source:* USDA 2007.

The geographic distribution of dairy farms in the United States (Figure 7.3) reflects the need to produce a bulky perishable product (fluid milk) near centers of population and consumption. Dairy pricing policy was initiated to support these localized markets. Major drivers of the dairy industry's westward shift have been the availability of less expensive land, favorable climate that permits large-scale operations with lower animal housing costs, local availability of high-quality feed at low cost, access to cheap hired labor, and proximity to major new markets for dairy products (USDA 2004b). Although milk production for fluid consumption remains concentrated near large population centers, production of milk for manufacturing purposes is increasingly located in low-cost areas of the West and Southwest. However, the current policy structure may lower the financial returns of some western dairy operations (USDA 2004b).

Increasing Dairy Cow Population Densities

The land area and number of dairy cows managed by US dairy farmers varies by region. In the traditional Northeast and Midwest dairy regions, farms tend to have smaller herds and a larger land base for forage and grain production compared to western US dairy farms. For example, in Wisconsin, the median ratio of lactating cows to land owned and operated by a dairy farmer was about 0.49 cows/ha in 2002, the time of the last national agricultural census (Figure 7.4). In contrast, in the Central Valley of California where dairy production has grown rapidly over the past 10 years, the median ratio was 8 cows/ha.

Changes in dairy cattle density have also varied regionally over the past decade. In Wisconsin, the number of dairy farms declined by 44%, yet the number of cows declined by only 18%. This resulted in an increase in the number of cows per unit land area on most farms, except those with very low or very high densities (lower panel of Figure 7.4). By contrast, in the Central Valley of California, there were significant increases over a wide range of cow/land densities by 2002. The overall number of farms decreased by 11%, whereas the number of cows increased by 31%. The average density of the densest 1% of reporting farms was 44.5 cows/ha in Wisconsin, compared to 955 cows/ha in the Central Valley. These averages declined by nearly one half in Wisconsin between 1992 and 2002, yet in California they increased by nearly 75% over the same period. This contrast in the land base reflects differences in the structure of dairy farming, including capital investment, relative costs and logistics of obtaining forages, local and statewide regulations, and availability of land for spreading manure. Isik (2004) found that dairy cow inventories per farm were lower in counties with more land suitable for agriculture, highlighting the fact that new, larger facilities are increasingly avoiding capital investment in land.

Changes in Forage Production

As herd sizes increase, dairy farmers seek to maximize forage yields per unit of cropland area. Over the past decade, the most significant shift in dairy diets has been the switch from alfalfa (lucerne, *Medicago sativa* L.) as a major source of fiber to corn (maize, *Zea mays* L.) silage. This change was driven by several more favorable economic characteristics of corn silage versus alfalfa (Klemme 1998):

- Higher dry mass yield, especially in the warmer environments where dairy has been expanding
- Higher energy content
- More uniform quality
- Fewer required harvests.

The statewide yield of corn silage in California has averaged 56 tonnes/ha since 1990, up from 42 tonnes/ha in the mid-1970s. In the cooler environment of Wisconsin, corn silage yield increased only 8% between the mid-1970s and 1990, but has since increased 30% to 38 tonnes/ha, in part due to genetic improvements. By

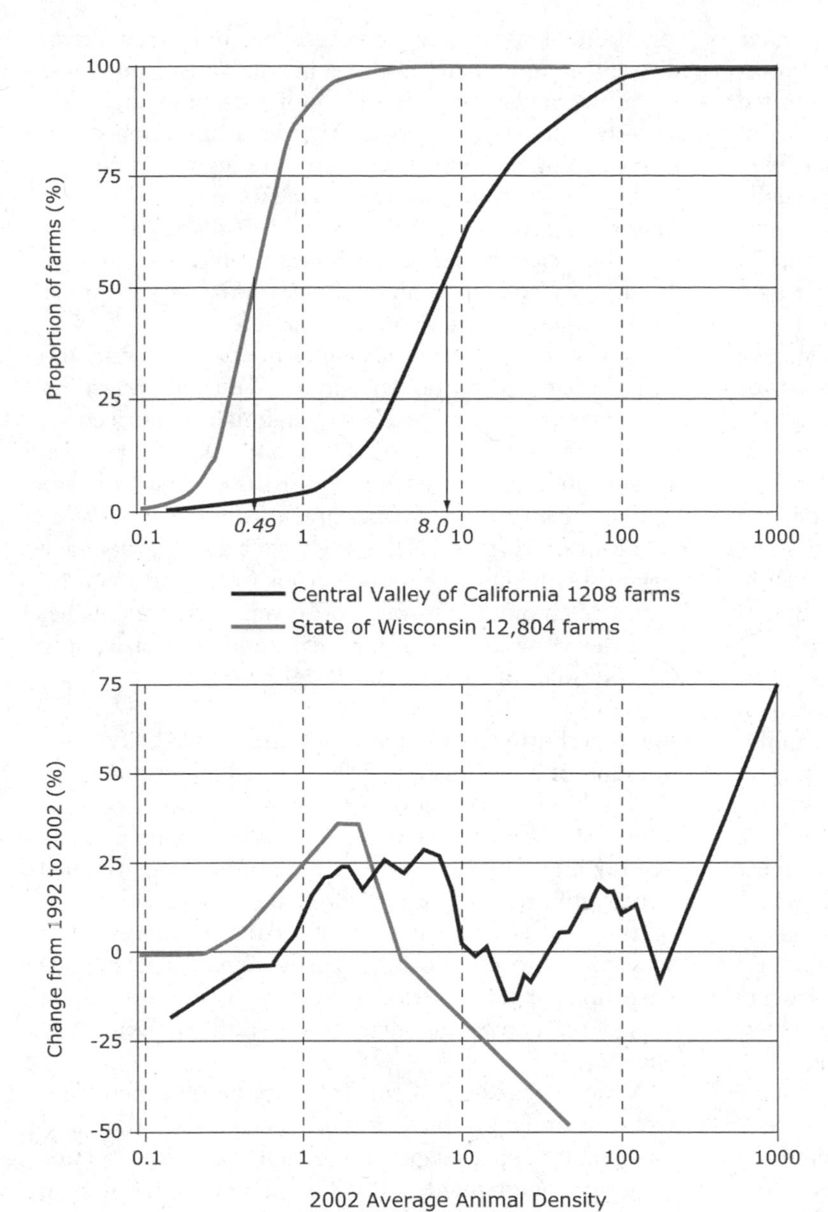

Figure 7.4. Cumulative frequency diagram of dairy cow densities in 2002, and the relative change in density from 1992 to 2002.
Source: NASS 2008.

contrast, for alfalfa the yields of new cultivars are not markedly higher than older cultivars except in stressful environments, despite improvements in disease resistance (Lamb et al., 2006). Statewide average yields of alfalfa hay have stagnated during the past 25 years, at 15.2 tonnes/ha in California and 5.4 tonnes/ha in Wisconsin. The yield advantage for corn silage is very significant.

Dairy Production Systems

The dairy industry is characterized by very diverse production systems, each with different cost structures, capital, and labor requirements. Once their production system is established, many dairy farmers find themselves less able to change and diversify as a strategy for managing risks. The principal aim of US dairy policy has been to stabilize milk prices and profits, because dairy farmers are more dependent on income from the farm business than other farm types. Three general dairy farm types can be distinguished in the United States (USDA 2004b): confinement feeding systems, pasture-based dairy farms, and dry-lot dairy operations.

Confinement feeding systems (where cows are housed in barns) are the predominant dairy system type in the United States. Small to intermediate-size confinement dairy farms grow most of their own feed, and labor is supplied by the farm family. Large confinement operations make extensive use of purchased feed and hired labor. Over the past 15 years or so, many confinement dairy farms have converted from traditional stanchion barns (each lactating cow held and fed in an individual stall) to free-stall barns (cows move freely among stalls and are fed in alleyways that separate group stalls).

In the US Midwest and Northeast, dairy farmers have been following a fairly standard confinement system for producing milk. Cows and replacement heifers are fed primarily farm-grown feed, from crop rotations comprising alfalfa, corn, and soybean (soya, *Glycine max* (L.) Merr.). Protein and mineral supplements are usually purchased to supplement dairy rations. However, during an economic crisis in the dairy industry during the 1980s, some farmers began to pasture their cattle as a means of reducing costs.

Pasture-based dairy farms in the United States have been modeled on western European, New Zealand, or Australian dairy systems, where pastures are grazed for short periods, then left for several weeks of regrowth. Although definitive statistics are not available, about 10 to 20% of the dairy farms in the US Northeast depend on pasture as a major feed source. This practice is less prevalent elsewhere except in a small area of the South, where dairy farmers traditionally have depended on winter grazing of Italian ryegrass (*Lolium multiflorum*).

These pasture systems have lower labor costs than confinement operations (because cows harvest their own feed), and smaller investments in machinery and buildings. Pasture or grazing-based dairy systems have several other advantages, including lower fuel and veterinary costs. Milk production per cow and per unit area is lower on grazing-based dairy farms, and farmers must store feed for times of inadequate pasture availability.

A noteworthy characteristic of pasture systems is that farmers are willing to mentor others, share their ideas and experiences, and open their account ledgers to see if others have ideas for improving profit. Average net farm income per cow is higher for grazing- than for confinement-based dairy herds (Fischer et al., 2005), although financial management is key in both dairy system types. The switch from confinement to grazing dairy production systems was possible because milk producers could retain their marketing arrangements, unlike the swine and poultry sectors, which are more vertically integrated (Hinrichs and Welsh 2003).

On pasture-based dairy farms, there is a continuing interest in improving efficiency, for example, by the following:

- Improved pasture production and utilization (through fertilization, inter-seeding other pasture species, altered grazing management, etc.)
- Stockpiling feed in place for autumn and winter grazing
- Determining optimum supplementation of feed rations to improve profit
- Finding low-cost feed during drought.

In the few studies that have been conducted to date, pasture-based dairy systems appear to conserve soil nutrients better than the average confinement-based feeding operation. This is likely because pasture-based farms have lower animal units per unit land area, and most of the nutrients contained in both feces and urine are deposited directly on pastures by grazing animals. Nitrate leaching losses from pastures are higher on coarse-texture soils in humid environments, especially with shallow-rooted perennial ryegrass (*Lolium perenne*) and white clover (*Trifolium repens*) mixtures, than on finer-texture soils in subhumid environments with more deeply rooted species (Rotz et al., 2005).

Drylot dairy operations, which are found in arid and semiarid regions (particularly in the West), are relatively new. These producers raise a large number of dairy cows, rely heavily on purchased feed, and make intensive, rather than extensive, use of land. As is evident from the high cow densities in the example of the Central Valley of California (Figure 7.4), many drylot dairy farms have no land on which to produce crops or spread manure. These are among the lowest-cost production systems because their low capital requirements and large size allow for economies of size (USDA 2004b).

Changing Farm Ownership and Labor Availability

An influx of dairy farmers to the United States and Canada has been occurring over the past two decades from Europe and other countries. The primary reason farmers gave for immigration was to escape milk quota systems in their countries of origin (Brolsma 2004). The availability of good opportunities for dairy producers was the second most important reason noted. The biggest constraints to immigration were legal issues, and these focused on converting temporary visas to permanent residency.

Major expansion of smaller dairy herds depends on increasing milk production while decreasing labor and management expenses per unit of milk produced. Major changes in the manager's tasks are also required during farm expansion, including a shift in focus from crops and herd to managing labor and finances; finding animals, land, and feed; meeting environmental regulations; and engaging in public relations (Hadley et al., 2002). Human resource management is another frequently listed priority and includes finding full-time workers, inexperience in communicating with employees, and developing fair criteria for evaluation.

Working conditions and relationships for immigrant laborers with management are not uniformly positive in the United States. Managers often cite language barriers with immigrant dairy farm workers as challenging. Spanish is the primary language of most immigrant dairy farm workers (Wilber et al., 2006). In a survey of 14 farms in East Central Wisconsin, immigrants were characterized as being more willing than US citizen workers to work additional hours. They received farm-supplied housing and utilities more frequently, and had pay rates, health insurance, and vacation leave similar to their US counterparts, but they were not in management positions. On most of the farms, both the manager and Spanish-speaking employees were attempting to learn the other language, but bilingual employees were often relied upon for translation. Managers frequently allowed immigrant employees to assist in recruiting, hiring, and training new employees.

The great structural change in the US dairy industry has raised concerns about the economic and social effects of different production systems. As large industrial-type dairy farms have gained in importance, concerns about their impact on the environment have grown. Increasing concentration of ownership has also raised concerns about competition in dairy markets and the viability of small farms. Dairy farm expansion in the West has been facilitated by federal grain support pricing, by less stringent environmental regulation in some states, and by milk support prices that disproportionately benefited large farms. Both pricing and regulation have become divisive issues among states.

Drivers of Change in the US Dairy Industry

A principle driver of change in the US dairy industry has been associated with regional shifts in human populations, which increased the demand for milk and therefore the number of dairy cows in the Southwest region (Figure 7.3). Economies of scale in milk production have also encouraged more large and fewer small dairy farms. Recent historic increases in the price of energy, feeds, and fertilizer, and the rising demand for grain and other biomass for ethanol production are putting new additional pressures on the US dairy industry. Although the steep price increases of 2007 and 2008 have subsided, it remains uncertain how input prices will impact change over the next few years. Environmental regulations will require greater investments in manure storage, energy conversion, and land application technologies. This will continue to add economic pressures to dairy farms with smaller herd sizes.

Economies of Scale Favor Large Dairy Farms

The ongoing trend of fewer and larger dairy farms in the United States can be attributed to their higher financial returns relative to the costs of production. In 2005 large dairy farms (> 1000 milk cows) had 15% lower production costs per unit of milk produced than farms with 500 to 599 cows, and 25 to 35% lower costs than farms with 100 to 499 cows (Table 7.2). The greatest financial advantage of the larger dairy farms was their ability to use capital and labor far more intensively than smaller dairy operations (MacDonald et al., 2007).

Changing prices have been an important driver of production and changes in farm size. Milk prices usually rise and fall in a pattern that reflects classic supply and demand economics. In the United States, fluid milk prices have recently reached historic highs (Figure 7.5). Under normal supply–demand conditions, higher milk prices have led to increased production and increased profitability for large-scale dairy farms, which have lower costs per unit of production,. This has enabled larger farms to expand herd size, add new buildings, buy new machinery, and increase market share. However, price rises from 2006 onward have been related primarily to escalating costs of major feed grains (Figure 7.5) and fuel. Hence net returns are not increasing as before, and many farmers may not increase their production as they would normally. This may help to keep milk prices higher over a longer period than usual.

Milk Pricing Policy

Dairy pricing policy in the United States was designed initially to stabilize farmers' incomes by influencing milk prices. Price support programs were designed to assure minimum prices for all farmers, regardless of herd size. These programs have had varied effects, some of which have not been neutral to farm size. For example, because of the great differences in production costs between small and large farms (Table 7.2), milk prices that may cover costs on midsize farms would, on large farms,

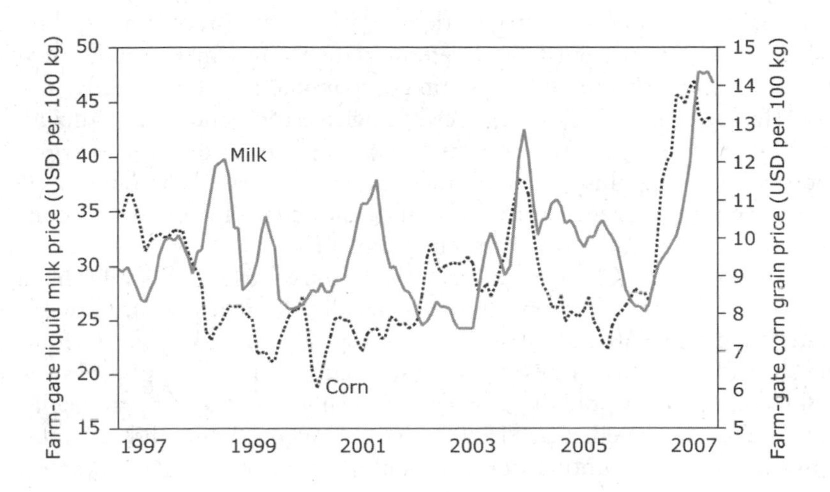

Figure 7.5. Trends in liquid milk and corn grain prices, United States, 1997–2007. *Source:* USDA 2007.

yield large profits, and even provide very strong incentives for expansion (MacDonald et al., 2007). To address this disparity, the Milk Income Loss Contract (MILC) program promulgated in 2002 put limits on farmer payments. Under this program, farmers are paid premiums only up to the first 1.1 million kilograms of production—the annual production from about 120 cows. Payments under MILC begin when milk prices fall below a reference level, and during periods of low prices they provide stronger revenue support to small operations and to regions where such farms predominate (Midwest and East; Figure 7.3). MILC provides targeted assistance to small farms during market downswings, and this helps to prevent foreclosures. Despite this, the powerful cost advantages of large dairy farms are likely to sustain the ongoing trend toward fewer small dairy farms in the United States (MacDonald et al., 2007).

Environmental Regulations

Federal, state, and local regulations have been implemented to minimize the environmental impacts of animal agriculture. Due to recent court decisions, federal regulations in the United States have been relaxed (EPA 2008) so that only those CAFOs that discharge or propose to discharge manure nutrients to navigable waters need to obtain a permit. State and local environmental regulations are often much more stringent than federal regulations, and this has impacted the regional distribution and consolidation of dairy farms in the United States. Fewer cows are kept and the greatest reduction in cow numbers occurs in those areas where local environmental regulations are most stringent (Isik 2004). Dairy farm operations may move into or expand within areas that have less stringent environmental regulations. However, their arrival is often followed by an increase in environmental problems and public concern, and promulgation of more regulation in those areas. The pressure from growing rural and exurban populations plays a role in this pattern. Human populations tend to increase faster in counties with greatest infrastructural change due to milk production, such as improved roads (McBride 1997), and this eventually leads to subsequent declines in dairy cow inventories as residential building permits increase (Isik 2004). Dairy farm size, regardless of its social ties, is a strong predictor of the level of complaints from neighbors (Jackson-Smith and Gillespie 2005) Nevertheless, new large-scale dairy farms are being established in the Midwest by operators who do address environmental concerns and engage in public relations to improve communication.

Price and Use of Feed Grain

Federal government feed grain policy has contributed to the rise of large-scale dairy farms and to the shift of dairy production to the West (USDA 2004b). Subsidies for feed grains have encouraged abundant supplies. The abundance of cheap corn transformed dairy cows from harvesters of pasture grasses and legumes grown on marginal cropland and in locations with marginal crop-growing weather, into consumers of energy in the form of grains grown on prime cereal cropland.

The recent escalation in corn grain prices (Figure 7.5) was fueled in part by recent government mandates to increase ethanol production (RFA 2006). It is uncertain how ethanol production will impact feed costs and overall dairy farm profits in the future, because it increases demand for corn grain, but also produces coproducts—wet and dry distillers' grains—that are used in ruminant rations. However, for the foreseeable future, large dairy farms will continue to have substantial capital and labor cost advantages over small dairy operations (Table 7.2), and will continue to increase their production. This will continue to place downward pressure on industrywide costs and prices, thereby offsetting some of the impact of any long-term increases in feed costs.

Price and Use of Fertilizer

As we describe in the consequences section, excessive use of nutrients in feed production and dairy farming can lead to significant environmental problems, especially air and water pollution. There are many reasons for the excess use. Nutrients, whether for land producing feedstuffs or for the animals themselves, have been relatively inexpensive in the United States. Efficiencies of scale in fertilizer manufacture and delivery have helped reduce prices relative to inflation. When expressed in constant dollars adjusted for changes in the gross national product (GNP), fertilizer costs declined by 20 to 50% between 1960 and 2002, with marked cycles of cost swings (Figure 7.6). The rapid rise in fertilizer cost in 2007 and 2008 exceeds the price spike in the mid-1970s; both were related to increases in the cost of fuel. Nevertheless, on a constant dollar basis, urea was less expensive in early 2008 than in 1960.

Because of the low prices, nutrients were generally applied in excess of crop and livestock requirements. Fertilizers and manure have been applied to the land in amounts that maximize economic returns of cropland. In addition, nutrient-rich diets have been recommended by nutritionists and veterinarians to maximize animal production and maintain good animal health and reproduction. However, overfeeding livestock and overfertilizing crops has exacerbated the potential for on-farm nutrient surpluses. These practices have increased the buildup of nutrients in the soil, and subsequent losses to and contamination of the environment.

Changes in overall fertilizer use nationally have varied among the main fertilizers (Figure 7.7). Between 1960 and 1980, use of nitrogen increased about fivefold. Slower rates of increase occurred in use of phosphorus and potassium. Use of phosphorus and potassium plateaued after 1980, whereas nitrogen use did not plateau until 1995. Although there are no national data on

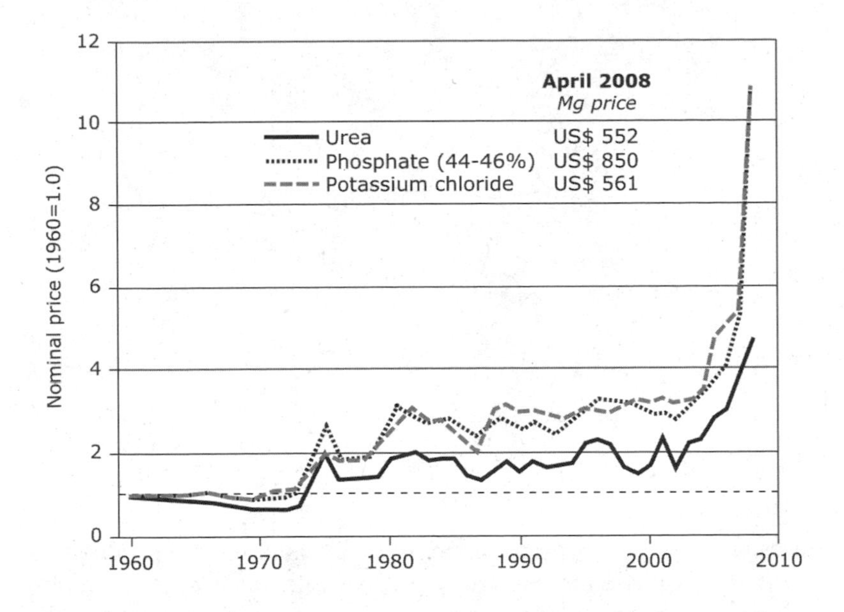

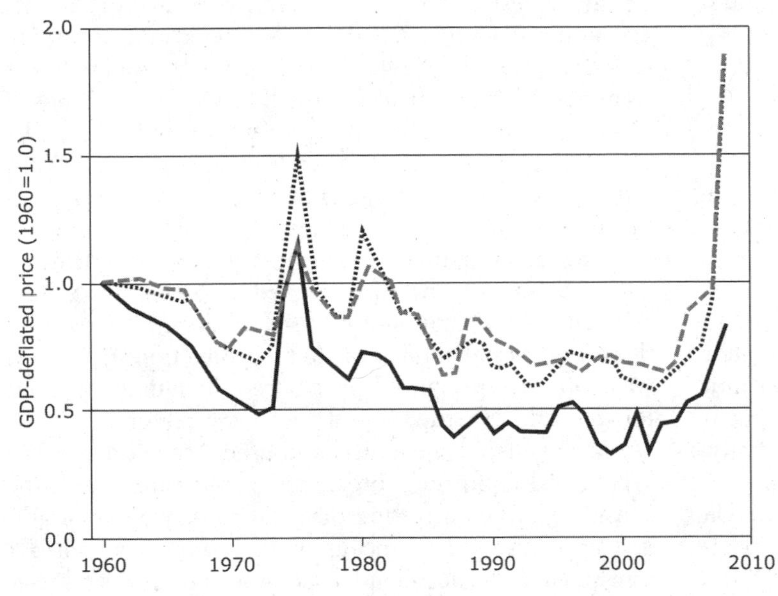

Figure 7.6. Trends in fertilizer prices in the USA, 1960-2008.
Source: NASS 2008.

fertilizer use specifically on dairy farms, if we assume that nutrient use for all farms reflects nutrient use on dairy farms, it appears that fertilizer use has been increasing while dairy manure production has been decreasing. It is unclear how fertilizer use will be affected by high prices. Anecdotally, dairy farmers have expressed increased interest in optimizing their use of manure nutrients. This is in contrast to earlier surveys that indicated poor farmer recognition of manure's nutrient value (Schmitt et al., 1999, Gassman et al., 2002).

In addition to nutrients being relatively inexpensive, farmers also apply additional nutrients to avoid the risks of nutrient undersupply, which could lead to adverse impacts on production and the environment. However, some of the high nutrient use in agriculture can be associated with inevitable biological inefficiencies with which nutrients are incorporated into crop and livestock products. For example, of total feed protein and minerals consumed by livestock, general averages of 60% for poultry, 50% for swine, 30% for lactating dairy, and 20% for beef steers, respectively, are incorporated into animal products; the remainder is excreted in manure (Kornegay 1996). Excessive dietary protein (Wu and Satter 2000, Olmos Colmenero and Broderick 2006) and phosphorus (Satter et al., 2005) is fed to dairy cows in the range of 20 to 30% above recommended levels. Field crops incorporate only a general average range of 30 to 60% of applied fertilizer and manure N and P into grain and other crop products.

Because of inevitable inefficiencies of nutrient use, most feed N and P for dairy cattle will always be excreted in manure, and after land application, manure N and P losses are inevitable. A continuous general challenge facing animal agriculture is to apply nutrients in recommended amounts in order to minimize nutrient loss and the resulting environmental contamination

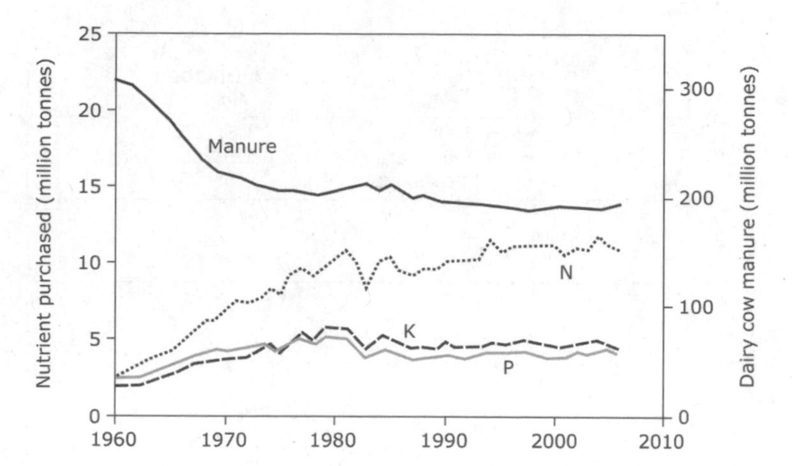

Figure 7.7. National fertilizer nutrient use (all farms) and estimated dairy cow manure production in the United States, 1960–2007. *Source:* NASS 2008.

through good management. The recent great increases in feed (Figure 7.5) and fertilizer prices (Figure 7.6) created new opportunities for dairy farmers and their feed and crop consultants to devise improved strategies to optimize overall nutrient use.

Consequences of Change

The combined trends of separate crop and livestock production and geographical concentration, and excess use of feed and fertilizer nutrients, has various consequences, including greater export of nutrients to the wider environment. In the United States, the N from animal wastes that is transferred to surface waters or is volatilized to the atmosphere as ammonia may be the single largest source of N that moves from agricultural operations into coastal waters (Howarth et al., 2002). Balancing nutrient inputs and outputs through proper animal density, feed, fertilizer, and manure management has become a major environmental challenge facing not only the US dairy industry but animal agriculture in most industrialized countries (Steinfeld et al., 2006).

Unlinking Crop and Livestock Production

The changes that have taken place in agricultural production since the mid-1900s can best be summed up in the term *industrialization* (Lanyon and Thompson 1996), encompassing specialized production techniques, geographic concentration of crops and livestock, increasingly specialized management functions, and substitution of capital for labor. Industrial agriculture has radically changed the relationship of livestock production to land resources and the environment. Before industrialization, crops and livestock were closely linked: agricultural production depended on on-site recycling of nutrients from animal manure or from biological N fixation by legumes. Since industrialization, inexpensive fertilizer and low transport costs have allowed crop and livestock production to be unlinked. Today crops can be grown in one location and fed to livestock in other locations, while human populations live in distant urban centers.

Close proximity of livestock and manure production to farms where crops are grown is fundamental to making the fullest and most effective use of manure for its agronomic benefits. The more livestock and crops are separated, the less likely the manure will be used to boost fertility, and the more likely it will be wasted or disposed of in ways that lead to environmental problems. In the United States, livestock specialization separated from crop production is most pronounced in the vertically integrated feedlot cattle, swine, and poultry industries. In dairy production, many dairy operations in the Northeast and Midwest regions of the United States continue to be associated with crop and pasture production. However, these traditional modes of dairy production are giving way to more specialized production, including irrigated forage in the Northwest and Southwest regions.

Many environmentalists and others contend that the "ecological footprint" of animal agriculture should be considered when assessing the total consequences of animal production. This means that environmental impact assessments should include not only the nutrient losses and resulting pollution that is generated on-farm, but also runoff, emissions, and pollutants generated during the production of the feed that farmers import onto their farms. For the purpose of this chapter, we consider only implications of on-farm nutrient use and how this may impact the environmental performance of dairy operations.

Environmental Problems of Excess Use of Feed and Fertilizer Nutrients

Only 20 to 30% of the N (crude protein) fed to dairy cows is converted into milk. The remaining feed nitrogen (N) is excreted about equally in urine and feces at moderate N intake, but at higher intakes more of the excess protein is excreted in urine than in feces (Figure 7.8). Urinary N is much more susceptible to environmental loss than fecal N through its rapid conversion to ammonia gas or to nitrate, which can be leached and denitrify in soils.

Excreted N follows several different pathways into the atmosphere and aquatic environment. About three

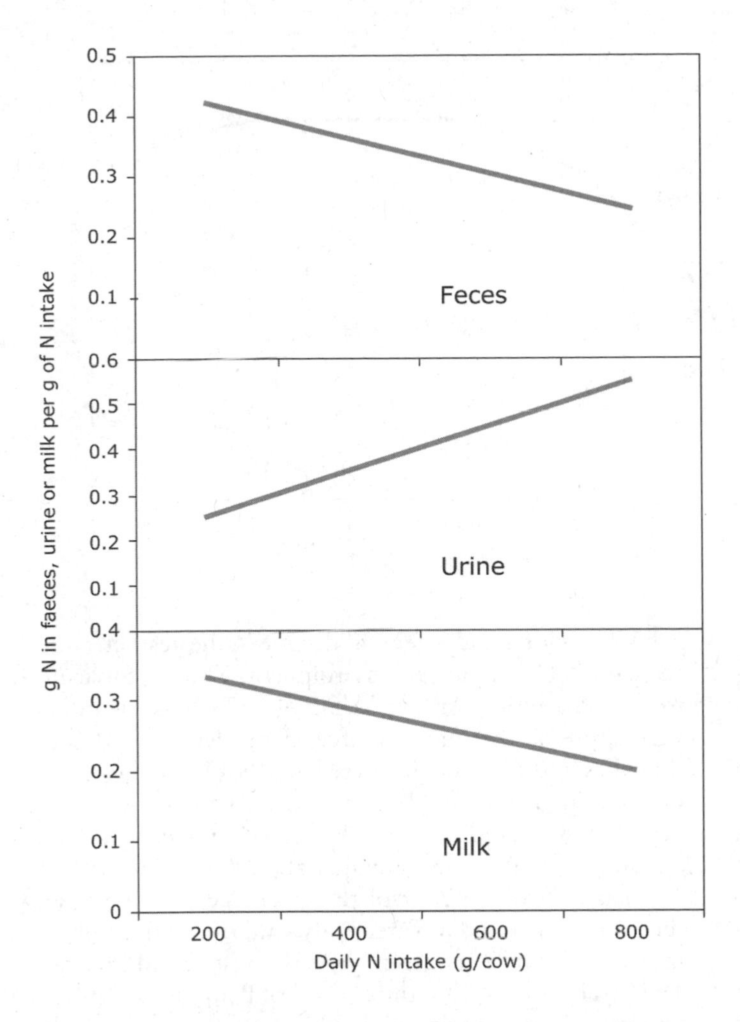

Figure 7.8. Generalized effects of increasing feed N intake by dairy cows on N levels in feces, urine, and milk.
Source: Adapted from Castillo et al., 2000.

fourths of the N contained in urine is in the form of urea. Urease enzymes, present in feces and soil, rapidly convert urea to ammonium. Ammonium, in turn, can be transformed quickly into ammonia gas and lost to the atmosphere.

Of the total amount of manure N excreted by dairy cows, approximately 30 to 40% is often lost as ammonia gas from barns, manure storage, and after land application. After release, ammonia gas combines with other chemicals in the atmosphere to form ammonium-containing dust particles that adversely affect human health. Ammonia is also redeposited as acid rain and nitrates that can be detrimental to natural ecosystems, especially aquatic ones. The ammonia produced by dairy farms in the Midwest may be a major contributor to the N loading of the Mississippi River and the hypoxia ("dead") zone in the Gulf of Mexico (Burkart and James 1999).

When N (as fertilizer, manure, legume N, and other organic sources) is applied to cropland and pastures in excess of agronomic requirements, nitrate leaching can increase (Figure 7.9). High nitrate leaching can contaminate ground and surface water and increase losses of N in gaseous form via denitrification. Although denitrification may constitute only a small fraction (2 to 5%) of applied N, production of nitrous oxide contributes to global warming and ozone depletion. In addition, application of manure slurries with low dry matter to artificially drained soils can rapidly contaminate surface and ground waters with pathogens, excess nutrients, and organic compounds that increase biological oxygen demand.

A similar picture applies for the mineral phosphorus (P) in dairy cow feed and P in fertilizer, which have also been used in excess due to their relatively low cost. Many dairy farms have accumulated P in soils over time, because imports of P in the form of feed and fertilizer exceed exports in the form of milk, cattle, and surplus grain or hay (Table 7.3).

The excessive dietary P supplements fed to dairy cows increase total and soluble P in manure (Figure 7.10). When the manure is applied to land, this greatly increases the potential for runoff of P into streams and lakes, where it promotes algae growth and eutrophication of surface waters (Ebeling et al., 2002). Excessive dietary P also decreases the N:P ratio of manure relative to N:P requirements of most crops (Powell et al., 2001). When such manure is applied to cropland in amounts sufficient to meet crop N demand, crops will be unable to take up all the P. Thus soil P increases much more quickly than when manure is derived from cows fed appropriate amounts of P.

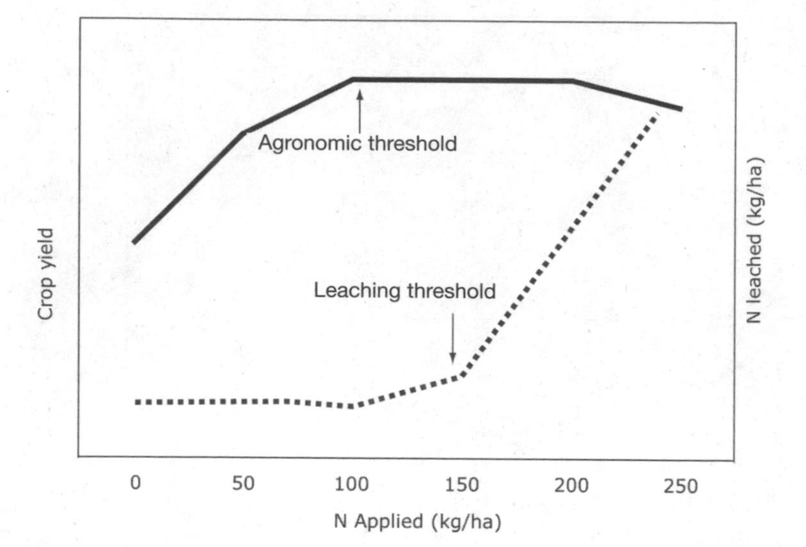

Figure 7.9. Relationships between N applications, relative crop yield, and N leaching.

Note: Agronomic threshold for crop yields is lower than leaching threshold. This means that N application in excess of agronomic threshold increases the risk of loss via nitrate leaching.

Source: Lord and Mitchell 1998.

In many areas of intensive livestock production the amount of P in manure often exceeds local crop requirements (Kingery et al., 1994, Sharpley et al., 1993), because manure has been applied at rates determined by disposal needs rather than agronomic requirements. In most dairy regions of the United States, soil test P levels are in the high plant availability range (Figure 7.11). When P levels rise above agronomic recommended levels the risk of P in runoff increases greatly (Figure 7.12).

The result of these excessive N and P inputs for lakes and streams has been to accelerate eutrophication and impair water quality. Excessive P runoff into surface waters increases growth of weeds and algae. When these decompose, dissolved oxygen levels are depleted, leading to fish kills, odors, and a general decline in the aesthetic and recreational value of the environment. The US Environmental Protection Agency (USEPA 1996) has identified agriculture as the major source of nutrients in 50% of the lakes and 60% of the river length of impaired water quality. Environmental pollution deriving from livestock production, including dairy, is highly significant among agricultural sources (Steinfeld et al., 2006).

It is difficult to control the exchange of N between the atmosphere and a water body, and the fixation of atmospheric N by blue-green algae (Krogstad and Lovstad 1991). This means that the control of P inputs is of prime importance in reducing eutrophication (Sharpley et al., 1994). Management aimed at reducing P losses to the

Table 7.3. Annual mass phosphorus balance for dairy farms, New York, USA

	Size of dairy, cows/farm			
Item	45	85	320	500
	kg P/year			
Inputs				
Purchased feed	907	1,542	7,619	12,880
Purchased fertilizer	1,088	816	1,814	9,070
Purchased animals	0	0	27	0
Outputs				
Milk	363	617	3,477	4,988
Meat	45	91	453	453
Crops sold	18	54	0	0
Remainder				
Tons	1,569	1,596	5,530	16,509
% of Inputs	79	68	59	75

Source: Klausner 1995.

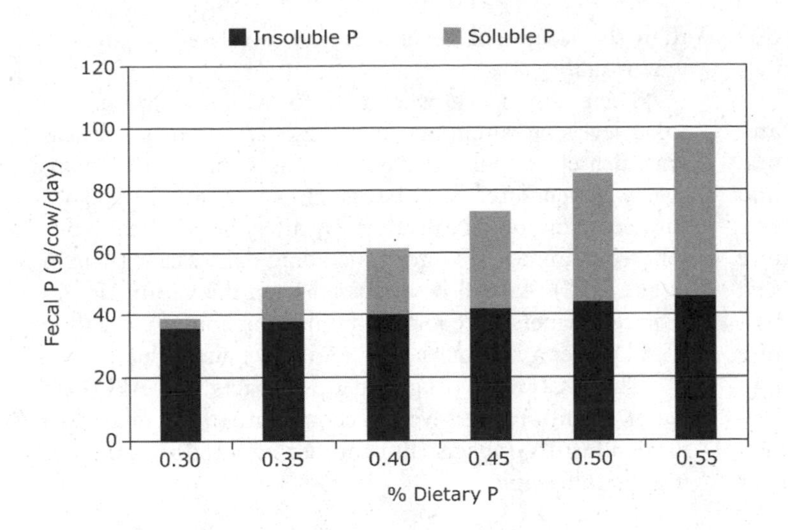

Figure 7.10. Effect of increasing feed P intake on P levels in feces of lactating dairy cows. *Source:* Satter et al., 2005.

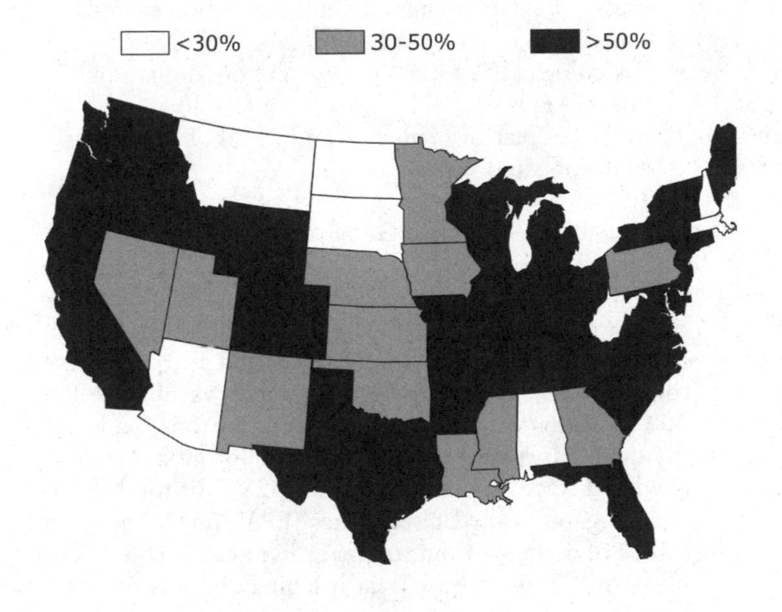

Figure 7.11. Percent of soils testing high or greater for P. In most states of the continental United States, soil test P levels are in the agronomic high or greater availability range. *Source:* Fixen 1998.

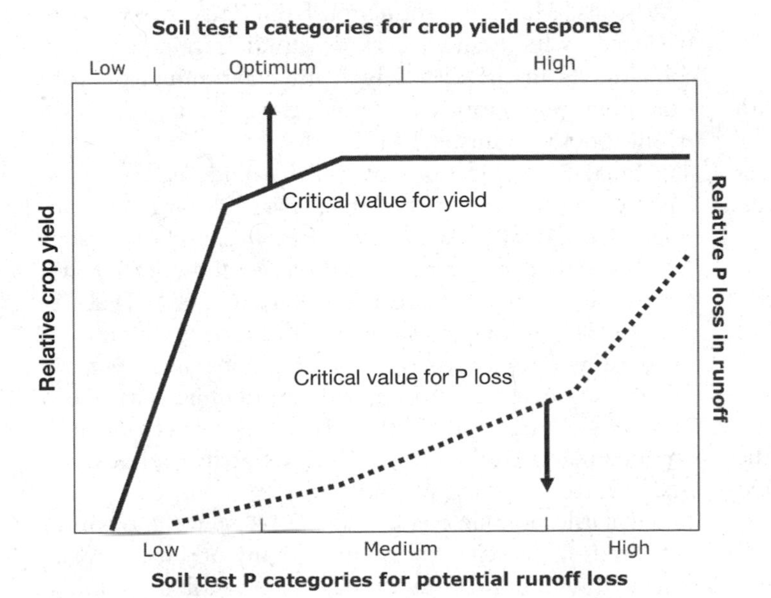

Figure 7.12. Relationships between soil test P levels, relative crop yield, and P loss in runoff.

Note: Critical soil test P level for optimum crop yield is lower than critical soil test P level for P losses in runoff. Because most US agricultural soils have high soil test P levels (Figure 7.11), additional P application will not increase crop yields but increases the risk of P runoff.

Source: Kleinman et al., 2000.

environment must therefore try to minimize P imports onto the farm (Table 7.3), while also controlling surface runoff and erosion (Sharpley and Withers 1994).

Other adverse impacts of dairy farming on water quality have been attributed to lack of appropriate manure management, generally in relation to the rate and timing of nutrients applied. Barn flush water systems used in the western United States produce dilute manure that is often used for irrigation after solids have been removed. Conventional management in the Central Valley of California has been to apply a blend of manure pond water with irrigation water during the spring and the fall. Commercial fertilizer is applied to corn in summer and sometimes to small grains in winter, and excess pond water is disposed of on fields in the winter (Harter et al., 2001). This overapplication of both N and water has resulted in nitrate contamination of shallow groundwater. Such losses can be reduced by a simple accounting scheme for total N applications (Campbell-Mathews 2007). In this scheme, farmers install water meters to quantify pond water application rates. They are taught to calculate the total N application rate in pond water, and learn to manage both pond water and fertilizer rates to more closely match crop N needs.

Impacts on Human Health

Concentrated animal feeding operations (CAFOs) can potentially have significant impacts on human health. The international scientific conference Environmental Health Impacts of Concentrated Animal Feeding Operations: Anticipating Hazards—Searching for Solutions (Thorne 2007) identified several major concerns associated with all types of CAFOs, including the following:

- Air and water contamination
- The rise of antibiotic-resistant bacteria in livestock
- The potential of influenza outbreaks arising from siting industrialized poultry and swine production in proximity to each other and to humans.

However, there is very little information on public health hazards associated specifically with dairy farms. The National Commission on Industrial Farm Animal Production (NCIFAP), including representatives from veterinary medicine, agriculture, public health, business, government, rural advocacy, and animal welfare, was established in 2005 (NCIFAP 2007). The principal mandate of NCIFAP has been to conduct an assessment of the industry's impact on public health, the environment, farm communities, and animal health and well-being. Scientific workshops and public meetings have been held to help inform NCIFAP commissioners and the public about the major environmental health issues associated with large, industrial-style livestock production facilities. However, the NCIFAP workgroups have listed few direct public health hazards associated with dairy farms, other than the long-held general concerns related to air and water quality.

There are concerns related to worker exposure to toxic levels of ammonia, but these are more associated with densely populated poultry houses than with expansive, well-ventilated dairy barns. However, fine particulates formed on ammonia can adversely affect human health distant from ammonia sources, including dairy farms (Asman et al., 1998). As will be discussed later, the Clean Air Act requires farmers to report ammonia emissions greater than 45.5 kg over a 24-hour period (Aillery et al., 2006a).

The NCIFAP workgroup on impacts of CAFOs on water quality made several recommendations related to human health impacts (Burkholder et al., 2007), including the following:

- Monitoring whole watersheds to understand the effects of extreme emission and deposition events on human and ecosystems health
- Toxicological assessment of water contamination from CAFOs
- Studies of primary effluents and metabolites in soils, sediments, and water.

Impacts on Greenhouse Gas Emissions

The trend toward fewer and larger livestock farms in the United States has increased public concern that livestock operations emit pollutants that adversely affect human health and soil, air, and water quality and also contribute to greenhouse gas emissions and global warming (NRC 2003). Methane, carbon dioxide, and nitrous oxide are the three gases held most responsible for global climate change. Livestock contribute about 28% of total methane emission in the United States (EPA 2005). The main source of methane from ruminant livestock is enteric fermentation (most released via belching and less by flatulence), which contributes about 75% of total livestock emissions. Methane production from dairy cows can be reduced by increasing starch or rapidly fermentable carbohydrates in the diet, which impact ruminal pH and microbial populations and regulate methane production (Johnson and Johnson 1995).

Small concentrations of nitrous oxide in the atmosphere are thought to contribute over 6% of the warming effect of all greenhouse gases because of nitrous oxide's extended atmospheric lifetime (about 150 years) and high thermal absorptivity (Godish 2004, Dalal et al., 2003). Although ammonia is the main pollutant gas emitted from dairy barns, small emissions of nitrous oxide, which originate primarily from manure, have been detected (Zhang et al., 2005). Nitrate derived from land-applied manure and fertilizers can denitrify and be emitted as nitrous oxide from soil.

Volatile organic compounds (VOCs) are a group of hundreds of reactive compounds, many of which are associated with odor. Some VOCs lead to global warming, others to ozone depletion, smog, and decreased visibility. Effects of VOCs are mostly associated with odor-producing compounds and their effects on human health (Schiffman et al., 1995). The majority of VOC emissions from dairy farms come directly from the cow, with smaller amounts emitted from fresh manure two to three hours after excretion (Mitloehner 2006). Additionally, ensiled feedstuffs (silages) are a significant source of VOCs (Rovner 2006). Different types of silage have different VOC emission potentials, but those low in sugar will have the least VOC emissions because of reduced fermentation rates.

Responses and Remedies

The regional shifts and intensification of dairy farming in the United States have elicited a wide range of responses from federal, state, and local governments, dairy supply and service sectors, and producers. Federal milk price support programs are being reevaluated for their impact on the profitability and viability of small- and medium-sized dairy farms. Environmental regulations are being modified to account for new public demands for cleaner air and water, as well as federal government response to pollution litigation. Feed and fertilizer dealers and veterinarians continue to revise their nutrient use recommendations in efforts to enhance the environmental performance of dairy operations. Improved manure handling, storage, and land-application strategies are being developed to maximize manure nutrient recycling through crops and pasture. Socioeconomic research is being incorporated into technology development so that recommended practices align more closely with producer resources including management skills.

Milk Pricing Policy

The notable trend toward larger dairy farms (Figure 7.2) has led to recent evaluations and proposed revisions of milk pricing policy. The Farm Security and Rural Investment Act of 2002 called for an evaluation of the economic impacts of all dairy policy programs. The resulting report *Economic Effects of U.S. Dairy Policy and Alternative Approaches to Milk Pricing* (USDA 2004c) provides a comprehensive description of 70 years of dairy support programs, and analyses of their impacts on farms, rural economies, and markets for dairy products. The report concluded that dairy support programs have had only modest benefits to producers. They raised milk prices by only about 1%, and total producer revenues (returns plus government payments) by 3% over a 5-year period. Short-term effects can be significantly higher, however, and impacts are more pronounced during years of low milk prices.

Because of this modest influence on milk prices and returns, dairy support programs have had limited impacts on the profitability and viability of US dairy farms. For example, the dairy support programs could be associated with only a 5% increase in the ability of midcost and high-cost dairy farms (usually the smaller-scale producers) to meet expenses (USDA 2004c). The support programs increase returns and allow some high-cost dairy operations to stay in business over the short or medium term. But in the longer term, higher milk prices improve the profitability of low-cost, large-scale dairy farms, which historically has enabled them to expand production and increase market share.

The USDA report (2004c) provided several other conclusions about federal policy and dairy support. Overall, dairy support programs have raised consumer costs and increased government expenditures. There are also program conflicts. For example, price support programs established a safety net for milk prices, which would allow milk prices to fall to certain levels to induce a correction in oversupply or underconsumption. When the milk price falls toward the price support safety net, however, the MILC program, which provides production-linked payments, may encourage production and retard the supply adjustment. The result is that milk prices may stay lower for longer periods and raise government costs to maintain the programs. Non-MILC dairy programs alone raise the all-milk price by 4%, but when MILC is included, all-milk price is raised by only about 1% (USDA 2004c).

Water and Air Quality Legislation

Current political concerns focus on pollution of lakes, streams, and groundwater, and on air emissions, especially from farms close to environmentally sensitive areas (e.g., forests and other natural habitats or shallow groundwater) and urban centers. The first federal law in the United States to stem pollution of surface and ground waters was passed in 1948 and focused almost exclusively on point sources of sewage. The trend toward fewer and larger livestock farms led to heightened public concern about pollution from animal agriculture. In the 1970s the US Environmental Protection Agency (EPA) created two rules under the Clean Water Act that affected animal agriculture: (1) The National Pollutant Discharge Elimination System (NPDES); and (2) Effluent Limitations Guidelines (ELGs). In 2003, the EPA implemented pollution standards for all CAFOs. The rules were recently relaxed (EPA 2008) to include only those CAFOs that currently discharge or plan to discharge pollutants to US navigable waters.

Under the NPDES, CAFOs are required to follow individualized Comprehensive Nutrient Management Plans (CNMPs) designed to protect surface and ground water (Moody and Burns 2006). Adherence to CAFO-based manure management regulations that meet both water and air quality standards would be most costly to the hog and beef cattle industries because these animal production facilities usually lack land for manure spreading (Aillery et al., 2006b). Large dairy facilities typical of the western United States face similar costs for compliance because they have insufficient land (Figure 7.4). By contrast, most dairy farms in the traditional Northeast and Midwest regions of the United States are land based. Most forage and grain is grown on-farm, and farmers have adequate land for manure spreading (Powell et al., 2002, Saam et al., 2005).

For animal agriculture in the United States, environmental regulations have focused mainly on the amount and timing of manure application to cropland. The current regulatory focus is on large livestock operations, based on the assumption that they produce the most manure and therefore pose the greatest environmental risk. However, it is becoming increasingly evident that farms of any size can generate negative environmental impacts. Indeed it has been suggested that economies of scale, more modern technologies, and potentially higher management skills associated with large-scale operations may make these operations less likely to pollute compared to smaller, older facilities (Norris and Batie 2000). For example, stanchion or tie-stall barns are the most common housing types on dairy farms that have small to medium herd sizes, mostly in the US Midwest and Northeast (USDA 2004c). Cows are confined to stalls, and manure is collected in a gutter behind the cows. Cows also have access to small exercise lots, or may be allowed access to a pasture to graze for part of the day. These farms face particular challenges in managing manure in outside confinement areas. On Wisconsin dairy farms, relatively less manure is collected on farms that manage tie-stall barns than from those that manage free-stall barns, and manure collection per animal is relatively lower on farms having small to medium herd sizes than on farms having large herds kept in free-stall housing (Table 7.4). The current regulatory focus on large farms, therefore, may not address all significant sources of pollution from dairy operations.

Table 7.4. Housing type and herd class differences in manure collection on Wisconsin dairy farms

Category	Subcategory	Manure Collection (% of total manure mass)
Housing type	Freestall (13)[1]	89 (16.5)[2] a[3]
	Stanchion (34)	66 (18.9) b
Herd class	< 50 cows (20)	57 (12.6) c
	50–99 (24)	76 (18.2) b
	100–199 (6)	95 (5.1) a
	200+ (4)	100 (0) a

[1] Number of farms sampled in parentheses.

[2] Mean, standard deviation in parentheses.

[3] Within a category, subcategory means followed by different letters are significantly different ($P < 0.05$).

Source: Adapted from Powell et al., 2005.

States have widely differing regulations regarding water quality protection, and these regulations often vary even among local units of government. In response to widespread nitrate and salt contamination of groundwater and assessments of sources on dairy farms (Chang et al., 2005, Harter and Menke 2004), the Regional Water Quality Control Board (RWQCB) of California's Central Valley has ordered new waste discharge requirements for dairy farms. For those dairy farms covered by this discharge order (about 1600 facilities when the order was published in 2007), all domestic and agricultural supply wells and subsurface soil drainage systems in the production and/or manure land application areas must be sampled annually to verify that ground and surface water quality goals are being met. In addition, these farmers must develop whole-farm nutrient balances, follow a waste management plan targeted at various areas of the farm (fields, manure storage ponds, loafing areas, etc.), and file detailed annual reports (California RWQCB, Central Valley Region 2007). To help farmers comply with the discharge order, sampling procedures for water, manure, soil and plant tissue have been developed (California RWQCB, Central Valley Region 2008). This level of monitoring apparently is the first of its kind to be required of livestock farmers in the United States.

Current environmental regulations in the United States related to animal agriculture are generally based on the number of livestock per farm, but various other environmental performance indicators have been proposed. Whole-farm nutrient balances, or the difference between nutrients imported and exported from farms, provide general indicators of a farm's risk for nutrient buildup, loss, and environmental contamination (Beegle and Lanyon 1994, Koelsch 2005). Online guides are available to help producers and their advisers make these calculations (Harrison and White 2008). Animal-to-cropland ratios relate livestock numbers and the manure they produce to the cropland area available for manure application (Westphal et al., 1989, Saam et al., 2005). There is a direct relationship between a farm's ability to grow feed for its livestock and its ability to recycle manure nutrients through cropland. For example, dairy farms in Wisconsin having approximately 0.91 ha per lactating cow (1.1 cows/ha, 92% of the farms surveyed in Figure 7.4) are self-sufficient in forage and grain production and have adequate cropland area for recycling manure N and P (Powell et al., 2002, Saam et al., 2005). Only 5% of the surveyed farms in the Central Valley of California meet this criterion.

Air quality legislation targeted at animal agriculture is now being promulgated by the EPA. The Comprehensive

Environmental Response, Compensation and Liability Act (CERCLA) enacted in 1980 aims to control the release of hazardous substances that might endanger public health. The Clean Air Act amendments of 1990 required the EPA to establish National Ambient Air Quality Standards for pollutants considered harmful to human health. Of principal concern are fine particles in the atmosphere, referred to as PM 2.5, or particles less than 2.5 micrometers in size. Ammonia is a major precursor for fine particulates (NRC 2003). CERCLA requires the reporting of the release of a hazardous substance in excess of threshold levels (e.g., 45.5 kg of ammonia over a 24-h period). Although CERCLA is focused mainly on emissions of hazardous wastes from industrial plants, the increased size and geographic consolidation of animal feeding operations make their ammonia emissions subject to the notification provisions (Aillery et al., 2006a).

A major challenge facing environmental policy related to animal agriculture is to devise practices that simultaneously protect both air and water quality. Legislation and on-farm practices aimed at controlling air emissions may actually exacerbate the potential for water pollution (Table 7.5). For example, manure injection into soil may reduce ammonia loss (and improve air quality), but it may also increase nitrate leaching (thus reducing groundwater quality), denitrification, and the production of nitrous oxide. Thus technologies to enhance manure recycling must address potential impacts at all stages of the production chain. Tillage practices recommended for decreasing N losses will also have to consider possible impacts on manure P losses in runoff.

Enhanced Feed Management

More precise feeding of protein and mineral supplements can reduce feed costs and imports, concentrations of N and P in manure, and therefore risks of environmental pollution (Table 7.6). Feed use efficiencies (the relative amount of feed nutrients converted into product) provide an indirect basis for evaluating feed management impacts on nutrient concentrations in manure. On dairy farms, management methods can have a large impact on the amount of feed N and P that is transformed into milk and excreted in manure. In Wisconsin, milk production and feed N use efficiency are highest on farms that use total mixed rations, balance rations four times per year, and milk three times per day (Table7.7). Feed P use efficiency is higher on farms that balance rations at least four times per year. These practices transform relatively more feed nutrients into product (milk), and less into manure.

Although some dairy farmers formulate their own dairy cow rations, most rations are formulated by dairy nutrition consultants, many of whom sell feed and may have an economic conflict of interest pushing them to promote overfeeding. Many dairy cows continue to be fed protein and P in excess of the requirements for the milk levels they produce, despite the fact that the relationship between feed levels, manure, and environmental risks is becoming more apparent. Reductions in manure N and P excretion can be obtained simply by feeding closer to the recommendations of the National Research Council's Subcommittee on Dairy Cattle Nutrition (NRC 2003). In Wisconsin, about 40% of 98 surveyed dairy farms had a positive P balance (Powell et al., 2002).

Table 7.5. Comparisons of major N loss pathways for manure application under various management regimes and environmental conditions

Manure Management		Soil Drainage	Nitrogen Loss Pathway		
Rate	Placement		Ammonia	Denitrification	Leaching
Placement Comparisons				*Relative loss*	
Medium	Surface	Well-drained	High	Low	Medium
Medium	Incorporated	Well-drained	Low	Medium	Medium
Medium	Injected	Well-drained	Low	Medium	Medium
Soil Drainage Comparisons					
Medium	Incorporated	Excessively drained	Low	Low	High
Medium	Incorporated	Poorly drained	Low	High	Medium
Application Rate Comparisons					
Low	Incorporated	Poorly drained	Low	Low	Low
Medium	Incorporated	Poorly drained	Low	Medium	Medium
High	Incorporated	Poorly drained	Low	Medium	High

Source: Adapted from Meisinger and Thompson 1996.

Table 7.6. Dietary strategies that reduce the mass and nitrogen (N) and phosphorus (P) content of dairy manure

Feed Management Strategy	Principal Effect of Manure and the Environment
Feed protein in relation to milk production	Reduces total N, urine N, and ammonia production
Refine mineral supplementation	Reduces total P, water-soluble P, and runoff P
Increase feed intake and improve forage quality	Reduces mass and N content per unit milk output

The simple practice of adopting the National Research Council's dietary P recommendations (that is, 0.38% P in the diet for high-producing dairy cows) would reduce the number of farms and amount of land in positive P balance by approximately two thirds.

Dietary P levels on dairy farms in the United States appear to be declining. Regional and national surveys indicate that the average dairy diet P content recommended by consultants and the feed industry in 1999 was 0.48% of ration dry matter. Yet by 2003, feed analysis of over 300 dairy total mixed rations submitted for testing to a commercial laboratory showed a P content of about 0.44 to 0.45% (Satter, unpublished information). Surveys in Wisconsin (CVTC 2004, Powell et al., 2002, 2006) and anecdotal evidence from nutritionists and feed companies confirm that dietary P levels have been reduced.

The decline in dietary P levels can be attributed to two causes: (1) the need to conform to P-based manure land application regulations; and (2) improved confidence that reducing dietary P will not decrease reproductive performance or milk production or quality. Many states have adopted nutrient management regulations to protect surface water quality based on topsoil P levels and risk of runoff (e.g., the Wisconsin soil test P index http://wpindex.soils.wisc.edu/). Full lactation trials (Satter et al., 2005) and related research have engendered higher confidence among feed consultants and producers that dietary P levels could be safely reduced. Lower manure P concentrations resulting from reduced P concentration in the ration helps farmers meet land application limits, thus improving farm profitability and reducing negative environmental impacts of manure.

In addition, water conservation strategies can reduce manure mass, thereby making manure more transportable for land application. The use of water as part of barn cleaning systems can impact nutrient losses in housing facilities, and the amount of manure and waste produced. Low-labor alternatives to water-dependent barn flush systems may be needed to reduce manure mass and facilitate manure handling, storage, and land application. In some locations, the price of water has risen because of water shortages and labor and transportation costs for manure handling, compelling some farmers to drastically reduce water use during barn cleaning, manure handling, and storage.

Improved Manure Handling and Storage

Improved manure handling and storage offers another valuable approach to meeting environmental challenges. The management of animal manure includes collection, handling, storage, treatment, and land application. The specifics of these activities differ depending on the operational

Table 7.7. Impact of feed management and milking frequency on milk production, and feed N (FNUE), and feed P (FPUE) use efficiencies on 54 Wisconsin dairy farms

Practice	Practice Use	Milk Production	FNUE[1]	FPUE[1]
		kg/cow/d	%	%
Use of total mixed rations (TMRs)	Yes	33.5a[2]	27.0a	28.9
	No	26.1b	24.1b	29.0
Balance rations 4x/y	Yes	30.6a	26.5a	30.0a
	No	24.7b	21.0b	24.8b
Milk thrice daily	Yes	40.2a	32.6a	34.6
	No	28.8b	24.9b	28.7
Use Posilac	Yes	37.1a	29.0a	28.7
	No	27.7b	24.6b	29.1

[1] Percentages of feed N and P transformed into milk.

[2] Within a practice, means followed by different letters differ significantly ($P < 0.05$).

Source: Adapted from Powell et al., 2006.

features of a dairy farm, such as housing (Table 7.4) and the presence or absence of manure storage. The recommended expansion of manure storage during the 1980s and '90s was premised on labor efficiency, the notion that storage would facilitate calculation of manure nutrients available, and also allow for land application during favorable weather conditions and close to crop nutrient demands. The appropriateness of manure storage depends, however, on cost levels and on the farmer's ability to spread the costs over many animal units. Most small-scale dairy farms are not able to afford long-term manure storage. Small-scale operations need low-cost alternatives to current practices, such as filter strips, or cement pads with retaining walls for stacking manure. These are also technologies that do not put additional burdens on family labor. Small-scale dairy farms rely almost exclusively on family labor, and the frequent removal and land spreading of manure are compatible with their labor supply. These frequent applications of small quantities of manure are not usually incorporated in the soil, are not uniform with regard to rate over a field, and are subject to volatilization losses of ammonia and runoff of nutrients and other constituents. It is difficult to predict nutrient supply in these systems, so farmers often compensate by ignoring manure nutrient credits.

Manure's impact on air and water pollution can also be reduced by using it as an energy source. Covered lagoons, complete mix digester systems, and plug flow digester systems capture methane, which can be converted into energy and used for gas or electricity production, heating, and cooling. Methane generation, recovery, and energy conversion is becoming increasingly attractive in areas where dairy farm concentration, and therefore the supply of manure and other organic sources, is high. Under these conditions it can produce energy that is competitive with classical energy sources. Community-scale, multiple dairy farm anaerobic digesters are being marketed where 2500 cow-equivalents are available for economic, steady biogas production (Bunting 2007). Starting in 2008, a private company, BioEnergy Solutions, began producing methane from manure digestion for the regional gas and electric company in central and southern California (PG&E 2008), After solid/liquid separation, methane produced by the liquid fraction is cleaned and transported by pipe to local storage, to replace natural gas or to fuel turbines producing electricity.

Enhanced Manure Recycling through Crops

Nationwide, US livestock producers have not been making full use of the nutrients contained in manure. Improvements in this situation are likely only under conditions of more intense regulation and price pressures (Schmitt et al., 1999). Tighter manure management regulations and rapid increases in fertilizer prices now have stimulated a growing interest in using manure in place of fertilizer. The potential here is considerable. As excreted, dairy manure contains about 1.1 million metric tons of N in the United States—a significant amount, when compared with an annual average of around 12 million nutrient tonnes of N applied to plants between 1992 and 2006 (USDA 2008). However, a significant share of manure is still not collected and managed, which may lead to point sources of water contamination (Harter et al., 2001, Powell et al., 2005, Russelle et al., 2007a).

Many dairy farmers now appear to be looking more favorably on manure to reduce fertilizer expenditures. Farm surveys approximately a decade ago (Nowak et al., 1998, Russelle, 1999) determined that Wisconsin and Minnesota dairy farmers were not allowing for applied manure nutrients when calculating the fertilizer requirements of their crops. However, Powell et al. (2007) recently found that most Wisconsin dairy farmers are now integrating fertilizer–manure–legume–N management, resulting in N and P application rates closer to agronomic recommendations. Increased use of manure as fertilizer promises a reduction in overall pollution risks.

Environmentally sound manure application strategies depend on the following:

- Land type (slope, soil texture, nutrient attenuation potential)
- Amounts and method of manure application (surface applied or incorporated
- Timing of application relative to crop growth
- The nutrient demands of the subsequent crops.

Strategies for reducing nutrient losses from manure are therefore necessarily site-specific (Table 7.5). For example, if the potential for nitrate leaching to drinking water aquifers is high, then N management should be a priority consideration. If runoff and erosion potentials to public surface water bodies are high, then P should be the main focus of management. Manure management based on site susceptibilities to N and P losses should aim to mitigate the excessive buildup and loss of soil P, and at the same time lower the risk for nitrate leaching to ground water. Manure land-application strategies need to be based on what pollutants are contributing to a problem (e.g., sediment, nutrients, bacteria), where pollutants are being transported (surface water, groundwater, air), and how the pollutants are being delivered.

Many considerations affect farmers' decisions about where to apply manure, including the amount of manure actually collected, the presence of manure storage, labor availability and machinery capacity for manure spreading, variations in the number of days manure can be spread given regional differences in weather and soil conditions, and the distances between the sites where manure is produced and the fields where it can be applied (Nowak et al., 1998). Manure spreading is also related to landownership—as the percentage of owned cropland operated by livestock farmers increases, so does the percentage of operated cropland that receives manure (Saam et al., 2005). More than half of all farmed land in the United States is rented by farmers, and this land is less likely to receive manure or other improvements, such as drainage.

Technology: No One Size Fits All

Numerous technologies have proven effective in minimizing pollution from livestock operations. However, farmer adoption of manure management technologies is closely linked to need, capability, and cost. Cost depends on farm size, or farmer ability to spread costs over many animal units. Most small-scale dairy farmers do not have additional resources to invest in the housing, manure collection, storage, and land-spreading options that are being promoted to improve manure management.

It is often assumed that pollution is simply a matter of choice, and that policy should "examine the question of how to induce farmers who cause water quality damages through their choice of production practices to adopt pollution prevention and pollution control practices that are consistent with societal environmental quality objectives" (Horan and Shortle 2001). Most livestock producers, however, do not actively choose to adopt practices that pollute, but rather may find themselves in environments that limit their management choices. Differences in soil type may hinder farmers in one geographic area in using as much of their cropland as possible for manure application (McCrory et al., 2004). For example, dairy farmers in the southwest part of Wisconsin, a region characterized by silt loam soils of relatively high permeability and drier field conditions in the spring and fall, have approximately 28% more days during the fall period (September–November) for surface application of manure and 60% more days available for fall tillage and manure injection operations than northeastern Wisconsin, a region of more finely textured and less permeable clayey and red loam soils (Figure 7.13). Flexible manure management regulations are therefore needed to reflect the diverse biophysical conditions farmers face. Advances in geographic information systems and weather forecasting are enhancing our ability to devise manure land-application options that minimize risks of nutrient runoff.

Although it is technologically possible to achieve significant improvements in the environmental performance of dairy farms, most advances will depend on producers voluntarily changing their behaviors. The socioeconomic literature suggests a number of possible reasons why farmers often fail to follow "best management practices," including individual characteristics of farmers and characteristics of the technologies.

Individual Characteristics of Farmers

It is often assumed that many farmers seek only maximum production and are less concerned about environmental issues (Horan et al., 2001). If farmers are unconcerned about environmental impact, the argument goes, one might therefore expect them to be reluctant to change management practices, or to make significant investments that would enhance the environmental performance of their farms. However, farmers are more aware of environmental concerns than is often appreciated. Most farmers agree that manure management is a critical issue in the industry, that they must do a better job of protecting the environment, and that there is room for improvement. Most attitudinal surveys have documented that levels of environmental concern are higher if the questions focus on local, regional, or national level problems, and lower if the question asks farmers whether they were concerned about environmental impacts of their own farm operation. This latter attitude may stem from a desire to avoid self-incrimination and/or a lack of recognition about deficiencies in their own practices. Awareness and concern about environmental problems are only partial prerequisites for change. These first must be personalized, but then knowing what to do, being able to do it, and a willingness to act are required to achieve behavioral changes that affect environmental outcomes.

Characteristics of the Technologies

Many studies have shown that the costs of some environmentally sound technologies may outweigh the benefits farmers expect to receive. For example, the added costs and large labor input required to handle, store, transport, and land-spread manure—with little confidence of an economic return—deters many from managing manure more effectively (Nowak et al., 1998). Historically, commercial fertilizers have been relatively inexpensive (Figure 7.6), and can be more easily handled and supply plant nutrients more readily than manure. Perhaps the biggest challenge facing efforts to improve manure management is therefore to create more meaningful incentives.

The compatibility of new agricultural technologies with existing farm management, land, labor, and capital resources is another prime determinant of adoption patterns (Nowak 1987). Although lined and covered manure storage is obligatory in some European counties, this technology has been adopted by only the largest dairy operations in the United States (USDA 2004b). The adoption of lined manure storage depends on the ability of farmers to spread costs over many animal units. Thus, even when public funds are available to subsidize the construction of manure storage, larger operations will continue to be more likely than smaller farms to invest in such structures. There may also be different adoption rates depending on farmer age. Because significant capital investments are required for manure storage, this technology is likely to have a relatively long payback period. It may not make economic sense, therefore, for dairy farmers nearing the end of their career to invest in manure storage facilities.

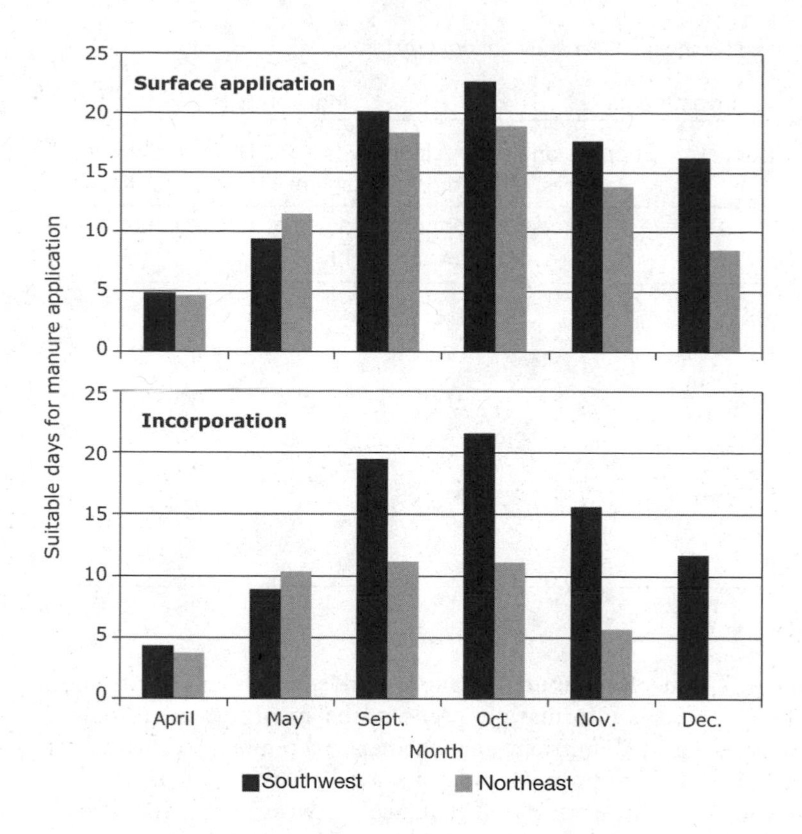

Figure 7.13. Regional differences in manure surface application and incorporation, Wisconsin.
Source: McCrory et al., 2004.

On farms that rely solely on family labor, improved manure management may be constrained by seasonal labor bottlenecks. In the US Northeast and Midwest, spreading large volumes of dairy manure in the spring can be a monumental task when farmers are already working long hours preparing and planting fields for a relatively short growing season. In these regions, timely planting is critical to high yields. Manure spreading within such seasonal labor constraints is best done in small installments, on a year-round basis, as and when labor is available. One promising alternative to frequent spreading of manure is corralling dairy cattle directly on cropland, which can improve N cycling and reduce gaseous and dissolved N losses (Powell and Russelle 2009).

Increasing herd size, greater animal-to-land ratios, and environmental regulation drive the need for regional integration of dairy farms with crop producers for manure sharing. This model has typified swine production in the United States and Canada for manure utilization, but dairy farms have greater potential to utilize the variety of feedstuffs produced by neighboring crop farms (Russelle et al., 2007b). This two-way flow of nutrients has the potential for improving sustainability of both crop and livestock enterprises (Steinfeld et al., 1997, Powell et al., 2004).

Partnerships to Enhance Nutrient Management on Dairy Farms

The development, dissemination, and adoption of technologies that enhance environmental performance are not straightforward processes. They necessitate partnerships consisting of key players, as well as policies that stimulate investments and inducements that integrate and improve nutrient management (Table 7.8). Hence attempts to improve nutrient management must engage dairy producers, their feed and fertilizer consultants, and policy makers in critical assessments of the real and perceived risks of nutrient utilization, the key factors that affect nutrient transformation and loss, and how these factors may be managed more effectively to enhance profitability and reduce environmental impacts. For example, nutrient supply dealers, such as representatives of the feed and fertilizer industries, need to be integral partners in any effort to reduce nutrient loads in manure through optimal feeding, or through land application techniques that combine manure and other nutrient sources (e.g., fertilizer, legume-N) to optimize plant nutrient use.

Involving the nutrient supply and service industries that affect farmer nutrient use behavior will be critical to achieving desired goals of improving regional and whole-farm nutrient balances. Pilot insurance policies are being tested to reduce risks to farmers of possible production loss due to reductions in feed or fertilizer nutrient use. Anecdotal evidence suggests that some policy makers would advocate disincentives, such as fines, to limit suppliers from selling feeds and fertilizers to farmers that exceed published recommended levels.

Improved manure collection, treatment, and storage technologies are expensive and will also require cost

Table 7.8. Possible investments and inducements to improve feed and manure management on dairy farms

Technology domain	Key players in technology implementation (order of importance)	Investments		Inducements	
		Capital	Supplies and Services	Incentives (e.g., cost share)	Disincentives (e.g., taxes or fines)
		Relative investment opportunity and use of inducements (1= low; 5 = high)			
Feeding strategies	Producer, feed industry, research, extension/outreach	1	4	2	1
Manure collection, treatment, and storage	Producer, extension/ outreach, research, policy	5	3	5	1
Manure land application	Producer, policy, custom manure haulers, extension/ outreach	3	5	3	3

sharing if farmers, especially small- to medium-size farms, are to adopt them. Conventional technologies may be financially out of reach of resource-limited dairy producers, and for them alternative low-cost technology options will likely be needed. Because of narrow profit margins, farms with small herd sizes are much less able than larger farms to afford technologies that improve environmental performance but not improve profits. Many current environmental technologies are cost effective for medium- and large-scale farms. Small farms having high pollution risks may require not only different technologies but also additional subsidies, which may include full cost subsidies.

Part of manure mismanagement can be attributed to shifts in educational messages. As dairy production has expanded and specialized, manure often has been viewed as an undesirable by-product. The widely adopted term *animal waste* has been counterproductive to the essential message that manure is a valuable source of fertilizer and energy. When *waste disposal* became an engineering term associated with industrial livestock systems, connotations of manure's intrinsic value were lost (Nowak et al., 1998). Alternative terminology to *waste* management should be sought when developing training materials aimed at affecting farmer behavior and environmental impact.

Conclusions

The US dairy industry has been undergoing great change. More cows are being kept on smaller land areas, and more feed is being purchased rather than homegrown. Greater cow numbers, supported by importation of relatively inexpensive feed and fertilizer, have increased the risk of on-farm nutrient surplus, soil nutrient buildup, nutrient loss, and environmental pollution. The dairy industry could benefit from a better understanding of the key factors that impact nutrient inputs and outputs and resource flow rates within subcomponents of the feed–animal–manure–soil–environment continuum. Such information needs to be incorporated into integrated nutrient management recommendations adaptable to prominent dairy system types.

Restoring the balance between livestock density and the nutrient adsorptive capacity of the surrounding environment will be central to any strategies aimed at improving the performance of any animal industry, including dairy. This will involve a series of different technological, financial, regulatory, and institutional tools. Technological tools encompass strategies such as optimal feeding to reduce the amount of manure nutrients produced, and therefore the land base required to recycle manure nutrients. Technology will also play a key role in enhancing manure collection, handling, storage, and land application to maximize manure nutrient recycling. Consensus needs to be sought on the comparative advantages of federal, state, and local governments in promulgating and enforcing environmental standards. In some locations, regulatory tools may be needed to strengthen zoning laws and regulations and to arrive at a better spatial distribution of crop/pasture and livestock production. Private institutions, especially different stakeholders in the feed and fertilizer industries, may need to change practices (such as commissions on sale of nutrients) to maximize the efficient use of agricultural nutrients. Associative institutions, such as cooperatives, may be able to facilitate areawide integration of specialized crop and dairy production and exert peer pressure to enhance environmental performance.

References

Aillery, M., N. Gollehon, R. Johansson, N. Key, and M. Ribaudo. 2006a. Managing manure to improve air and water quality. *Ag Nutrient Management: Part 2* 2(3): 13–20.

Aillery, M., N. Gollehon, R. Johansson, N. Key, and M. Ribaudo. 2006b. Managing manure to improve air and water quality. *Ag Nutrient Management: Part 1* 2(2): 14–19.

Asman, W. A. H., M. A. Sutton, and J. K. Schjorring. 1998. Ammonia: emission, atmospheric transport, and deposition. *New Phytologist* 139: 27–48.

Beegle, D. B. and L. E. Lanyon. 1994. Understanding the nutrient management process. *Journal of Soil and Water Conservation* 49(2): 23–30.

Brolsma, J. L. 2004. International immigration into the Wisconsin dairy industry: opportunities and constraints. Madison: Dept. of Rural Sociology, Univ. of Wisconsin. Cited 25 November 2008. Available at http://www.babcock.cals.wisc.edu/sites/default/files/documents/Intl_Immigration04.pdf.

Bunting, S. 2007. Unlocking manure's full potential. *Progressive Dairyman* 21(3): 34–38.

Burkart, M. R. and D. E. James. 1999. Agricultural nitrogen contributions to hypoxia in the Gulf of Mexico. *Journal of Environmental Quality* 28: 850–859.

Burkholder, J., B. Libra, P. Weyer, S. Heathcote, D. Kolpin, P. S. Thorne, and M. Wichman. 2007. Impacts of waste from concentrated animal feeding operations on water quality. *Environmental Health Perspectives* 115: 308–312.

California Regional Water Quality Control Board, Central Valley Region. 2007. Waste discharge requirements general order for existing milk cow dairies. Cited 25 November 2008. Available at http://www.waterboards.ca.gov/centralvalley/board_decisions/adopted_orders/general_orders/r5-2007-0035.pdf.

California Regional Water Quality Control Board, Central Valley Region. 2008. Approved sampling and analysis procedures for nutrient and groundwater monitoring at existing milk cow dairies. Cited 25 November 2008. Available at http://www.waterboards.ca.gov/centralvalley/water_issues/dairies/general_order_guidance/sampling_analysis/sampling_procedures_rev_23may08.pdf.

Campbell-Mathews, M. 2007. A checklist for using dairy lagoon water for crop production. University of California Cooperative Extension Service. Cited 25 November 2008. Available at http://groups.ucanr.org/LNM/Checklist_for_Using_Dairy_Lagoon_Water/.

CAST. 1999. *Animal Agriculture and Global Food Supply*. Task Force Report No. 135. Ames, Iowa: Council for Agricultural Science and Technology.

Castillo, A. R., E. Kebreab, D. E. Beever, and J. France. 2000. A review of efficiency of nitrogen utilization in lactating dairy cows and its relationship with environmental pollution. *Journal of Animal and Feed Science* 9: 1–32.

Chang, A., T. Harter, J. Letey, D. Meyer, R. D. Meyer, M. Campbell-Mathews, F. Mitloehner, S. Pettygrove, P. Robinson, and R. Zhang. 2005. Managing dairy manure in the Central Valley of California. Davis: University of California. Cited 25 November 2008. Available at http://groundwater.ucdavis.edu/Publications/uc-committee-of-experts-final-report%202006.pdf.

CVTC. 2004. *Whole-Farm Phosphorus Management*. Project Report. Eau Claire, Wisconsin: Chippewa Valley Technical College.

Dalal, R. C., W. J. Wang, G. P. Robertson, and W. J. Parton. 2003. Nitrous oxide emission from Australian agricultural lands and mitigation options: a review. *Australian Journal of Soil Research* 41: 165–195.

Ebeling, A. M., L. G. Bundy, J. M. Powell, and T. W. Andraski. 2002. Dairy diet phosphorus effects on phosphorus losses in runoff from land-applied manure. *Soil Science Society of America Journal* 66: 284–291.

EPA. 2005. *Inventory of U.S. Greenhouse Gas Emissions and Sinks: 1990–2003*. EPA 430-R-05-003. Washington, DC: US Environmental Protection Agency. Cited 25 November 2008. Available at http://www.epa.gov/climatechange/emissions/downloads06/05CR.pdf.

EPA. 2008. Concentrated animal feed operations final rule-making fact sheet. Cited 18 November 2008. Available at http://www.epa.gov/npdes/pubs/cafo_final_rule2008_fs.pdf.

Fischer, D. B., M. F. Hutjens, and E. N. Ballard. 2005. Pasture-based feeding programs for dairy cattle. Illini DairyNet Papers. Cited 25 November 2008. Available at http://www.livestocktrail.uiuc.edu/dairynet/paperDisplay.cfm?ContentID=7160.

Fixen, P. E. 1998. Soil test levels in North America. *Better Crops* 82: 16–18.

Gassman, P. W., E. Osei, A. Saley, and L. M. Hauck. 2002. Application of an environmental and economic modelling system for watershed assessments. *Journal of the American Water Resources Association* 38: 423–438.

Godish, T. 2004. *Air Quality*. 4th ed. Boca Raton, Florida: CRC Press.

Greene, C. 2007. U.S. organic farm sector continues to expand. *AmberWaves* May 2007. Cited 17 November 2008. Available at http://www.ers.usda.gov/AmberWaves/May07SpecialIssue/Findings/Organic.htm.

Hadley, G. L., S. B. Harsh, and C. A. Wolf. 2002. Managerial and financial implications of major dairy farm expansions in Michigan and Wisconsin. *Journal of Dairy Science* 85: 2053–2064.

Harper, D. 2000. Requiem for the small dairy: agricultural change in northern New York. In *Dairy Industry Restructuring: Research in Rural Sociology and Development,* Volume 8, ed. H. K Schwarsweller and A. P. Davidson, 13–45. New York: JAI.

Harrison, J. H. and R. A. White. 2008. Whole farm nutrient management: A dairy example. Livestock and Poultry Environmental Learning Center, eXtension Initiative. Cited 25 November 2008. Available at http://www.extension.org/pages/Whole_Farm_Nutrient_Management_-_A_Dairy_Example.

Harter, T., M. Mathews, and R. Meyer. 2001. Effects of dairy manure nutrient management on shallow groundwater nitrate: a case study. In ASAE Paper Number 01-2192. ASAE Annual International Meeting, July 30–August 1, Sacramento, CA.

Harter, T. and J. Menke. 2004. Cow numbers and water quality: is there a magic limit? In *A Groundwater Perspective*. Proceedings of the National Alfalfa Symposium, December 13–15, San Diego, CA.

Hinrichs, C. and R. Welsh. 2003. The effects of the industrialization of US livestock agriculture on promoting sustainable production practices. *Agricultural Human Values* 20:125–141.

Horan, R. D. and J. S. Shortle. 2001. Environmental instruments for agriculture. In *Environmental Policies for Agricultural Pollution Control*, ed. J. S. Shortle and D. Abler, 19–66. Wallingford, UK: CAB International.

Horan, R. D., M. Ribaudo, and D. G. Abler. 2001. Voluntary and indirect approaches for reducing externalities and satisfying multiple objectives. In *Environmental Policies for Agricultural Pollution Control*, ed. J. S. Shortle and D. Abler, 67–84. Wallingford, UK: CAB International.

Howarth, R. W., A. Sharpley, and D. Walker. 2002. Sources of nutrient pollution to coastal waters in the United States: implications for achieving coastal water quality goals. *Estuaries* 25: 656–676.

Huffman, M. 2008. Organic milk production: are you getting what you pay for? Cited 18 November 2008. Available at http://www.consumeraffairs.com/news04/2008/05/organic_milk.html.

Isik, M. 2004. Environmental regulation and the spatial structure of the U.S. dairy sector. *American Journal of Agricultural Economics* 86: 949–962.

Jackson-Smith, Douglas and G. W. Gillespie Jr. 2005. Impacts of farm structural change on farmer's social ties. *Society and Natural Resources* 18: 215–240.

Johnson, K. A. and D. E. Johnson. 1995. Methane emission from cattle. *Journal of Animal Science* 73: 2483–2492.

Kingery, W. L., C. W. Wood, D. P. Delaney, J. C. Williams, and G. L. Mullins. 1994. Impact of long-term land application of broiler litter on environmentally related soil properties. *Journal of Environmental Quality* 23: 139–147.

Klausner, S. 1995. Nutrient management planning. In *Animal Waste and the Land–Water Interphase*, ed. K Steele, 383–391. New York: Lewis Publishers.

Kleinman, P. J. A., R. B. Bryant, W. S. Reid, A. N. Sharpley, and D. Pimentel. 2000. Using soil phosphorus behavior to identify environmental thresholds. *Soil Science* 165: 943–950.

Klemme, R. M. 1998. The economics of forage production in a rapidly changing dairy sector. In *Midwest Dairy Management Conference Proceedings*, 17–18 August 1998, Kansas City, MO USA, 58–65. Madison, Wisconsin: University of Wisconsin.

Koelsch, R. 2005. Evaluating livestock system environmental performance with whole-farm nutrient balance. *Journal of Environmental Quality* 34: 149–155.

Kornegay, E. T. 1996. *Nutrient Management of Food Animals to Enhance and Protect the Environment.* Boca Raton: Lewis Publishers.

Krogstad, T. and O. Lovstad. 1991. Available soil phosphorus for planktonic blue-green algae in eutrophic lake water samples. *Arch. Hydrobiol.* 122: 117–128.

Lamb, J. F. S., C. C. Sheaffer, L. H. Rhodes, R. M. Sulc, D. J. Undersander, and E. C. Brummer. 2006. Five decades of alfalfa cultivar improvement: impact on forage yield, persistence, and nutritive value. *Crop Science* 46: 902–909.

Lanyon, L. and P. Thompson. 1996. Changing emphasis on farm production. In *Animal Agriculture and the Environment: Nutrients, Pathogens, and Community Relations*. Proceedings from the Animal Agriculture and the Environment North American Conference, Rochester, NY, December 11–13, 15–23. Ithaca, New York: Northeast Regional Agricultural Engineering Service.

Lord, E. and R. D. J. Mitchell. 1998. Effect of nitrogen inputs to cereals on nitrate leaching from sandy soils. *Soil Use and Management* 14: 78–83.

MacDonald, J. M., W. D. McBride, and E. J. O'Donoghue. 2007. Low costs drive production to large dairy farms: profits and losses lead to fewer small dairy farms and more large operations. *Amber Waves*, September 2007. Cited 18 November 2008. Available at http://www.ers.usda.gov/AmberWaves/September07/Features/DairyFarms.htm.

McBride, W. D. 1997. *Change in U.S. livestock production, 1969–1992*. Agricultural Economics Report No. 754. Washington, DC: U.S. Department of Agriculture, Economic Research Service.

McCrory, D. F., H. Saam, J. M. Powell, D. Jackson-Smith, and C. A. Rotz. 2004. Predicting the number of suitable days for manure spreading across Wisconsin. In *Agronomy Abstracts (CD), Proceedings of Annual Meeting*, American Society of Agronomy, Madison, Wisconsin.

Meisinger, J. J. and R. B. Thompson. 1996. Improving nutrient cycling in animal agriculture systems. In *Animal Agriculture and the Environment: Nutrients, Pathogens, and Community Relations*. Proceedings from the Animal Agriculture and the Environment North American Conference, Rochester, New York, December 11–13, 15–23. Ithaca, New York: Northeast Regional Agricultural Engineering Service.

Mitloehner, F. M. 2006. Volatile organic compound and greenhouse gas emissions from dairy cows, waste, and feed. In *Proceedings of Workshop on Agricultural Air Quality: State of the Science*, 5–8 June 2006, Potomac, Maryland, 921–922. Raleigh: North Carolina State University.

Moody, L. B. and R. T. Burns. 2006. Comprehensive nutrient management planning for your facility. *Ag Nutrient Management* 2(1): 11–13.

NASS. 2008. National Agricultural Statistics Service. Cited 14 November 2008. Available at http://www.nass.usda.gov.

NCIFAP. 2007. National Commission on Industrial Farm Animal Production. Cited 14 November 2008. Available at http://www.ncifap.org/.

Norris, P. E. and S. S. Batie. 2000. Setting the animal waste management policy in context. USDA Agricultural Outlook Forum, Washington, DC. Cited 25 November 2008. Available at http://ageconsearch.umn.edu/bitstream/33455/1/fo00no01.pdf.

Nowak, P. 1987. The adoption of agricultural conservation technologies: economic and diffusion explanations. *Rural Sociology* 52(2): 208–220.

Nowak, P., R. Shepard, and F. Madison. 1998. Farmers and manure management: a critical analysis. In *Animal Waste Utilization: Effective Use of Manure as a Soil Resource*, ed. J. Hatfield, B. Stewart, 1–32. Chelsea, MI: Ann Arbor Press.

NRC. 2003. *Air Emission from Animal Feeding Operations: Current Knowledge, Future Needs.* Washington, DC: National Research Council, National Academy of Sciences.

Olmos Colmenero, J. J. and G. A. Broderick. 2006. Effect of dietary crude protein concentration on milk production and nitrogen utilization in lactating dairy cows. *Journal of Dairy Science* 89: 1704–1712.

PG&E. 2008. PG&E and Bioenergy Solutions turn the valve on California's first "cow power" project. News release, 4 March 2008. Cited 25 November 2008. Available at http://www.pge.com/about/news/mediarelations/newsreleases/q1_2008/080304.shtml.

Powell, J. M., Z. Wu, and L. D. Satter. 2001. Dairy diet effects on phosphorus cycles of cropland. *Journal of Soil and Water Conservation* 56(1): 22–26.

Powell, J. M., D. Jackson-Smith, and L. Satter. 2002. Phosphorus feeding and manure nutrient recycling on Wisconsin dairy farms. *Nutrient Cycling in Agroecosystems* 62: 277–286.

Powell, J. M., R. Pearson, and P. Hiernaux. 2004. Crop–livestock interactions in the West African drylands, *Agronomy Journal* 96: 469–483.

Powell, J. M., D. McCrory, D. Jackson-Smith, and H. Saam. 2005. Manure collection and distribution on Wisconsin dairy farms. *Journal of Environmental Quality* 34: 2036–2044.

Powell, J. M., D. Jackson-Smith, D. McCrory, H. Saam, and M. Mariola. 2006. Validation of feed and manure data collected on Wisconsin dairy farms. *Journal of Dairy Science* 89: 2268–2278.

Powell, J. M., D. Jackson-Smith, D. McCrory, H. Saam, and M. Mariola. 2007. Nutrient management behavior on Wisconsin Dairy farms. *Agronomy Journal* 99: 211–219.

Powell, J. M. and M. P. Russelle. 2009. Dairy cattle management impacts manure collection and nitrogen cycling through crops. *Agriculture, Ecosystems and Environment* 131: 170–177.

RFA. 2006. From niche to nation: ethanol industry outlook 2006. Renewable Fuels Association. Cited 18 November 2008. Available at http://www.ethanolrfa.org/objects/pdf/outlook/outlook_2006.pdf.

Rotz, C. A., F. Taube, M. P. Russelle, J. Oenema, M. A. Sanderson, and M. Wachendorf. 2005. Whole-farm perspectives of nutrient flows in grassland agriculture. *Crop Science* 45: 2139–2159.

Rovner, S. 2006. Farm emissions control. *Chemistry and Engineering News* 84: 104–107.

Russelle, M. P. 1999. Survey results of forage nutrient management on Minnesota dairy farms. *Proceedings of the 23rd Forage Production and Use Symposium*, 26–27 January 1999, Appleton, Wisconsin, 30–38. Wisconsin Forage Council. Cited 25 November 2008. Available at: http://www.uwex.edu/ces/forage/wfc/Russelle.html.

Russelle, M. P., J. Lamb, N. Turyk, B. Shaw, and B. Pearson. 2007a. Managing N contaminated soils: benefits of N_2-fixing alfalfa. *Agronomy Journal* 99: 738–746.

Russelle, M., H. M. Entz, and A. J. Franzluebbers. 2007b. Reconsidering integrated crop–livestock systems in North America. *Agronomy Journal* 99: 325–334.

Saam, H., J. M. Powell, D. Jackson-Smith, W. Bland, and J. Posner. 2005. Use of animal density to estimate nutrient recycling ability of Wisconsin farms. *Agricultural Systems* 84: 343–357.

Satter, L. D., T. Klopfenstein, G. Erickson, and J. Powell. 2005. Phosphorus and dairy-beef nutrition. In *Phosphorus Agriculture and the Environment,* ASA-CSSA-SSSA Monograph No. 46, ed. A. N. Sharply, 587–606. Madison, Wisconsin: ASA-CSSA-SSSA.

Schiffman, S. S., E. A. Sattely-Miller, M. S. Suggs, and B. G. Graham. 1995. The effect of environmental odors emanating from commercial swine operations on the mood of nearby residents. *Brain Research Bulletin* 37: 369–375.

Schmitt, M. A., M. P. Russelle, G. W. Randall, and J. A. Lory. 1999. Manure nitrogen crediting and management in the USA: survey of university faculty. *Journal of Production Agriculture* 12: 419–422.

Sharpley, A. N., T. Daniel, and D. Edwards. 1993. Phosphorus movement in the landscape. *Journal of Production Agriculture* 6: 453–500.

Sharpley, A. N. and P. Withers. 1994. The environmentally-sound management of agricultural phosphorus. *Fertilizer Research* 39: 133–146.

Sharpley, A. N., S. Chapra, R. Wedepohl, J. Sims, T. Daniel, and K. Reddy. 1994. Managing agricultural phosphorus for protection of surface waters: issues and options. *Journal of Environmental Quality* 23: 437–451.

Steinfeld H., C. de Haan, and H. Blackburn. 1997. Livestock–environment interactions: issues and options. Suffolk, UK: WREN-media. Cited 18 November 2008. Available at http://www.fao.org/docrep/x5305e/x5305e00.htm.

Steinfeld H., P. Gerber, T. Wassenaar, V. Castel, M. Rosales, and C. deHaan. 2006. *Livestock's Long Shadow: Environmental Issues and Options*. Rome: FAO. Cited 25 November 2008. Available at http://www.fao.org/docrep/010/a0701e/a0701e00.htm.

Thorne, P. S. 2007. Environmental health impacts of concentrated animal feeding operations: anticipating hazards—searching for solutions. *Environmental Health Perspectives* 115: 296–297. Cited 18 November 2008. Available at http://www.ehponline.org/members/2006/8831/8831.pdf.

USDA. 2004a. *2002 Census of Agriculture*. Volume 1 Chapter 1: U.S. National Level Data. Cited 18 November 2008. Available at http://www.agcensus.usda.gov/Publications/2002/index.asp.

USDA. 2004b. *Dairy 2002: Nutrient Management and the U.S. Dairy Industry in 2002*. Washington, DC: USDA Animal and Plant Health Inspection Service. Cited 25 November 2008. Available at http://nahms.aphis.usda.gov/dairy/dairy02/Dairy02Nutrient_mgmt_rept.pdf.

USDA. 2004c. Economic Effects of U.S. Dairy Policy and Alternative Approaches to Milk Pricing. Cited 18 November 2008. Available at http://www.usda.gov/documents/NewsReleases/dairyreport1.pdf.

USDA. 2007. *USDA Agricultural Projections to 2016*. Washington, DC: Interagency Agricultural Projections Committee, USDA. Cited 18 November 2008. Available at http://www.usda.gov/oce/commodity/archive_projections/USDA%20Agricultural%20Projections%20to%202016.pdf.

USDA. 2008. U.S. consumption of nitrogen, phosphate, and potash for 1960–2006. Cited 26 March 2008. Available at http://www.ers.usda.gov/Data/FertilizerUse/.

USEPA. 1996. *Environmental Indicators of Water Quality in the U.S.* USEPA 841-R-96-002. USEPA, Office of Water (4503F). Washington, DC: US Gov. Print. Office.

Westphal, P. J., L. E. Lanyon, and E. J. Partenheimer. 1989. Plant nutrients management strategy and implications for optimal herd size and performance of a simulated dairy farm. *Agricultural Systems* 31: 381–394.

Wilber N., G. Hadley, G. Bonde, and T. Anderson. 2006. Producer perceptions: diverse workforce acceptance on Wisconsin dairy farms and farming communities—Shawano and Waupaca counties. Univ. of River Falls, UW-Extension, and Center for Dairy Profitability Report. Cited 25 November 2008. Available at http://www.cdp.wisc.edu/pdf/Shawano_and_Waupaca_County_Diverse_Workforce_Final_Report%5B1%5D.pdf.

Wu, Z. and L. D. Satter. 2000. Milk production during the complete lactation of dairy cows fed diets containing different amounts of protein. *Journal of Dairy Science* 83: 1042–1051.

Zhang, G., J. S. Strom, B. Li, H. B. Rom, S. Morsing, P. Dahl, and C. Wang. 2005. Emission of ammonia and other contaminant gases from naturally ventilated dairy cattle buildings. *Biological Systems Engineering* 92: 355–364.

8

Denmark–European Union

Reducing Nutrient Losses from Intensive Livestock Operations*

Søren A. Mikkelsen, Torben Moth Iversen, Brian H. Jacobsen, and Søren S. Kjær

Abstract

Since 1985, a number of action plans have been implemented in Denmark to reduce nitrate leaching from agriculture. This chapter summarizes the regulatory measures applied in this period, with a focus on nitrogen and nitrogen leaching, and the effect of these measures in agriculture and the aquatic environment. Measures have included area-related measures such as wetlands establishment and afforestation, and nutrient-related measures such as mandatory fertilizer plans and improved utilization of nitrogen in manure. To assess changes in nitrogen (N) losses from the agricultural system, three national indicators are defined: N surplus, N efficiency, and N leaching. For the period studied (1979 to 2004) nitrogen surplus has been strongly reduced, and nitrogen efficiency has been increased from 27% to 38%. The reduction in nitrate leaching, based on model calculations, is estimated at 48% by 2003. Environmental monitoring programs show a decrease in nitrogen concentration in water leaving the root zone, in rivers, and in coastal waters. The economic analyses carried out show that the annual costs involved in the action plans, including Action Plan III, are around €600 million, of which agricultural measures cover €340 million—the costs are evenly divided between the private agricultural sector and public funding. The most cost-effective measures in Action Plan II were catch crops, wetlands, increased utilization of animal manure, and improved feeding practices. The total annual cost for AP II is estimated at €70 million, or €2 per kg of reduction in nitrogen leaching. The Danish approach to regulating nutrient losses from agriculture, based on research programs and dialogue between authorities and the agricultural community, has proven successful. A more regional or local approach is believed to be necessary in the future to complement regulation on the national scale.

* This chapter is an updated version of a paper presented at the OECD Workshop on Evaluating Agri-Environmental Policies, Paris, December 2004. There have been additions to this chapter regarding the economic consequences.

Introduction

The European Union (EU) consists of 25 member states. Over the years, a comprehensive framework of common environmental policies and directives has been developed. Several of the directives are of great importance to livestock production and to agriculture in general. The Nitrates Directive from 1991, the Natura 2000 directives, and the Water Framework directive from 1999 all set certain standards for environmental protection that require changes in the livestock production system.

The Nitrates Directive in particular sets a number of standards focusing on livestock and the need to control the nutrient content in livestock manure. They include standards for the maximum annual application of nitrogen in livestock manure per hectare (170 kg N/ha per year), and a request to countries to develop national standards for good agricultural practices. Each individual member state will develop an action plan to implement the directive. Apart from the 170 kg limit, there is scope for different approaches to applying the rest of the provisions of the directive. The EU Commission decides if the national action plans are in compliance with the directive.

The Nitrates Directive is required to be fully implemented in national regulations. However, so far only a limited number of countries have done so. Denmark was the first member state (1998) to implement the EU Nitrates Directive correctly. More recently, Sweden, Finland, Netherlands, Germany, Austria, and Ireland have followed. This case study focuses on the Danish experience. However, livestock production systems and their environmental impact are similar in large parts of the European Union and countries like Switzerland, and to some extent also in certain states in the United States.

Table 8.1. Livestock production in Denmark in 1990, 2003, and 2007 ('000s)

	1990	2003	2007
Number of dairy cows (stock)	750	600	545
Number of pigs (stock)	9700	13,300	13,700
Production of pigs for slaughter per year in Denmark*	16,400	24,200	20,500

* Excluding export of live piglets (2 million in 2003 and 5 million in 2008).

Source: Jacobsen 2004 and Danish Statistics, 1992 and 2008.

Danish Agriculture and Livestock Production

The total numbers of livestock in Denmark in 1990, 2003, and 2007 are shown in Table 8.1. The table shows an increase in pig numbers and production, and a decline in the number of dairy cows. The total number of livestock units (LUs)[1] of all types has been constant from 1990 to 2007 at around 2.4 million. The average livestock intensity has remained stable from 1990 to 2007, at around 1 LU per hectare.

Danish farm structure is based on a mixed approach to agriculture, where livestock farms also include crop production. Even the larger livestock farms always have integrated production of livestock and crops. This is partly due to the historic tradition and development of the agricultural sector in Denmark and partly because the maximum livestock intensity rule sets a maximum standard for livestock manure per hectare. This forces livestock farmers to own or rent land, or at least to make agreements with crop farmers about manure utilization.

In recent decades, Danish agriculture has undergone significant and rapid structural change. The number of farms fell from 100,000 in 1980 to 40,000 in 2003. Half of the 2003 total consists of part-time farms. This consolidation will probably continue. The work force in the primary farm sector fell from over 120,000 in 1993 to 62,000 in 2004. This steep drop was due to economies of scale in the farm sector, technological development of farm practices, and employment opportunities in other sectors. These tendencies have continued since 2003 and seem likely to continue.

The intensification of agriculture during the last 50 years has had important environmental effects. It has disturbed the natural nitrogen cycle, causing significant losses through emissions of ammonia to the atmosphere and leaching of nitrates to water. The impacts on the aquatic environment include high concentrations of nitrates in groundwater and surface water, causing unacceptable drinking water quality (EEA 2003). High concentrations of nitrates may also cause eutrophication of lakes and coastal marine areas. The oxygen deficiency in Danish coastal waters in the early 1980s put nitrogen emission from Danish agriculture on the political agenda, and literally started the regulation of Danish agriculture in this context.

1. One livestock unit is equivalent to one adult dairy cow.

Political Action Plans and Regulatory Measures

Since 1985, a series of political action plans have been imposed, with remarkable effects on the agricultural N efficiency and N pollution (Table 8.2). Consequently, Denmark has been one of the most successful among the EU countries in reducing N surpluses and N losses (OECD 2001). Moreover, these effects have been achieved while still increasing animal production and the value of agricultural products.

A number of the action plans reaffirmed the original goal of reducing nitrate leaching by 49%. In 2003, the final evaluation of Action Plan II (Grant and Waagepetersen 2003) showed a model-calculated reduction of 48% for Denmark as a whole.

The measures in the Ammonia Action Plan were scheduled to be implemented in full in 2007. AP III was evaluated in 2008, and the preliminary findings indicate that it will be difficult to reach the target reduction in N leaching, whereas the reduction in P-surplus most likely will be reached.

The regulatory measures applied in this period can be divided into two groups:

- Area-related measures such as wetlands and afforestation
- Nutrient-related measures such as fertilizer plans and utilization of N in manure.

Nitrogen Surplus, Efficiency, and Leaching

In order to assess whether the measures to reduce nitrogen losses from the agricultural system have achieved their intended effects, we define three national indicators: N surplus, N efficiency, and N leaching. All three are measured in tonnes of nitrogen per year.

N surplus is defined as N imports minus N exports for the agricultural system. Annual values for N imports and N exports are derived from national agricultural statistics (Statistics Denmark 2005, according to the method of Kyllingsbæk 2005). Nitrogen imports include the following:

- N in commercial fertilizers and waste materials spread on fields
- N in imported concentrate fodders like soybean cakes, fodder urea, fish products, and so forth
- Atmospheric net N deposition
- Estimated N fixation via legumes and free-living microorganisms.

Table 8.2. An outline of the Danish action plans and measures imposed to reduce nutrient losses from agriculture*

Danish Policy Actions	Policy Measures
1985: NPo Action Plan to reduce N and P pollution Target: general reduction of N and P	• Minimum 6 months slurry storage capacity. • Ban on slurry spreading between harvest and 15 October on soil destined for spring crops. • Maximum stock density equivalent to 2 LU/ha. • Various measures to reduce runoff from silage clamps and manure heaps. • A floating barrier (natural crust or artificial cover) is mandatory on slurry tanks.
1987: Action Plan I for the Aquatic Environment (AP I) Target: 49% reduction of N leaching compared to mid-1980s	• Minimum 9 months slurry storage capacity. • Ban on slurry spreading from harvest to 1 November on soil destined for spring crops. • Mandatory fertilizer and crop rotation plans. • Minimum proportion of area to be planted with winter green crops. • Mandatory incorporation of manure within 12 hours after spreading.
1991: Action Plan for a Sustainable Agriculture N Target: as in AP I	• Ban on slurry spreading from harvest until 1 February, except on grass and winter rape. • Obligatory fertilizer budgets. • Maximum limits on the plant-available N applied to different crops, equal to the economic optimum. • Statutory norms for the utilization of manure N (Pig slurry: 60%, cattle slurry: 55%, deep litter: 25%, other types: 50%).
1998: Action Plan II for the Aquatic Environment (AP II) N Target: as in AP I	• Subsidies to establish wetlands. • Subsidies to enable reduced nutrient inputs to areas designated as environmentally vulnerable areas. • Improved animal feeding practices to improve utilization of feed. • Reduction of the stock density maximum to 1.7 LU/ha for cattle and 1.4 LU/ha for other species. • Subsidies to encourage conversion to organic agriculture. • The statutory norms for the utilization of manure N are increased as from 1999. (Pig slurry: 65%, cattle slurry: 60%, deep litter: 35%, other types: 55%) • Maximum limits on the application of plant-available N to crops reduced to 10% below the economic optimum. • Mandatory 6% of the area with cereals, legumes, and oil crops to be planted with catch crops. • Subsidies to encourage afforestation.
2000: AP II Midterm Enforcement N Target: as in AP I	• Increased economic incentives to establish wetlands. • The N assumed to be retained by catch crops must be included in the fertilizer plans. • Further tightening of the statutory norms for the utilization of N in manure. • Until 2001: pig slurry: 70%, cattle slurry: 65%, deep litter: 40%, other types: 60%. • From 2002 pig slurry: 75%, cattle slurry: 70%, deep litter: 45%, other types: 65%. • Reduced fertilization norms for grassland and restrictions on additional N application to bread wheat.
2001: Ammonia Action Plan	• Subsidies to encourage good manure handling in animal housing and improved housing design. • Mandatory covering of all dung heaps. • Ban on slurry application by broadcast spreader. • Slurry spread on bare soil must be incorporated within 6 hours. • Ban on the treatment of straw with ammonia to improve its quality as an animal feed. • Restrictions on agricultural expansion near sensitive ecosystems.

2004: Action Plan III for the Aquatic Environment (AP III) Targets: • 13% reduction of N leaching in 2015 compared to 2003 • P surplus in Danish agriculture to be halved by 2015	• Increased mandatory area with catch crops. • Further tightening of the statutory norms for utilization of manure N based on research results. • Establishment of further wetland areas. • Establishment of buffer zones along streams and around lakes before 2015 to reduce discharge of P. • Improved utilization of N and P in feed to reduce losses of N and agricultural surplus of P. • A tax of 4 Danish Kroner (DKR) per kilo mineral P in feed. • Protection zones of 300 m around ammonia-sensitive habitats such as raised bogs, lobelia lakes, and heaths larger than 10 ha. • Strengthening of organic farming. • Based on evaluations in 2008 and 2011 further initiatives will be considered.

* Measures in other sectors are not included in the table.

Source: Modified from Dalgaard et al., 2004.

Nitrogen export includes N that is incorporated in animal products in the form of eggs, milk, meat, live animals, or livestock received by offal destruction plants, and vegetable products in the form of cereals, seeds for manufacturing and sowing, beets for sugar production, potatoes, and other fruit and vegetable products.

N surplus indicates the potential N losses from farming to the environment. It covers several components. The largest N loss component is leaching of nitrates. N leaching is of special importance in relation to groundwater and surface water pollution. Other main N losses are ammonia emissions and atmospheric N emissions due to denitrification. Some of the N surplus may temporarily accumulate in the soil, but over time, as the system approaches steady state, N surplus and N loss will converge.

N efficiency is the proportion of nitrogen taken up in agricultural products in relation to the amount applied and is defined at the national level as N export per N import.

N leaching is the loss of N from soils and is modeled for the country as a whole based on statistical data and various modeling concepts (Grant and Waagepetersen 2003).

Developments in N imports and N exports for the last century are shown in Figure 8.1. The gap between N imports and N exports corresponds to the N surplus. With few exceptions during the two World Wars (1914–18, 1940–45), and the oil crisis (1972–74), the N surplus generally increased from 1900 until the mid-1980s, when the action plans toward agricultural N losses were politically initiated.

To follow the effects of these action plans, we assess developments in N surplus, N efficiency, and N leaching for the period 1979 to 2004 (Figure 8.2). Both N surplus and N leaching were reduced significantly over the period, whereas N efficiency was raised. N surplus decreased from about 490,000 tonnes N in 1985 to about 290,000 t N in 2004. N efficiency increased from 25% to 38% in the same

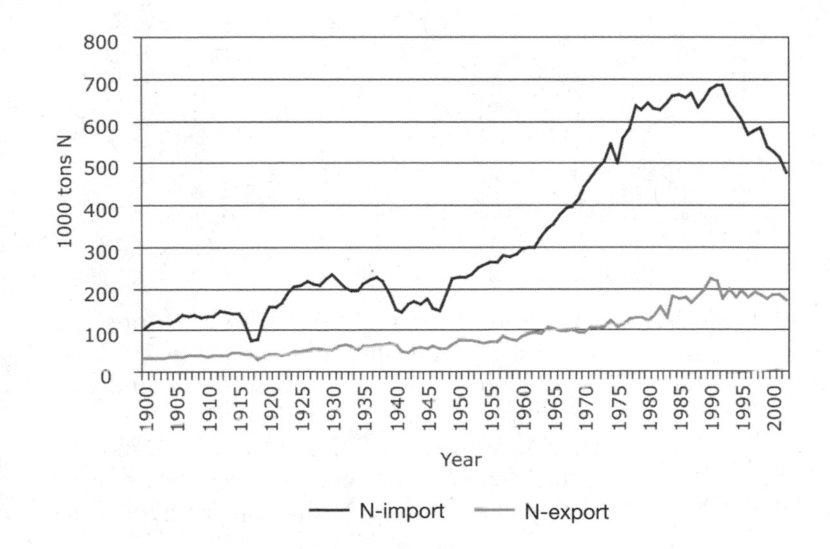

Figure 8.1. Developments in N imports to and N exports from Danish agriculture 1900–2002. *Source:* Modified from Dalgaard and Kylllingsbæk 2004.

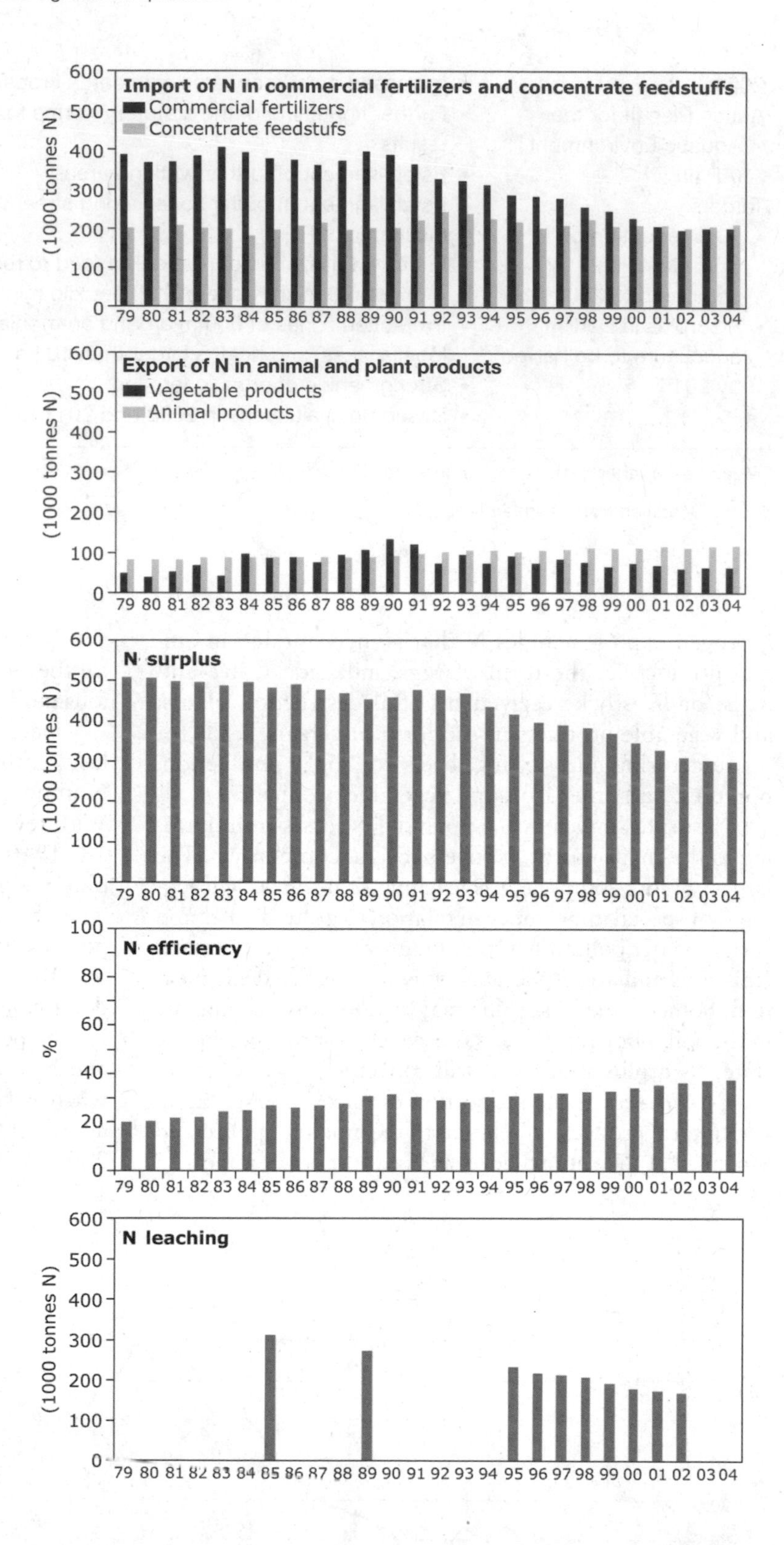

Figure 8.2. N import, N export, N surplus, N leaching, and N efficiency for Denmark 1979 to 2004.

period, whereas N leaching was reduced by 46% from about 311,000 t N in 1985 to about 168,000 t N in 2002.

Overall agricultural production has been relatively stable over the same period. From 1985 to 2004, the N in animal exports increased by 33%, whereas N in total plant production was reduced. However, in the same period the total agricultural area was also reduced by about 7% from 2.85 million ha to 2.65 million ha. Moreover, set-aside of land was introduced by the EU in 1992, so the utilized agricultural area without set-aside land was lower in the years after 1992.

The main instruments of N regulation in Denmark are mandatory fertilizer and crop rotation plans, with limits on the plant-available N that can be applied to different crops, and statutory norms for the utilization of nitrogen in animal manure. The norms reflect how much nitrogen in the manure is assumed to be plant available. This also sets a limit on how much mineral fertilizer each farmer can apply. Information on the amount of purchased N in mineral fertilizer is reported to the Ministry of Food every year. The application of N from animal manure and mineral fertilizer cannot exceed the total N-norm for that farm.

These two instruments have been strengthened several times, for example, with the 1991, 1998, and 2000 restrictions of the norms for the utilization of manure N (Table 8.2). In addition, improved feeding has had a significant effect on the efficiency of N utilization in animal feeds. Throughout the period, N regulations have been designed in close dialogue with researchers and farmers' associations. They have also been followed up with information campaigns, extension service, and education. Also, extensive strategic research programs have been supported. A major achievement of this bottom-up approach of continuous dialogue has been the ability to design N regulations in a way that minimizes negative impacts on crop and animal production.

With the Action Plan III for the Aquatic Environment (AP III), national Danish regulation of agriculture includes nitrogen, phosphorus, and odors. The timing of AP III coincides with the implementation of the Water Framework Directive, which will supplement national regulation with additional measures at water basin level if necessary to achieve good ecological quality.

Effect on the Aquatic Environment

In the first Action Plan on the Aquatic Environment (1987) it was decided to establish a coordinated national monitoring program (Miljøstyrelsen 1989) to document the implementation of the different measures in different sectors and the ecological consequences in the aquatic environment. The program was designed to cover groundwater, rivers, lakes, coastal and open marine waters, atmospheric deposition, point sources, and small agricultural watersheds. In a driving force–pressure–state–impact–response (DPSIR) context (Figure 8.3) the program covers agricultural driving forces or causes (D), pressures (in this case pollutants P), state or quality of environment (S), and impact on the health of the ecosystem (I) (EEA 1999).

The monitoring program has now run for about 15 years and costs approximately €30 million per year (2004 prices). By documenting how the measures agreed upon in the late 1980s were inadequate, it has significantly influenced the regulation of Danish agriculture (Grant et al., 1995). Our presentation will mainly be based on the results of this program.

In five agricultural watersheds, ranging in size from 5 to 15 km^2, annual information is collected on agricultural practices (Drivers) at field and farm level. Leaching of N (pressure) is frequently monitored, along with groundwater and river water N concentrations (state) (Grant et al., 1995).

The results show that nitrogen concentration in water leaving the root zone has decreased significantly since 1990, on clay as well as sandy soils (Figure 8.4). On sandy soils this has been accompanied by a decrease in N concentrations in upper groundwater. On clay soils no decrease has been detected yet, most likely because the major runoff on clay soils is surface and subsurface, so groundwater formation is small. More than 99% of the Danish drinking water is based on groundwater. On a national basis there are still significant problems with high

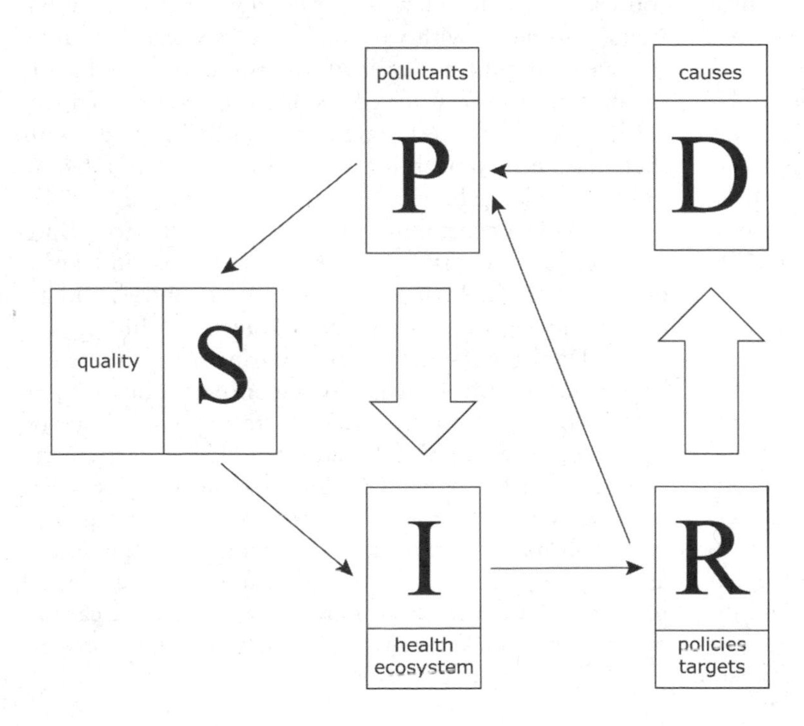

Figure 8.3. The DPSIR concept.

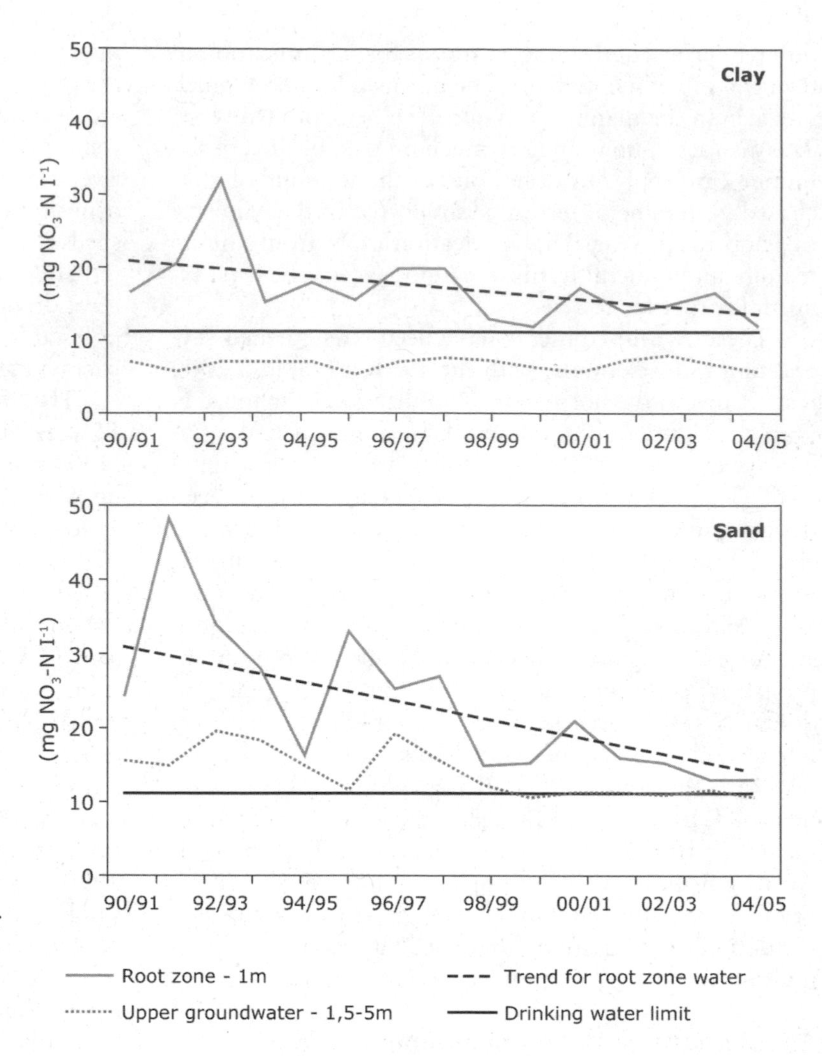

Figure 8.4. The development 1990/91–2002/03 of measured N concentration in root zone water and upper groundwater in two clayey and three sandy Danish watersheds.
Source: Grant et al., 2004.

nitrate concentrations in the groundwater. The age of this groundwater, however, is more than 20 years, and so no impact of the agricultural regulations can be seen yet.

Figure 8.5 shows the distribution of modeled N leaching by municipality in 1985 and 2002. The two patterns reflect the distribution of livestock in Denmark and show the widespread effect of the regulation. It is worth noticing that the N leaching in some water basins is still high, and further regional reduction may be needed under the Water Framework Directive to achieve good ecological quality in the respective coastal waters.

The concentration of N in rivers has decreased by 29 to 33% over the last 15 years, in watersheds with point sources such as urban wastewater plants, as well as in watersheds with agriculture and no point sources (Figure 8.6). In 50 out of 63 rivers in watersheds with agriculture and without point sources the decrease is statistically significant (Bøgestrand 2004).

In Danish coastal and open marine areas, nitrogen is the main limiting nutrient; hence additional N can stimulate growth of algae. From Figure 8.7 it can be seen that N transport to marine areas is dominated by diffuse sources (agriculture) and varies significantly between years, correlated with varying levels of water discharge. It is clear that diffuse agricultural pollution is by far the major contributor. After correcting for variations in water discharge, it is estimated that since the mid-1990s there has been an overall average decrease of about 43% (95% confidence limits: 33 to 61%) (Bøgestrand 2004).

In Danish coastal marine areas N inputs from Danish rivers are the major source of N, whereas in Danish open marine waters inputs from other countries and atmospheric deposition dominate (Bøgestrand 2004).

In Danish coastal and open marine waters there has been a significant decrease in N concentrations (Figure 8.8). In the open waters N concentration is much lower, but a decrease can also be detected. The ecosystem response to these changed N concentrations is less clear because there are significant year-to-year variations, and other factors such as insolation, grazing by zooplankton, and filtering mussels may be significant. When corrected for year-to-year variations in freshwater inputs, clear improvements can be seen in algal density and transparency (Ærtebjerg et al., 2004).

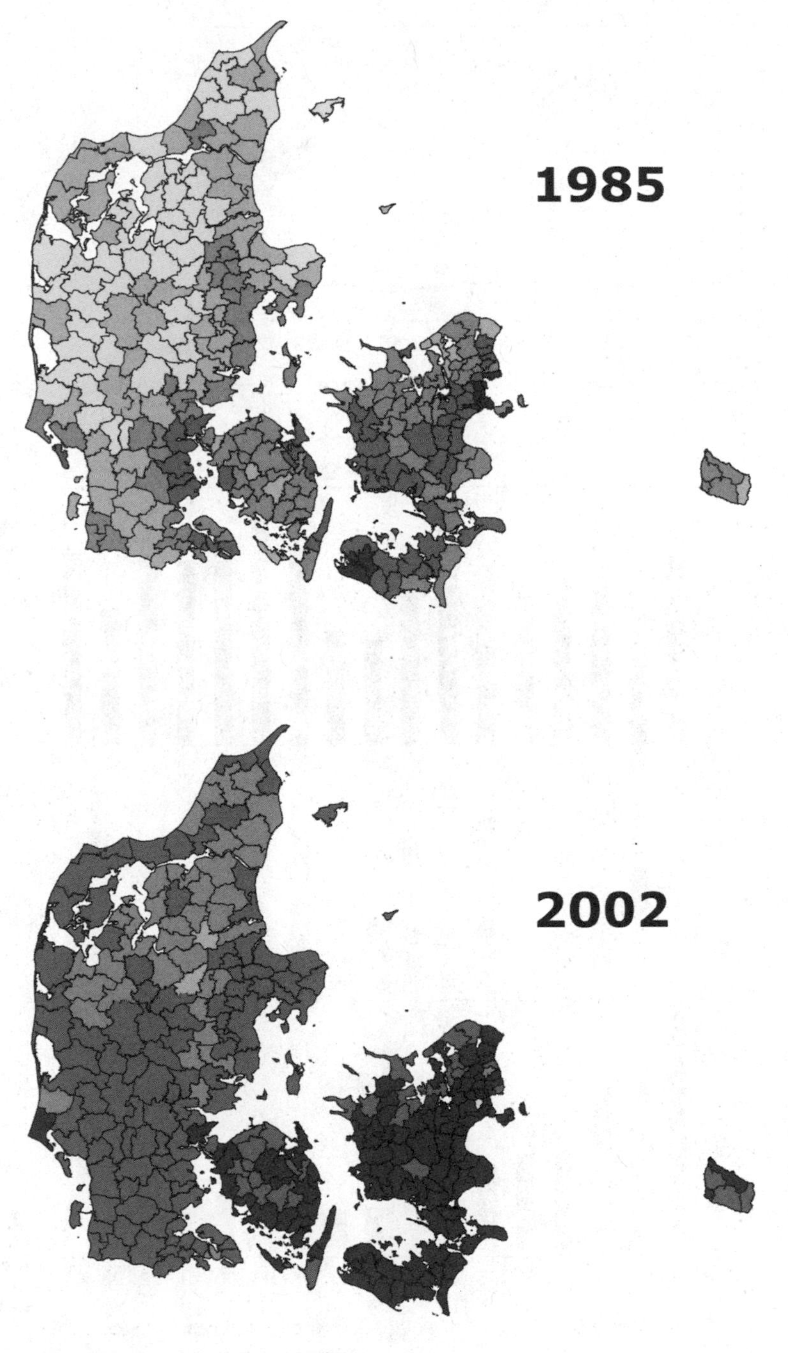

Figure 8.5. Geographic distribution of the modeled N leaching by municipalities 1985 and 2002.
Source: Dalgaard et al., 2004.

However, this has not yet translated into a measurable improvement in oxygen deficiency in Danish waters; indeed the worst case of oxygen deficiency ever seen occurred in 2002. Danish coastal waters are generally more sensitive to eutrophication than most other marine areas worldwide because the exchange of water with the open sea is generally small, and because stratification due to a combined thermo- and halocline limits the transport of oxygen to the bottom water (Ærtebjerg et al., 2004). The reason for this lack of response is not yet clearly understood.

Overall it can be concluded from the results of the monitoring program that the measures taken in agriculture have resulted in clear reductions in nutrient

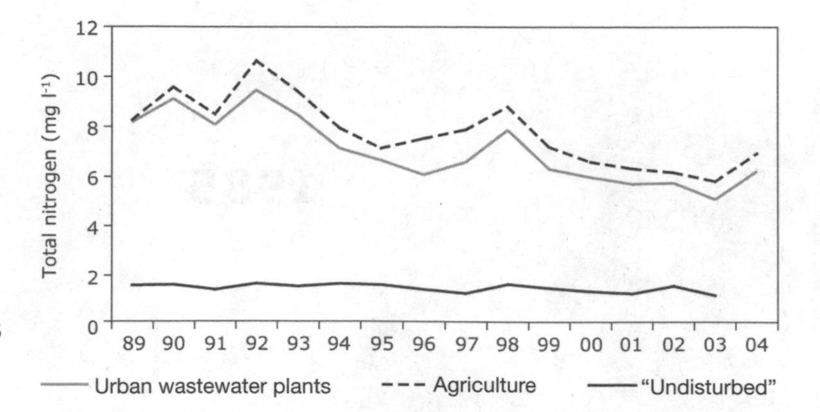

Figure 8.6. Nitrogen concentrations 1989–2003 in Danish rivers affected by different pressures. *Source:* Bøgestrand 2004.

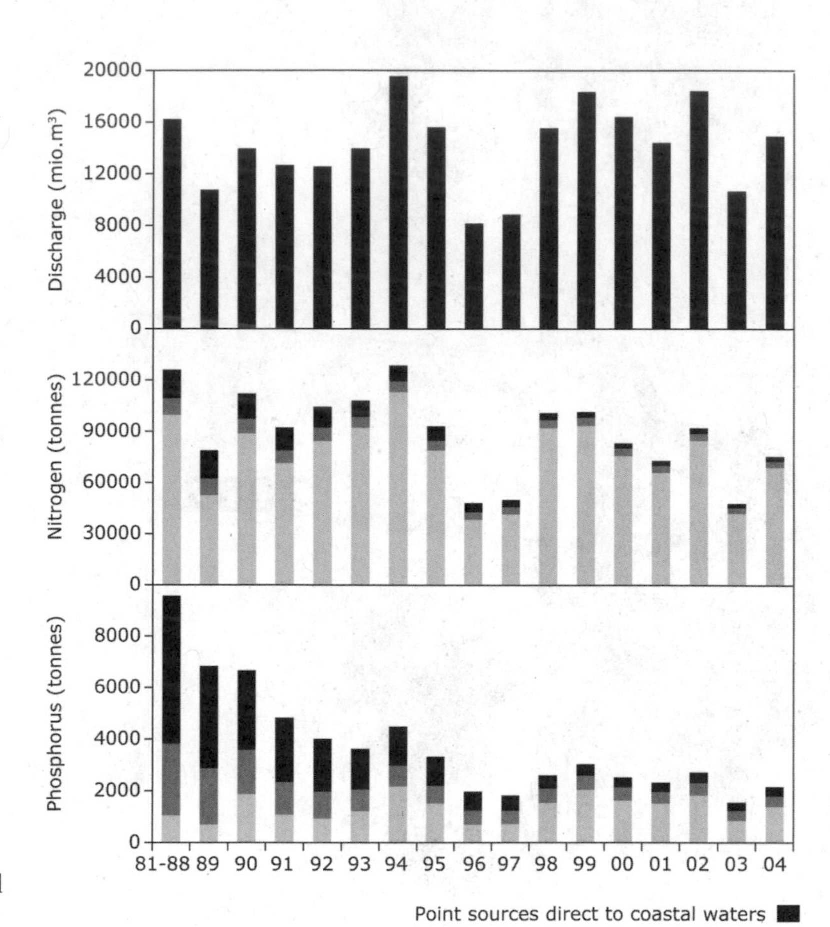

Figure 8.7. Discharges of water and nutrients from different Danish sources to Danish coastal areas, 1981–88 (mean) and from the period 1989 to 2003.
Source: Bøgestrand 2004.

concentrations. However, the biological response is still difficult to detect due to large year to-year variations and due to ecosystem complexity.

The Costs of Action Plans

The annual costs of all the action plans so far (including agricultural and nonagricultural measures) are around €600 million (2005 prices). Of these the annual cost related to agricultural measures is about €340 million, and the rest is related to industry and sewage treatment plants. The estimated costs related to the different action plans are described in Table 8.3.

The payment of the costs in AP II have been evenly divided between the agricultural sector and public funding. The administrative costs are not included. There seems to be a tendency for the agricultural sector to pay for farm-related measures (changed production), whereas the state pays for land taken out of production through measures that are cofinanced with the EU. In terms of farm types, the majority of costs, which relate

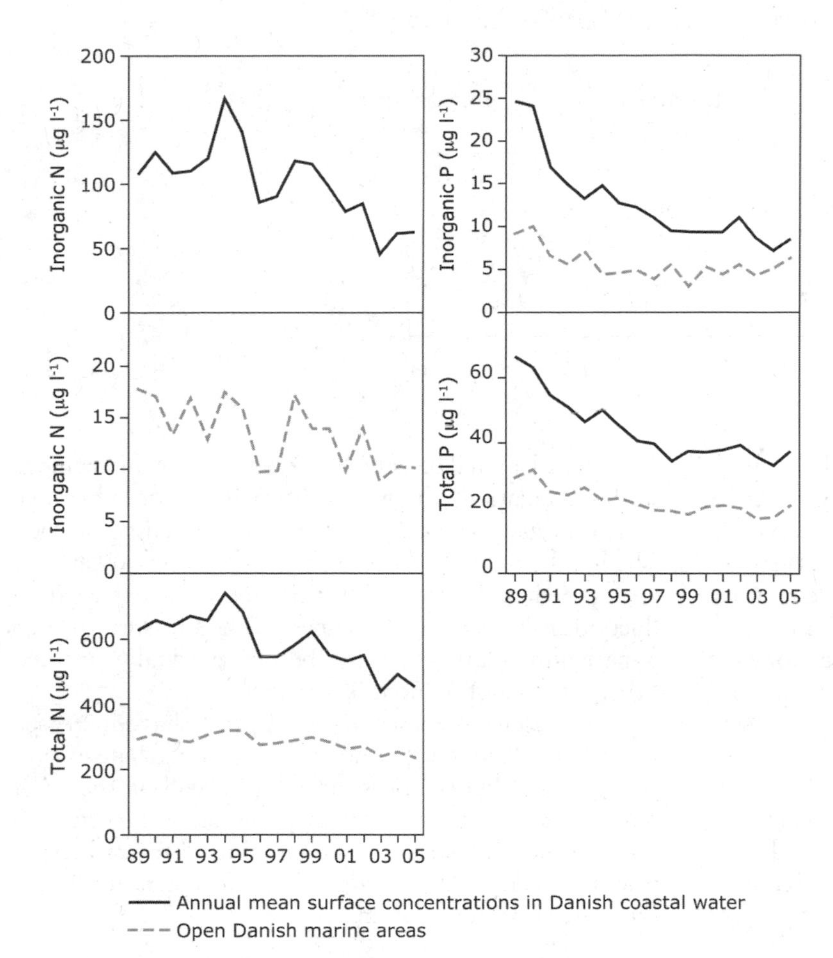

Figure 8.8. Annual mean surface concentrations in Danish coastal water and open Danish marine areas for inorganic and total nitrogen and inorganic and total phosphorus.
Source: Ærtebjerg et al., 2004

to measures like changes in feeding, lower livestock density, and increased utilization of N, are borne by animal farms, whereas other costs (lower N quota and catch crops) are more evenly divided among farms.

In general, measures were chosen partly on their cost-effectiveness, partly on other benefits that politicians wanted to promote. A significant reason for the success of the Danish policies is that, when designing the policies, efforts have been taken to reduce the costs to farmers. In Action Plan I increased slurry storage capacity was promoted with a state subsidy scheme in order to reduce the investment cost to farmers. The definitions of winter green crops were also changed, allowing winter wheat to be categorized as winter green crops. This lowered the cost (and the environmental effect) of this measure significantly because winter wheat has for some years been more profitable than spring crops. The large increase in manure utilization has been less costly than anticipated and has led to lower use of mineral fertilizer.

The requirements for storage capacity helped to increase the production of biogas based on centralized biogas plants owned by several farmers (Christensen 1999). The use of biogas for heating has lowered storage costs, decreased smells, and increased utilization of slurry. The 20 to 25 centralized biogas plants also ease the redistribution of animal manure to arable farms. Furthermore, they facilitate waste recycling from other sectors (e.g., animal fat). Despite this, no new plants have been built in recent years—biogas plants have become less viable than before because of lower subsidies, and it has been difficult to find suitable locations for the biogas plants.

As a final example of how the measures have reduced some costs for farmers, in implementing of fertilizer plans, great effort is made to ensure that the N quota is related to the crop rotation, soil type, and region of the individual farm. This has significantly lowered the costs of implementation.

Table 8.4 shows the costs and cost-effectiveness of various measures in AP II. The total costs are costs paid by the public (top four measures) or by the farmer (last five measures in the table). The assessment includes not only an analysis prior to implementation but also a mid-term and a final (ex-post) evaluation of the direct costs, which few other countries have carried out (Jacobsen 2004). The conclusion is that the total direct costs for farmers plus the state-related costs to AP II have been lower than were expected in 1998 at the beginning of the plan. The main reasons are that the implementation of new technologies has been cheaper than expected, and

Table 8.3. Estimated costs of agricultural measures in different action plans to reduce nitrogen leaching from agriculture (2005 prices)

	Ex-ante costs (mill. €/year)	Ex-post costs (mill. €/year)
Action plan for aquatic environment I—AP I (1987)	84	Not calculated
Action plan for a more sustainable agriculture (1991)	134	Not calculated
Action Plan II (1998–2003)	92	70
Action Plan III (2004–2015)	30	Not ready
Total	340	

Source: Jacobsen et al., 2004 and Jacobsen 2004.

the area involved has been lower than expected. The total annual cost for AP II was estimated at €92 million at the outset and is now estimated at €70 million, or €2 per kg of reduced nitrogen leaching (2004 prices). The costs are difficult to compare with results from other countries; for example, the Netherlands have found it difficult to analyze the costs and the environmental change for each measure, but some overall cost estimates for the Mineral Accounting System (MINAS) are available (Berentsen, personal communication, RIVM 2004).

Overall, the most cost-effective measures in AP II have been the requirements for catch crops and wetlands, increased utilization of animal manure, and improved feeding practices. The least cost-effective have been set-aside and increased areas with grass, as well as the requirement for reduced animal density. Other benefits are not included in the calculation, and this is the main reason why area-related measures generally have the lowest cost efficiency. The uptake of area-related measures has been lower than anticipated, and they have not produced the expected reduction in N leaching. As expected the measures related to agriculture have, over time, become gradually more expensive per kg of reduced N leaching.

The recent increases in cereal prices have increased the costs of reducing N-norms and taking land out of production. Higher prices on N and P will on the other hand increase the value of animal manure as farmers will try to reduce the purchase of mineral fertilizers. Recently, prices on cereals have dropped again, indicating larger fluctuations in future farm prices.

Table 8.4. Cost effectiveness for different measures in Action Plan II

	Reduction in N Leaching (tonnes N/year)	Total Cost (million €/year)	Cost-effectiveness (€/kg N)
Wetlands	800	0.7[3]	0.9
ESA-areas	700	7.7	10.9
Forestry	800	4.7[3]	5.9
Organic farming	3,700	14.0	3.8
Changed feeding	3,800	5.7	1.5
Lower livestock density[1]	140	1.5	10.4
Catch crops (6%)	3,000	6.4	2.1
Increased utilization of N in animal manure (15%)[1]	10,110	6.7	0.7
Reduced N quota (10%)[1]	12,850	22.8	1.8
Sum[2]	35,900	70.2	2.0

[1] In the technical evaluation of Action Plan II the effect of these measures was not estimated individually, hence the figures given here are approximate estimates.

[2] Changes in use of area and animal production as well as other matters are not included in the table.

[3] Annuity based on 4% interest and infinite lifetime.

Source: Jacobsen 2004.

Conclusions

In intensive Danish agriculture it has proven possible to reduce N leaching by almost 50% while maintaining crop yields and significantly increasing livestock production. This has been achieved by a strong focus on improving nitrogen efficiency facilitated by regulatory measures, intensive research efforts, and an innovative farming community.

Has Denmark performed well on a European, an OECD, or a global scale? We believe so. In the EU, Denmark in the late 1990s was unique in fulfilling the requirements of both the EU Urban Waste Water Directive and the EU Nitrates Directive. According to the OECD study by de Clercq et al. (OECD 2001) Denmark is one of the most successful EU countries in reducing N surpluses and N losses. These conclusions were based on data from 1997. At the time of writing (2006), we could find no more recent official benchmarking of the performance of different countries on N surpluses and N losses. There is a significant need for some European institution to publish regularly updated authoritative data on N surpluses and N losses.

With hindsight, it can be seen that some targets were not met within the planned time frame. The scientific foundation of the first action plan of 1987 was insufficient, and the target to reduce N leaching by almost 50% within five years proved unrealistic. Actually it took several action plans and 16 years to reach the 1987 target (Figure 8.9).

The lesson learned has been that environmental policies need to be developed in a dynamic way. Policies and regulations need to be continually readjusted in the light of experience, with new measures or tightening of existing measures. As part of this process continuous evaluation of progress is essential. The authorities should monitor developments in environmental impact or in the amount of emissions, so that failures in the implementation or wrong expectations about the environmental effect of certain instruments and policies can be observed and corrected.

The agricultural system and the environment are influenced by decisions taken by thousands of individual farmers who have to change their farming practice. In general, the training level and managerial capacity of Danish farmers are high. Currently a young farmer will go through a four-year training program before graduating as a trained farmer. The agricultural advisory system is well established and organized, with branches in every region of the country. It is run by the farmers' union, and local departments of the advisory system are guided by steering committees chaired by farmers. More than 80% of Danish farmers are supported by advisers from this system, and an even higher share of land and production is under this supervision.

Despite this level of competence, farmers do not always react in the way that the plans assume, nor do they necessarily have full awareness of the environmental problems that farming may cause for the rest of society. For example, in the first action plan it was expected that green cover of fields during winter would contribute substantially to reduce nitrate leaching. This was based on the assumption that farmers would meet requirements by cultivating grass catch crops after spring-sown barley and wheat. However, the farmers instead chose a strategy of shifting from crops planted in the spring—primarily barley—to winter-sown crops. Consequently the expected environmental effect was not realized, although the legal requirements were more than fulfilled.

In addition, when the action plans were initially introduced, the problem of nitrate losses and pollution was not generally recognized by the majority of farmers. For example, according to AP I farmers had to establish nine months' storage capacity for manure, so that manure could be spread on the fields in spring instead of autumn. However, the monitoring program showed that it took some years to build this storage capacity—and some additional years before the majority of the manure was spread in the spring.

Farmers and their organizations accepted the existence of environmental problems gradually. They also accepted new policies, although their opinion quite frequently is that measures should be based on voluntary approaches and that the time horizon should be longer. Given the high level of training and the comprehensive advisory system, there is no doubt that Danish farmers are capable of adopting new rules and technologies within a very few years.

Since 1987 the national monitoring program and several research programs have significantly increased

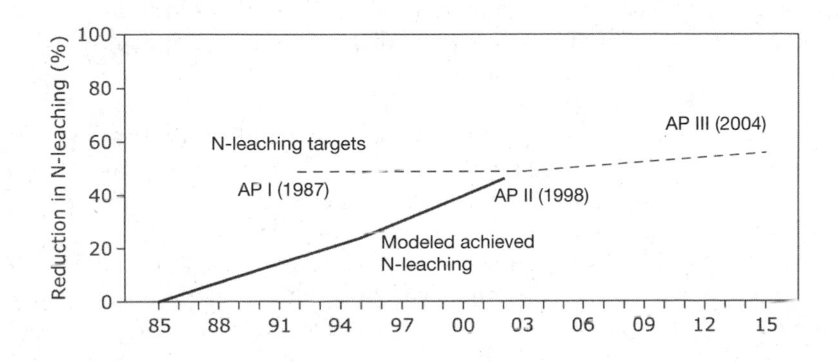

Figure 8.9. Comparison of N reduction targets in Danish Action Plans 1987, 1998, and 2004 with modeled actual N leaching from Danish agriculture.
Source: Data on leaching from Grant and Waagepetersen 2003.

our understanding of how agriculture interacts with the environment. This knowledge has been used in the preparation of AP II as well as in the midterm and final evaluations of AP II. However, assessing the environmental impact of changes in agriculture is not always easy. It is complicated by the fact that year-to-year variations in climate significantly influence the outcome. For example, it was extremely hard to explain to the public that Danish agriculture in 2002 was on the proper track and would fulfill its obligations in AP II, at the same time as the worst case of oxygen deficiency in Danish coastal waters ever took place.

It has been important that the legislation helped to promote the implementation of new technology and practices. Implementation is enhanced if new technologies and techniques offer gains to both farmer and environment. The large increase in the utilization of animal manure is a good example. However, even in situations where there is a common interest between economy and environment—for instance where resources are used more efficiently—it can sometimes be difficult to convince farmers to change current practices.

Finally, therefore, it is important to highlight the decisive importance of enforcement. Enforcement of environmental regulations is decisive in order to change behavior and to reduce environmental impact. Enforcement is, however, also about creating awareness and providing information about environmentally friendly technologies and practices. Looking at other countries with intensive livestock farming and with environmental policies in place, it appears that the Danish experiences with enforcement are among the best in the world. Efficient enforcement of environmental regulations ensures that things change at the farm level, so that major results can be achieved in a short time span. In contrast, environmental policy and regulation without enforcement are not normally productive.

In Denmark, enforcement of the environmental regulation of farms has been based on two approaches. The municipalities carry out physical inspection at the farm site. Their focus is on quality of buildings and storage facilities, and size of the herd in relation to the level allowed at the specific site. The important focus is on progress from one inspection to the next, rather than on sanctioning minor violations.

The second tier of enforcement is the responsibility of the Ministry of Food and Agriculture. All farmers are required to submit fertilizer accounts in which they report cropping patterns, cultivated area, use of commercial fertilizers, use of livestock manure, and livestock numbers. Each farmer must submit an annual report. Excess fertilization of nitrogen is punished with a fine of up to €3 per kg.

The "self-control" included in the Danish fertilizer system has ensured that farmers selling and buying animal manure have opposite interests, which ensures market pricing. This ensures that reduced application would lead to lower income on a given farm. In the Netherlands the MINAS system allowed farmers such high application levels that they could "sell" some of the quota without it reducing their income (Jacobsen et al., 2005, OECD 2007). This led to fictive transport of manure in the fertilizer accounts from one part to another part of the country. In these cases the buyer and the seller both had an interest in hiding the fact that the transportation was fictive because the buyer received money from the seller and the seller avoided paying the levy. Another effect in the Netherlands was the need for independent weighing facilities to determine the amount and content of exported animal manure. This was not required in Denmark.

The regulation of nitrogen in Action Plan III follows the same lines as previous action plans. In addition a ceiling for phosphorus surplus is included in the action plan, and phosphorus is regulated by means of a tax on mineral phosphorus in feed as a new instrument.

Future agroenvironmental initiatives will be based on a more holistic approach, integrating protection of the aquatic and terrestrial environment and natural habitats and linking national regulations to EC-directives and other international obligations. No doubt there will be a change from a national approach toward a water basin approach in order to meet environmental objectives for individual water bodies and natural habitats. For instance, wetlands will reduce N and P losses, and at the same time improve nature and reduce CO_2 emissions.

In conclusion, the Danish experience with regulation of nitrogen losses to the aquatic environment can be summarized as follows:

- The measures applied have had a significant effect in improving N utilization and reducing N surplus and N leaching.
- The measures applied in the agricultural production system have focused on improved utilization of animal manure and feedstuffs, fertilizer and crop rotation plans, and limitations on total N application.
- Regulation of the agricultural production system is complicated, involving the behavior of many individual farms and also involving a large effort in public control of plans and procedures.
- The knowledge-based approach, with intensive research programs and dialogue with the agricultural sector, has been successful.
- There is a considerable effect on the environment, but with several years' delay. This delay is due to the time needed for farmers to change their behavior, and to the time-delay before effects are visible in the agroecosystem and the water cycle.
- N balances are good indicators for agricultural systems. N balances can be applied at national, regional, and local scales and at the farm level.

- N balances should be the basis for further distribution of N losses on various components, including N leaching.
- The costs of new technology have been lower than expected. The average direct cost was about €2 per kg of reduced N leaching.

References

Ærtebjerg, G., et al. 2004. Marine områder 2003: Miljøtilstand og udvikling. NOVA 2003. Faglig rapport fra DMU nr. 513. Danmarks Miljøundersøgelser. Available at http://www2.dmu.dk/1_viden/2_publikationer/3_fagrapporter/rapporter/FR513.pdf.

Bøgestrand, J., ed. 2004. Vandløb 2003. NOVA 2003. Faglig rapport fra DMU nr. 516. Danmarks Miljøundersøgelser. Available at http://www2.dmu.dk/1_viden/2_publikationer/3_fagrapporter/rapporter/FR516.pdf.

Christensen, J. 1999. *Centralized Biogas Plants: Integrated Energy Production*. Copenhagen: Danish Institute of Agricultural and Fisheries Economics.

Dalgaard, T., C. D. Børgesen, J. F. Hansen, N. J. Hutchings, U. Jørgensen, and A. Kyllingsbæk. 2004. How to halve N-losses, improve N-efficiencies and maintain yields? The Danish Case. Paper for the 3rd International Nitrogen Conference, Nanjing, China, 12–16 October 2004. Cited 16 March 2009. Available at: http://www.lr.dk/planteavl/informationsserier/info-planter/plk06_96_1_t_dalgaard.pdf.

EEA. 1999. Environmental indicators: typology and overview. European Environment Agency Technical report No 25. Available at http://reports.eea.europa.eu/TEC25/en.

EEA. 2003. *Europe's Environment: The Third Assessment*. Copenhagen: European Environment Agency.

Grant, R. and J. Waagepetersen. 2003. *The Second Action Plan for the Aquatic Environment: Final Evaluation (in Danish)*. Foulum: Danish Institute of Agricultural Sciences.

Grant, R., G. Blicher-Mathiesen, H. E. Andersen, P. Berg, P. G. Jensen, and A. R. Laubel. 1995. *Landovervågningsoplande: Vandmiljøplanens Overvågningsprogram 1994*. Faglig rapport fra DMU nr. 141. Silkeborg: Danmarks Miljøundersøgelser.

Grant, R., G. Blicher-Mathiesen, P. G. Jensen, B. Clausen, M. L. Pedersen, and P. Rasmussen. 2004. *Landovervågningsoplande 2003. NOVA 2003*. Danmarks Miljøundersøgelser. Faglig rapport fra DMU nr. 514. Available at http://www2.dmu.dk/1_viden/2_publikationer/3_fagrapporter/rapporter/FR514.pdf.

Jacobsen, B. H. 2004. *Final Economic Evaluation of the Action Plan for the Aquatic Environment II*. Report no. 169. Copenhagen: Danish Research Institute of Food Economics.

Jacobsen, B. H., J. Abildtrup, M. Andersen, T. Christensen, B. Hasler, Z. B. Hussain, H. Huusom, J. D. Jensen, J. S. Schou, and J. E. Ørum. 2004. *Costs of Reducing Nutrient Losses from Agriculture: Work Related to Action Plan III*. Report no. 167. Copenhagen: Danish Research Institute of Food Economics.

Jacobsen, B. H., J. Abildtrup, and J. E. Ørum. 2005. Reducing nutrient losses in Europe—in light of the Water Framework Directive. Paper for the 15th IFMA Congress, Campinas, Brazil, 14–19 August 2005.

Kyllingsbæk, A. 2005. *Nutrient Balances and Nutrient Surpluses in Danish Agriculture 1979–2002*. DJF Report no. 116 (in Danish with English summary). Foulum: Danish Institute of Agricultural Sciences.

Miljøstyrelsen. 1989. *Vandmiljøplanens overvågningsprogram*. Miljøprojekt nr. 115. Copenhagen: Miljømimisteriet.

OECD. 2001. OECD nitrogen balance database. Available at http://www.oecd.org/document/29/0,3343,en_2649_33793_1890205_1_1_1_1,00.html.

OECD. 2007. *Instrument Mixes Addressing Non-point Sources of Water Pollution*. Paris: OECD.

RIVM. 2004. *Mineralen beter geregeld: Evaluatie van de werking van de Meststoffenwet 1998–2003*. Bilthoven, Netherlands: RIVM.

Statistics Denmark. 2008. *Agricultural Statistics*. Copenhagen: Statistics Denmark.

9

Nestlé

Responses of the Food Industry*

Manfred Noll and Hans Jöhr

Abstract

For Nestlé, milk represents by far the most important raw material of animal origin, with an annual spending of over $2 billion to buy fresh milk from farmers and cooperatives.

This chapter describes Nestlé's milk procurement and the objectives and achievements related to the environmental performance of Nestlé factories. We present different stakeholders interested or involved in livestock production, their roles and objectives, and current drivers in the livestock sector and their impacts.

We examine the responses of the food industry in general and Nestlé in particular to these ongoing developments. Various tools used by the food industry in their backward integration with farming allow for improvements in quality control and food safety, such as farm assurance schemes, supplier selection and development, traceability systems, and supplier education. These tools and others are also beginning to be used in relation to environmental goals such as soil and water conservation, reduction of water pollution, reforestation, energy conservation, and reduction of greenhouse gas emissions.

We describe Nestlé's responses in more detail through case examples from Nestlé's milk procurement operations worldwide. The examples illustrate key objectives, including provision of stable markets for agricultural products, ensuring food safety, quality and regulatory compliance, control of animal diseases, environmental sustainability of livestock production, and productivity improvement.

Possible gaps in current responses regarding environmental sustainability of livestock farming are addressed. An ongoing initiative is the SAI Platform (Sustainable Agriculture Initiative Platform), an association of the food industry to address the economic, environmental, and social aspects of agricultural production.

* This chapter presents the view of the authors but not necessarily those of the editors. The authors are working at the Agricultural Service Department in Nestlé's International Head Office in Switzerland.

Introduction

Founded in 1866 in Switzerland, Nestlé grew from a small company producing powdered milk products for infants into a global food company with a large portfolio of products. Over the decades, Nestlé has diversified into dairy, coffee, culinary, confectionery, bottled water, ice cream, and pet food. Some of its strongest brands include Nescafé, Nespresso, Milo, Nesquick, Nido, Nespray, Bear Brand, Nan, Lactogen, Maggi, Purina, and Friskies. By 2006, Nestlé had annual sales of $73 billion, with 487 factories in 83 countries and 253,000 employees worldwide (Nestlé 2007a).

Nutrition, health, and wellness are at the core of Nestlé's company policy; Henri Nestlé's infant food was created to combat infant mortality due to malnutrition (Heer 1966).

During the decades after its foundation, as Nestlé's sales expanded to many countries outside of Switzerland, the increasing fragmentation of markets and the growth of demand required local milk sourcing and processing to ensure availability and competitive costs of Nestlé products. By 1905 when it merged with the Anglo-Swiss Condensed Milk Company, Nestlé already had operating factories and milk districts in six countries: Switzerland, the United Kingdom, Germany, Norway, Spain, and the United States. As early as 1921, Nestlé began setting up dairy factories and milk collection from farmers in Brazil, followed by Jamaica, Panama, Mexico, Venezuela, Peru, and Colombia (Heer 1966). The first milk collection from farmers in Asia began in 1961 for Moga Factory in India, followed by factories in Sri Lanka, Indonesia, Pakistan, China, and Uzbekistan. All these Asian factories are collecting milk from tens of thousands of small-scale farmers (Goldberg and Herman 2005).

Often, competitive conditions necessitated setting up milk collection centers and factories in more remote rural regions, where low-cost dairy production systems relying on grassland or crop residue for fodder made the district economically and socially viable. This meant Nestlé frequently worked in areas where conditions were less developed, and where prior to Nestlé's arrival the local population had limited access to markets for their agricultural products. When a milk district was built, these constituents reaped benefits far beyond the selling and production of milk, including infrastructure, banking, training, and financing support to farmers (Goldberg and Herman 2005). In some countries, Nestlé is buying fresh milk directly from dairy farmers, whereas in other countries (e.g., Thailand, Indonesia, Morocco, and Dominican Republic) Nestlé is sourcing from farmer cooperatives active in milk procurement and sales.

The sustainable use of natural resources is a core element of Nestlé Corporate Business Principles (Nestlé 2004), which state the following policy related to protection of the environment:

> Since its early days Nestlé has been committed to environmentally sound business practices throughout the world and continues to make substantial environmental investments. In this way Nestlé contributes to sustainable development by meeting the needs of the present without compromising the ability of future generations to meet their needs. The Nestlé Policy on the Environment underlines this need. . . . Nestlé therefore:
>
> Supports a precautionary approach to environmental challenges;
>
> Undertakes initiatives to promote greater environmental responsibility;
>
> Encourages the development and diffusion of environmentally friendly technologies.

Nestlé's environmental performance is driven by the Nestlé Environmental Management System (NEMS), which conforms to the International Organization for Standardization's ISO 14001 standards and is mandatory for all operating companies and plants (Nestlé 2000, 2006). ISO 14001 sets out the requirements for environmental management systems, for organizations wishing to operate in an environmentally sustainable manner.

Focus on its environmental impact has allowed Nestlé to boost industrial productivity while at the same time considerably reducing the consumption of natural resources within its own operations. Applying the principles of its environmental policy, Nestlé has reduced its water consumption by 34.6% per tonne of manufactured product from 2002 to 2006. In the same period, Nestlé has reduced its energy consumption by 28.4% and its greenhouse gas emissions by 31.9% per tonne of product. In addition, Nestlé invests an average of US$24 million per year to expand wastewater treatment facilities in its factories and $32 million to reduce air emissions (Nestlé 2007b).

Nestlé extends its environmental policy not only to its own operations but also to its supply chain of raw materials. The Nestlé Corporate Business Principles state that Nestlé aims to contribute to sustainable development by ensuring environmental performance throughout the supply chain, supporting sustainable farming practices and sustainable usage of water in agriculture, as well as ensuring that key suppliers comply with Nestlé Corporate Business Principle (Nestlé 2004).

Drivers and Consequences

Roles and Objectives of Stakeholders

There are a large number of different stakeholders involved or interested in livestock production and in the food chain of livestock products. These stakeholders include consumers, varied types of producers (small- and larger-scale farmers, industrial livestock producers), the farm input industry, traders of livestock and livestock products, food industry and retailers, various

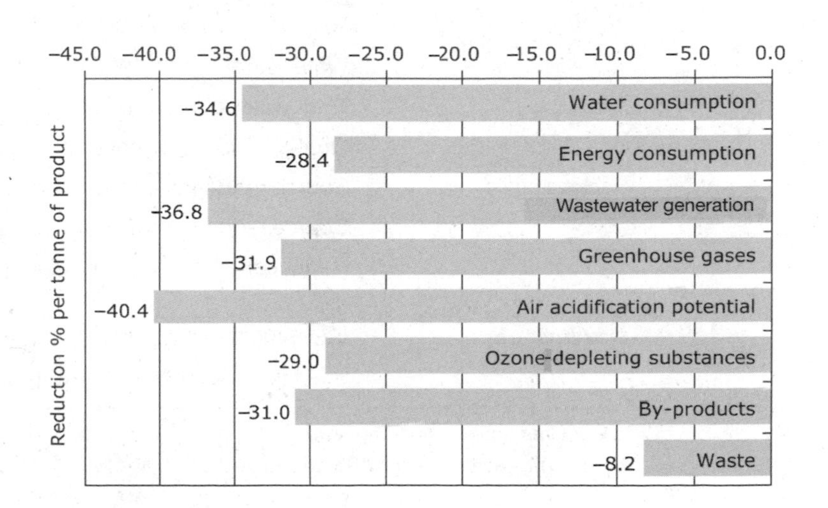

Figure 9.1. Nestlé environmental performance indicators (2002–2006).
Source: The Nestlé Water Management Report, Nestlé S.A. 2007b.

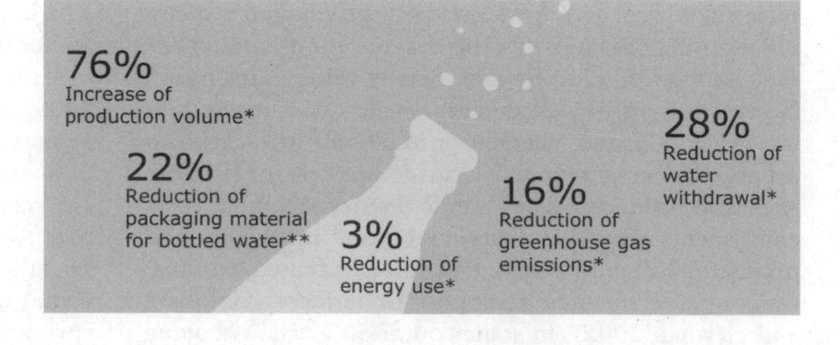

Figure 9.2. Environmental impact of Nestlé operations (developments 1998–2007).
* From 1998 to 2007.
** From 2002 to 2007.
Source: The Nestlé Creating Shared Value Report, Nestlé S.A. 2008.

nongovernmental organizations (NGOs), governmental institutions, as well as research institutions and international organizations. Each of these types of stakeholder has a different set of objectives, activities, and impacts on animal husbandry and markets for livestock products.

Role of the Food Industry

Responsible operators in the food industry, in addition to their commercial targets, usually have the following objectives:

- Ensuring food safety and regulatory compliance
- Ensuring quality and consumer satisfaction
- Meeting consumer needs and expectations
- Achieving competitiveness in terms of cost, volumes, and quality
- Ensuring continuous supply at quantities required
- Applying the principles of corporate social responsibility
- Promoting sustainable production methods.

Table 9.1. Nestlé creating shared value performance indicators

	Total Group Sales (CHF billion)	107.6
	Total shareholder return: 1 January 1997–31 December 2007	342.5%
Materials	Raw materials and ingredients (except water, million tonnes)	20.48
	Packaging materials (million tonnes)	4.08
	By-products (for recycling, million tonnes)	1.07
	Reduction of by-products since 1998 (per tonne of product)	58%
	Waste (for final disposal, million tonnes)	0.372
	Reduction of waste since 1998 (per tonne of product)	58%
Energy	Direct energy consumption (peta-joules)	85.3
	Energy saved since 1998 (per tonne of product)	45%
Greenhouse gases	Direct CO_2 emissions (million tonnes)	4.1
	Reduction of direct CO_2 emissions since 1998 (per tonne of product)	53%
Water	Total water withdrawal (million m^3)	157
	Water saved since 1998 (per tonne of product)	59%
	Total water discharge (million m^3)	101
	Quality of water discharged (average mg COD/L)	62
Packaging	Source reduction (thousand tonnes): 1991–2007	326.3
	Source reduction (CHF million): 1991–2007	583.7
	Reduction of packaging weight (per L of product) Nestlé Waters: 2002–2007	22%
Governance	ISO 14001/OHSAS 18001-certified sites (number of certificates)	171
	Sites audited through CARE program	403

COD = Chemical oxygen demand
Source: The Nestlé Creating Shared Value Report, Nestlé S.A. 2008.

Businesses have social and environmental impacts—both positive and negative—through their operations. The activities of the food industry create a number of important benefits, such as opportunities for local development through local sourcing, employment, and training as well as safe and nutritional products at affordable prices. But each step in the value chain also has the potential for harmful consequences. Without sustainable growing practices, farms can deplete natural resources, and farmers can be marginalized. Livestock production and processing can cause air and water pollution and contamination of food products. Consumer products can create health risks if good practices are not in place in the supply chain and manufacturing operations (Nestlé 2006).

To meet its responsibilities, the food industry must not only create the inherent benefits already mentioned, it must work to maximize these benefits and to eliminate possible negative side-effects from its activities (Nestlé 2006).

Role of the Public Sector

The objectives of government institutions, in many ways, coincide with the interests of consumers and the food industry, though the focus may be different. In the interests of a thriving economy, it is one of the roles of the public sector to enable businesses to operate and to generate optimum and balanced benefits for the stakeholders. But the public sector also bears responsibility for setting the rules and regulations, ensuring equal and fair application, and providing incentives for businesses to maximize positive impacts while minimizing negative ones.

Public authorities have to define food safety standards and regulatory norms in accordance with the needs and circumstances of their country, and to enforce them. In developing countries, there is a particular challenge in ensuring that all players in the food sector, including the informal sector, adhere to minimum quality standards.

The public sector needs to ensure the necessary framework conditions, legal structures, and infrastructure to favor investments in procurement and processing of local agricultural products. Other possible activities of the public sector include helping to introduce innovations, disseminating know-how, and helping local farmer communities to develop producer associations or to gain market access.

The farming sector and the food industry need to join efforts in order to prevent or control animal diseases and to mitigate their impact. However, the individual farm and the individual food company have limited impact on the general epidemiological situation in a country. It is the public sector that needs to define and enforce plans for veterinary inspection and disease prevention, as well as contingency plans and sanitary measures in case of outbreaks. The future of the livestock sector in many countries will depend on effective disease prevention measures, particularly where export production is concerned.

Authorities are also in charge of ensuring the safety and quality of animal feeds on the market. This is a very challenging task, because in developing and developed countries alike, many contamination risks and incidents are caused by inadequate standards or lack of control in feed production. In developing countries, the task is all the more difficult because good facilities for storage of feedstuffs are not always available and supply chains may be complex. This results in higher contamination risks, mainly for mycotoxins but also for pesticides used during storage. Additionally, because higher-quality grains are often selected for human consumption, potentially contaminated products are more likely to be used in animal feeds.

Overview of the Drivers and Consequences of Livestock Production

One major driver of current changes in livestock production is the quickly growing demand for livestock products, particularly in the urban centers of developing countries. Growing demand is caused by growth of available incomes, changing food preferences, increasing urbanization, and population growth (Steinfeld et al., 2006).

Growing demand is what drives the present growth in production of meat, milk, and eggs. This growth in production is achieved by increasing animal numbers, but also by increasing productivity per unit through intensification of livestock production systems (Steinfeld et al., 2006). Very often this results in an industrialization of livestock production systems around big cities, where mainly chicken, eggs, and pork are produced based on purchased feed. Farms increase productivity through improved breeds, feeding, housing, and management but also through higher usage of farm inputs such as veterinary drugs, pesticides, fertilizers, and concentrated feeds. This may lead to higher contamination risks for agricultural products, while the knowledge and tools to control and manage such risks are not yet fully developed at all levels in some developing countries. Issues can arise relating to food safety and quality as well as to regulatory compliance.

Growth and intensification of production also have important environmental impacts and socioeconomic consequences. Livestock and feed production use land, water, and energy resources. Farm effluents may pollute water resources, and emissions may impact air quality.

The growth in demand for livestock products creates enormous opportunities for improving farm incomes through value-added products. But it also exerts pressure on existing livestock production systems. There are opportunities for nonindustrial livestock producers, including small-scale ones, to participate in and to benefit from this so-called livestock revolution. However, it may be more difficult for them, due to restricted access to land, capital, know-how, and markets.

All these developments require appropriate responses from the public sector and from industry to minimize risks and to deliver the full benefits of the livestock revolution for society and local communities.

For a company like Nestlé, processing agricultural raw materials, the main drivers include the following:

- Increasing demands on the supply chain regarding chemical and biological contaminants, risk management, and traceability of raw materials, even as supply chains are becoming international and more complex
- Emerging animal diseases and the potential trade impact of these diseases
- Increasing focus on the sustainability of supply of agricultural raw materials
- Food companies moving away from primary processing and direct procurement of agricultural raw materials (e.g., fresh milk) toward purchasing of semiprocessed materials (e.g., milk powder) and comanufacturing agreements
- Changing consumer expectations about the functional and emotional benefits of products, such as the environmental and social impact of raw material production, or animal welfare
- Changes in consumer dietary habits and preferences about the composition and ingredients of food products (e.g., increasing demand for fat-reduced dairy products).

Role of Technology and Farm Size

Farm sizes and productivity are evolving. In many countries, smallholders are moving away from pure subsistence production, integrating into market economies and developing more intensive farming systems. Intensification can create higher returns to farm labor, ensuring that agricultural production in the local community remains an attractive way of earning a livelihood.

Larger livestock farms have advantages but also disadvantages compared to small-scale farms. For food companies, economies of scale can encourage a focus on larger suppliers because of their potential advantages regarding food safety, quality assurance, and procurement costs. The need for intensive contaminant monitoring and supplier development in developing countries favors larger farms versus small-scale traditional farmers. It is easier for a food company or raw material purchaser to control and to educate a limited number of larger farms than large numbers of small-scale suppliers. Reducing the complexity of the upstream supply chain, in itself, reduces the risks for food safety and quality as well as the cost of collection and procurement. On the other hand, small-scale dairy farms often have advantages regarding cost of production, availability of labor, and flexibility to adapt to changing market conditions.

The challenges for small-scale producers and the food industry are to aggregate supply and to ensure quality standards that are comparable to the larger professional farms. Farmer associations can help in accomplishing this task.

Both intensive and extensive livestock production systems are found in small-scale as well as in larger-scale farms. Small-scale farms often show features of industrial systems, such as market-oriented production and high usage of farm inputs, including purchased feed. However, larger farm size facilitates the introduction of modern production technologies, easier access to capital and know-how, as well as improved labor efficiency.

In some cases, intensive small-scale farming systems concentrated in relatively small areas can exert a pollution pressure on soil, water, and air that is similar to large-scale industrial livestock production systems. In large-scale systems, however, it may be easier to intervene with standardized solutions and to enforce environmental legislation.

On the other hand, small-scale livestock production systems offer more employment opportunities than the larger-scale industrialized farming sector.

Differentiation Based on Stages of Economic Development

The drivers and consequences of livestock production vary significantly between countries and require different responses and priorities depending on the local situation. Countries or rural areas can be categorized as developed economies, emerging economies (e.g., China and India), and least-developed economies. The typical situations are different for each of these categories.

In *developed economies*, government control systems and regulatory environment are generally in place and uniformly applied. Supplying farms tend to be larger-scale, professional, adhering to legal requirements, and applying good farming practices. All of these factors reduce the risk of food quality and safety issues. On the other hand, there is widespread consumer awareness and concern about contamination risks and environmental and social impacts, as well as about agricultural practices such as animal welfare.

The attention of authorities is focused, but media coverage is also high, which increases the risk of unexpected issues emerging. In case of complex intercontinental supply chains, the regulatory requirements and consumer preferences of importing countries may substantially differ from the ones of exporting countries. In terms of responses of the food industry, it is very important to ensure that all supplying farms are traceable, and to apply the necessary standards and practices. Early warning systems at company level help to identify and resolve new or unexpected issues in a timely fashion.

In *emerging economies*, livestock production is rapidly increasing along with the demand for livestock products. This situation triggers increased usage of agrochemicals, even while effective control systems are not always in place. The problem is complicated by the fact that many agricultural raw materials originate from small-scale farms, which makes supplier assessment and development difficult and costly. Also, there can be situations where regulatory requirements are not uniformly enforced.

Sometimes it becomes difficult for responsible food processing companies to buy raw materials with the right quality specifications because some competitors are ready to buy regardless of quality. This increases risks of issues arising from raw materials and can even result in situations where consumers distrust local products and show a preference for imports.

In response, the food industry will often focus on intensive monitoring schemes for contaminants in raw materials, possibly in close collaboration with the authorities. Supplier education and development are very important because farmers generally have less access to know-how and capital than in developed countries.

In the *least-developed economies* or in underdeveloped rural areas in other countries, the purchasing power of most of the population is very limited, resulting in reduced demand for processed food products. The farming sector is based on small-scale farmers, often producing mainly for subsistence and sometimes for the unregulated informal market. Infrastructure conditions make it difficult for the industry to set up procurement systems for larger volumes of livestock products. All these factors hamper the integration of the farming community into the formal market economy. This is particularly the case for nomadic/transhumant livestock production. Usage of agrochemicals is generally limited, hence contamination risks are lower. On the other hand, there may be higher mycotoxin and microbiological contamination risks due to inappropriate storage conditions or higher incidence of animal diseases.

Even in the situation of least-developed economies it is possible, given sufficient expertise and investment, to set up systems for successful procurement and processing of livestock products. This is illustrated by a number of successful Nestlé milk districts, for example, the recent development of a new milk district for a Nestlé Factory in Namangan in the Fergana Valley of Uzbekistan (Nestlé 2005).

Table 9.2. Change in average milk delivery of direct Nestlé suppliers 1998 versus 2006

	Kg milk per farm per day		
Country	1998	2006	Change (%)
France	487	649	33
Spain	194	571	194
Switzerland	177	256	45
China*	33	43	29
India*	8	10	26
Morocco*	11	9	–14
Pakistan*	4	7	93
South Africa	821	1274	55
Sri Lanka	9	9	0
Argentina	3196	5041	58
Brazil	119	527	342
Chile	229	777	239
Colombia	106	107	1
Dominican Republic	64	66	4
Ecuador	64	121	90
Jamaica	151	321	113
Mexico	117	385	230
Nicaragua*	91	67	–26
Trinidad	36	69	91
United States	10,612	19,404	83
Venezuela*	32	346	980

* In these countries, the number of supplying farms increased a lot between 1998 and 2006. Because the new suppliers are generally smaller than the existing suppliers, this depresses the average, or at least diminishes the increase (except for Venezuela where new suppliers have been larger scale).

Source: Nestlé data.

Responses of Private Sector

This section summarizes the responses of the food industry to these developments, based on the case of Nestlé and its milk district model.

The Nestlé Milk District Model

In order to operate successfully in any country, Nestlé needs to satisfy consumer expectations. Consumer satisfaction can only be achieved with products of high quality standards that provide value for money. Careful sourcing of fresh milk for processing in Nestlé factories is very important in achieving these targets. Milk needs to meet the following criteria:

- Available in the right quantities at the right time
- High in quality and free of contaminants
- Sourced at a competitive cost so that finished products remain affordable
- Sourced in a way that meets social and consumer expectations.

Achieving the quality requirements is more challenging in the case of fresh milk purchased from small-scale farmers in developing countries than in the European environment. But Nestlé has been successfully collecting and processing large quantities of fresh milk from small-scale producers in many countries for decades (Goldberg and Herman 2005).

Key to this success is an integrated approach throughout the supply chain. Managing and aligning each level of the chain in a professional way ensures quality, volume, and low cost. Farmers are all aware of the importance of high quality to succeed in the milk business.

This awareness did not come by itself. For many years, Nestlé field officers have been providing technical assistance and management advice to farmers, not only on milk quality and hygiene practices but also on a range of other topics that are essential for an efficient dairy business. Animal husbandry, feeding practices, fodder production, animal health, and breeding as well as economic aspects are all part of this integrated approach (Nestlé 2006).

The following factors are equally important for the success of the system:

- Direct contacts between Nestlé field staff and farmers (e.g., in the form of regular meetings for training and to discuss questions related to the business. This creates transparency in communication and mutual trust.
- Presence of Nestlé's collection infrastructure at the village level to receive milk on a daily basis and to preserve it in good quality.
- Continuous purchasing of milk without interruption.
- Regular cash flow delivered by the milk payments. This enables small-scale farmers to invest in continually improving and growing the business (Goldberg and Herman 2005).

In such a system, it is exceedingly important to educate and motivate small-scale suppliers to produce and deliver milk of a competitive quality. This is not easy in an environment where such requirements and quality standards did not exist previously. It requires time and communication, as well as a quality-based milk payment system under which farmers who deliver milk of high bacteriological quality, normal composition, and free from contamination will achieve their business objectives in terms of milk price and economic benefits. This transparent and professional business approach has been well understood and accepted by the farmers, and they appreciate the company's support in achieving their business objectives. Through this backward integration, the farmers become active partners of Nestlé in ensuring raw materials and products of high quality. A robust quality-monitoring scheme allowing feedback on milk quality to the farmers is a precondition for the success of such a system.

The following components of the Nestlé model for setting up and developing milk districts help to explain its success:

- Initial selection of locations with good agricultural potential for sustainable dairy farming, able to continuously increase milk production over time
- Provision of income with very high reliability through regular milk purchases
- Provision or facilitation of continuous technical and financial assistance for farmers (Goldberg and Herman 2005).

There are some situations where the milk district model faces limitations. High-yielding dairy cattle are not well adapted to humid tropical climates and to some of the cattle diseases prevailing there. In these environments, cows tend to have lower productivity and hence lower economic attractiveness for farmers than in temperate countries. In such environments, local producers may have difficulty in competing with imported dairy products, especially if they are exposed to world market prices. Therefore, in some tropical countries where there was little or no dairy tradition (e.g., parts of Southeast Asia, Africa, and the Caribbean), dairy farming has sometimes stagnated at low production levels. In some of these countries it may be better for public development efforts to allocate available resources to the development of alternative crops and income sources for which the country has competitive cost advantages, rather than to the development of dairy farming.

In recent years, Nestlé has reduced its procurement of milk for export production in some of the European countries. One of the reasons is the fact that Nestlé has been processing more and more milk from local farmers in Latin America and Asia instead of importing finished products from Europe. Some of the European factories have instead moved to special products requiring specialized technologies and know-how but less fresh milk.

Overview of Instruments and Processes Used by the Food Industry

Food safety and regulatory compliance are primary objectives of the food industry. The industry relies on a range of instruments and processes to ensure these requirements are met not only in its manufacturing processes but also in the upstream supply chain of its factories.

Food Safety and Contaminant Monitoring Schemes

For each type of raw material used by Nestlé factories, monitoring schemes define which laboratory tests have to be performed, at what time intervals, and for what purpose. Monitoring schemes are elaborated based on risk assessments of the specific production and supply chains. Contaminant monitoring usually includes heavy metals, mycotoxins, antibiotic residues, pesticide residues, and microbiological contamination. If a raw material does not fulfill the specifications (e.g., absence of antibiotic residues), then it will not be used in production and actions are taken to resume safe sourcing of raw materials. Contaminant monitoring schemes are implemented for both agricultural raw materials and finished products. This monitoring is often more intensive and more frequent in developing countries than in developed countries due to the higher risk situation.

Farm Quality Assurance Schemes
Farm quality assurance schemes are a comprehensive set of good farming practices, which, if consistently applied by farmers, ensure absence of contaminants and high quality of farm products. Farm assurance schemes describe, for example, the measures to be taken at the farm to ensure that milk from cows treated with antibiotics does not enter the food chain (e.g., marking treated cows and milking them separately).

The Global Partnership for Good Agricultural Practice (GLOBALGAP) is an example of such a farm quality assurance scheme, which sets voluntary standards for the certification of agricultural products around the globe. There is a range of other schemes adapted to specific crops and supply chains. The International Dairy Federation and FAO, for example, have jointly published a Guide to Good Dairy Farming Practice (IDF and FAO 2004), which is commonly applied by the dairy industry and dairy farming sector as a basis for elaborating farm quality assurance schemes (Nestlé 2005).

Assessment and Selection of Suppliers
Food companies assess and select their supplying farms based on specific criteria relating to application of good farming practices and implementation of farm quality assurance schemes. Farmers are first informed and educated about the requirements. Farm verifications, audits, and certification may follow at a later stage. The implementation of individual supplier assessment faces practical limitations in some developing countries where a dairy factory may be supplied by tens of thousands of small-scale milk producers.

Education and Development of Suppliers
In order to ensure the supply of high-quality raw materials, food companies often embark on programs to educate and develop supplying farmers. Such programs may focus on food quality and safety, but they frequently include additional topics such as farm management, animal feeding, disease prevention, productivity improvement, and environmental education of suppliers.

Traceability in the Food Chain
Some food companies use traceability systems that exceed the legal requirements. This is driven by the need to trace possible quality problems back to the root cause or forward to the finished product that may have been impacted. For traceability purposes, records and samples are stored for specified periods.

The foregoing description of instruments and processes makes it clear that the food industry is very much involved in ensuring good farming practices, as well as in assessing and developing supplier farms with regard to raw material quality and safety. This is particularly true for sensitive products, or in situations with an elevated risk of contamination problems.

The food industry will resort to the same set of instruments and processes when it wants farms to meet additional requirements—for example, relating to environmental impacts, farm labor conditions, animal welfare, prevention of animal diseases, or sustainability of production. The industry is adding these additional requirements to the farm assurance schemes and to supplier education.

The cost of contaminant monitoring and supplier development is high, particularly in developing countries, where many small-scale suppliers are involved. Some food companies may be reluctant to invest substantial financial resources and efforts in supplier development. It is therefore sometimes possible that importing agricultural raw materials becomes more attractive than sourcing from local suppliers.

On the other hand, there may be companies in developing countries that are sourcing local raw materials without having the necessary quality systems in place. This increases food safety risks and discourages sustainable production and sourcing methods. Therefore public authorities in developing countries should create a legal basis or even legal requirements to use instruments such as contaminant monitoring, farm assurance schemes, and farm audits.

Review of Responses—Nestlé Case Examples

The following section describes the responses in more detail related to the key objectives of Nestlé and illustrates these responses through case examples from Nestlé's milk procurement operations worldwide.

Provision of Stable Markets for Agricultural Products

Providing farmers with a stable market outlet for their products is in the interest of both food industry and farmers. The food industry needs to build up a sustainable and growing supply of raw materials as a base for business success and growth. On the farming side, stable market outlets for agricultural products are an extremely important stimulus to increase agricultural production and to develop a viable farming sector (Goldberg and Herman 2005). This is perhaps one of the most important, most beneficial and successful roles that primary processors and the food industry play in agricultural development.

Through its procurement practices for fresh milk, Nestlé helps to improve the economic and professional level of many farmers around the world. This is particularly true in the case of small-scale farmers who may otherwise face difficulties in finding a permanent and reliable market outlet for their daily milk production. Around half of Nestlé factories are located in developing countries, and investment in local manufacturing for local consumption is the general approach (Nestlé 2006).

Case Study: Nestlé Milk Districts in Pakistan

Nestlé has two dairy factories in the Punjab Province of Pakistan. Expanding continually since it started business here in 1988, the company now operates milk delivery points for farmers in over 3000 villages. This huge and complex operation covers an area of more than 70,000 km^2 and ensures an average daily supply of over 1200 tonnes of good-quality milk to the Nestlé factories, based on farms that on average deliver less than 10 kg of milk per day each. Every week, more than 130,000 dairy farmers receive their milk payments (Nestlé 2005).

Each supplier delivers milk to a collection point twice per day, directly after milking cows or buffaloes. The milk supplied by each farmer must undergo a basic and rapid quality check before being accepted. From the delivery points, Nestlé transports the milk in churns to centers equipped with milk cooling tanks and electricity generators—Nestlé Pakistan is operating more than 1500 milk cooling centers, and new ones are being added continually. After the milk is cooled to a safe temperature of 4° C, it is transported to 27 main centers equipped with large storage tanks. Onward transport to the factories is carried out by large-capacity milk tanker trucks.

The results of this approach are impressive. Every year Nestlé purchases milk from Punjab's farmers to a total value exceeding US$100 million. In addition to the participating farmers, more than 2000 people are directly employed or contracted by Nestlé for milk procurement and milk collection (Nestlé 2005). During the last few years, there has been a strong increase of the milk quantity sold per farmer, and in March 2007, Nestlé opened a new milk processing plant in Pakistan, making Kabirwala one of the biggest milk processing factories of Nestlé worldwide.

Ensuring Food Safety and Regulatory Compliance

Potential chemical contaminants in livestock products include residues of antibiotics and other veterinary drugs, mycotoxins and heavy metals originating from cattle feeds, and residues of pesticides used in production and storage of feed crops and in animal production.

Another potential food safety hazard is microbiological contamination of livestock products. In the case of milk products, the microbiological hazard is eliminated through thermization of milk and through good manufacturing practices at factories. But it is still essential to ensure the lowest possible microbiological contamination of raw milk so as to optimize the quality and shelf life of finished products.

As described earlier, Nestlé has implemented the following tools to ensure that contamination risks are under control:

- Quality control schemes, including monitoring of contaminants
- Specifications for acceptance and rejection of raw materials
- Farm quality assurance schemes
- Supplier assessment and selection
- Supplier education and development.

It is generally the role of the *authorities* to define food safety standards and regulatory norms and to enforce them. It is best if local authorities fix maximum residue limits for the relevant contaminants and to enforce the adherence of all food companies and products to these limits.

It is the responsibility of the *food industry* to ensure that food safety standards and regulatory norms are met in raw materials and finished products, and that the necessary control schemes are in place.

In some cases, Nestlé's norms need to be stricter than local requirements in order to ensure quality of sensitive products and to avoid multiple standards in different countries. For specific contaminants where there is no national legislation, Nestlé applies the maximum residue limits of the Codex Alimentarius as a minimum standard wherever possible. This practice goes beyond what many local competitors do, but Nestlé believes that it meets the expectations of consumers by implementing the highest food safety requirements. By communicating and ensuring its strict standards and through active lobbying for general high standards, Nestlé strengthens its position and contributes to raising food safety requirements in the countries where it operates.

Even where monitoring programs and farm assurance schemes are in place, unexpected food safety issues may still emerge. Therefore, Nestlé set up an early warning team of scientists and quality specialists to identify and address any emerging challenges (Nestlé 2006).

Nestlé is also in constant dialogue with national and international regulatory authorities and organizations such as WHO, FAO, Codex Alimentarius, and the International Dairy Federation. Nestlé experts actively participate in working groups and task forces on specific food safety issues that are set up by these organizations (Nestlé 2006).

The Nestlé Research Center also studies food contaminants as well as verification or development of new testing methods. Nestlé scientists publish research papers and frequently interact with external institutes. Cooperation with regulatory authorities and Nestlé's own research contribute to the development of better and more precise food safety regulations and analysis techniques.

Case Study: Nestlé Morocco—Elimination of Antibiotic Residues

Nestlé Maroc buys 60,000 tonnes of fresh milk annually from about 15,000 small-scale farmers who deliver milk twice daily to 100 collection centers (Nestlé 2005). Up to 200 supplying farmers are needed to fill one milk cooling tank of 2 tonnes capacity. Thus, if a single farmer delivers milk from a cow recently treated with antibiotics, there is a high risk that all the milk in the cooling tank and consequently all the milk in a 25-tonne tanker truck will be contaminated above the maximum residue limit.

For Nestlé, the risk of contamination with antibiotic residues was unacceptable. In its campaign against antibiotic residues, Nestlé Morocco had the following objectives:

- To eliminate quality risks for fresh milk by ensuring good practices at the farm level and at milk collection centers
- To promote prudent use of antibiotics in animal husbandry
- To avoid milk losses for farmers, cooperatives, and the Nestlé factory
- To show supplying farmers that Nestlé cares about the quality and safety of the raw material.

Nestlé took the following initial steps to assess the situation and to involve stakeholders (Nestlé 2005):

- Nestlé worked with agricultural and veterinary institutes to identify the antibiotics most frequently used in the milk district and the main reasons why antibiotics enter the milk supply chain.
- Nestlé collaborated with private veterinarians active in the milk district to educate farmers about withholding periods[1] of veterinary drugs.
- Nestlé involved farmers' cooperatives in the effort to eliminate residues.

In the year 2000, Nestlé's milk sourcing department intensified its campaign to educate farmers on how to keep antibiotic residues out of the cooling tanks. A poster with explanations was elaborated and displayed at all milk collection centers and village meeting places to raise awareness about the importance of not delivering milk from antibiotic-treated animals until the withholding period has elapsed.

Red-colored churns were placed at milk collection centers—farmers were expected to discard milk from recently treated cows not in the milk cooling tank but in the red churn. Visits of farmers to the Nestlé factory were organized to give them a better understanding of milk quality requirements. Milk positive for antibiotics declared by the farmers and poured in the red churn is destroyed, but the cooperative pays the farmers for it. This was certainly not best practice but was necessary to resolve the problem.

Additionally, the driver of the Nestlé milk tanker tests the milk in each cooling tank for antibiotic residues by means of a rapid test and refuses to load milk from a tank that tests positive. In such cases, the cooperative is responsible to resolve the problem and to cover the monetary loss. Nestlé has introduced and pushed through these stringent measures at high cost, in spite of severe competition in the milk district from other companies.

The project has achieved its targeted results (Nestlé 2005):

- It increased awareness among farmers and cooperatives about the need to avoid antibiotic residues in the fresh milk sold.
- Antibiotic residues have been completely eliminated in raw milk arriving at the Nestlé factory.
- It has contributed to sustainable milk production practices by working closely with veterinarians.
- These actions boosted the image of Nestlé as a company that considers quality as not negotiable.
- Cooperative associations and key government representatives have reacted positively.

Ensuring High Quality of Products

Consumers want a product that is not only safe but also nutritious and of high quality and good taste. Nestlé has built its business on the basis of product quality. A Nestlé brand name on the product aims to be a promise to the consumer that it is safe to consume, complies with all regulations, and meets high standards of quality.

The quality of finished products starts with the quality of the agricultural raw materials. Therefore Nestlé's purchasing guidelines and supplier contracts stipulate specific quality parameters, supported by adequate production, storage, and procurement methods. Nestlé employs hundreds of agronomists and purchasers who help to achieve Nestlé's quality objectives for raw materials (Nestlé 2006).

In its sourcing operations, Nestlé regularly takes raw material samples from each supplying farmer. These samples are analyzed for relevant quality parameters, and the results are recorded. The quality-based payment system pays farmers premiums above or deductions from the basic raw material price, according to quality results. This system helps to motivate farmers to deliver raw materials of the highest quality. Many companies that source raw materials from farmers in developing countries do not yet have such individual sampling and quality-based payment systems in place.

Buying high-quality raw materials from farms is not in itself enough to ensure high quality of raw materials arriving at Nestlé factories. This is particularly true in the case of raw milk procured from small-scale farmers. Raw milk is a very precarious raw material. It has to be cooled to around 4°C not later than two hours after milking. In developed countries, this is achieved by on-farm cooling in milk cooling tanks.

1. Withholding period is the minimum time that must elapse, following administration of an antibiotic, before products from the affected animal can be used.

In many developing countries, milk quantities per farm are much too low to enable on-farm cooling. In India, Pakistan, Sri Lanka, Uzbekistan, and Peru, Nestlé has set up cooling centers in local villages. Farmers deliver their milk to the village cooling center immediately after milking. Here the quality of the milk is checked, the quantity is measured, and the milk is poured into the cooling tank. These cooling centers are often set up and operated through contractors paid by Nestlé. There are hundreds of Nestlé cooling centers collecting the milk for one single Nestlé factory. Such a collection and cooling infrastructure is costly, but without it, it would be impossible to collect the raw material in a quality that is acceptable for Nestlé products. Many farmers in more remote villages would not have market access without this infrastructure. Despite this, many local competitors and milk traders still collect uncooled milk that reaches their factories only after several hours' delay. In many countries Nestlé's example has stimulated the local dairy industry to improve their collection systems and milk quality.

In other countries where average farm sizes reach the necessary daily milk quantities, Nestlé has supported farmers who wish to take the next step in quality improvement and to invest in farm-based cooling tanks. Such a scheme was started by Nestlé Brazil in 1996. The project was finalized in 2002: all milk is now cooled on-farm and achieves even higher quality than in the previous system. Some production units, however, were not able to participate in the scheme and many discontinued their activity.

Case Study: The Farm Assurance Manual of Nestlé Chile

Dairy farming in Chile has developed rapidly during the last two decades toward larger-scale professional farms. Nestlé Chile buys 350,000 tonnes of fresh milk annually from over 1500 dairy farms located in the south of the country. This quantity represents over 20% of Chilean production (Nestlé 2005).

In the past, the approach to assure the quality of fresh milk for the factories focused on quality control. Before releasing it for production, factory staff analyze the quality of the milk in each incoming tanker with a set of tests prescribed by the Nestlé Quality Monitoring Scheme. This testing will continue in order to guarantee the quality of the raw material, but the focus is now shifting to increased quality assurance at the farm level. Under the new procedure, farmers commit themselves to apply a comprehensive set of good dairy farming practices to guarantee the quality and safety of the milk produced on their farms. Through its field staff Nestlé supports the supplying farmers in implementing good practices.

The necessary guidelines are described in farm assurance manuals adapted to the specific country situation. By 1998 Nestlé had already elaborated and applied such manuals in several places—for example, in France and Australia. Based on discussions and surveys carried out with Chilean farmers, the Agricultural Service Department of Nestlé Chile developed a farm assurance manual adapted to the situation on local farms. The *Manual De Aseguramiento De Calidad De Leche Fresca* was published in the form of a folder in October 2004. To our knowledge, it was the first farm assurance manual for dairy in Latin America and in the Spanish-speaking world. Not only does it include good farming practices aimed at a safe and high-quality product, but it also promotes practices for environmentally sustainable dairy farming, such as water conservation, soil nutrient management, and protection of biodiversity.

Reaction in Chile to the guidelines was considerable. The Chilean Ministry for Agriculture and Nestlé jointly introduced the manual to the public through a press release. Then it was presented to Nestlé's milk suppliers during ceremonies at the three factories of Los Angeles, Osorno, and Llanquihue. The guidelines are now under implementation at the farm level in collaboration between Nestlé and official organizations. The farms were visited in 2005 and 2006 in order to assess existing gaps and to support farmers in establishing action plans to overcome the gaps. Farm visits and assessments continue on a regular basis (Nestlé 2005). Within a few years, Nestlé Chile will receive all its milk from suppliers certified to have the mandatory requirements in place. This work will further contribute to promoting the high quality of Nestlé products and the sustainability of dairy farming in Chile.

Animal Diseases and Animal Welfare

On principle, livestock products must come from healthy animals that are kept under adequate conditions. Animal diseases could have the following impacts on Nestlé supply chains for livestock products:

- Risk that products derived from sick animals enter the supply chain, which is of great concern, particularly in case of zoonotic diseases
- Risk that residues of veterinary drugs enter the supply chain
- Disruption of supplies or demand (e.g., through reduced production, sanitary, and export restrictions, or loss of consumer confidence).

In developing countries, the incidence of epidemic and zoonotic diseases is generally higher than in developed countries, partly because governmental control and eradication programs are not always in place. This situation includes diseases such as foot and mouth disease (FMD), brucellosis, and tuberculosis (TB). Other notifiable diseases are equally present in developed countries.

In its milk collection from farmers, Nestlé has been confronted with the problem of animal diseases for a long time. The critical control point for avoiding any zoonosis risk for dairy products is thermization of the raw milk under controlled conditions, which is ensured for all Nestlé products. However, Nestlé goes much further and considers implementation of good dairy farming practices, including disease prevention and control plans, as the decisive step in controlling disease incidence on farms. In countries like India and Pakistan, Nestlé field officers, who frequently have a veterinary background, carry out various measures from farmer education to actively conducting vaccination campaigns for cattle.

Recently, Nestlé has started a project in collaboration with a specialized consulting company (SAFOSO) to develop guidelines for its supply chains for managing animal disease risks. These procedures include risk assessments for individual diseases; prevention and contingency plans; supplier assessment; and adequate procedures for farming, transport, and primary processing. The procedures will standardize Nestlé's approach toward managing animal disease risks in its supply chains for animal products and will complement measures already taken. Standard requirements are the following:

- Requiring each dairy factory to carry out a risk assessment for different animal diseases and develop a prevention/contingency plan for the main disease risks
- Supporting vaccination programs against FMD (where applicable)
- Educating farmers on infection risks from different sources
- Paying premiums for farmers participating in official TB/brucellosis control programs (where necessary and applicable).

Sustainability of Livestock Production

The Nestlé Supplier Code states the following regarding sustainability (Nestec 2008):

> Nestlé supports and encourages operating practices, farming practices and agricultural production systems that are sustainable. This is an integral part of Nestlé's supply strategy and supplier development. Nestlé expects the supplier to Nestlé to continuously strive towards improving the efficiency and sustainability of its operations, which will include water conservation programs.

Sustainable farming practices are in the long-term interest of all stakeholders. Nestlé's engagement in sustainable agriculture aims at all three pillars of sustainability: economically, socially, and environmentally sustainable farming practices. The main goals for Nestlé are as follows:

- Protect the company's long-term sourcing requirements for agricultural raw materials
- Ensure quality, availability, and competitive prices
- Ensure the sustainable use of natural resources as part of corporate social responsibility
- Meet the expectations of society and of ethical consumers and investors who give preference to sustainable brands
- Promote social sustainability, which includes requirements for appropriate occupational safety measures and appropriate labor conditions on farms—for example, absence of child labor or forced labor (Nestlé 2004).

Nestlé's sourcing principles and practices as well as voluntary investments (e.g., in technologies reducing water usage) contribute to such sustainability objectives:

- Farm assurance schemes include requirements not only for food quality and safety but also for sustainable agricultural practices.
- Education of farmers through Nestlé field departments enables Nestlé to meet its quality and brand requirements while helping to develop and disseminate sustainable farming practices.
- Partnerships within the industry such as SAI Platform[2] aim to transform individual company policies and efforts toward a global initiative promoting proven sustainable agriculture practices, tools, and indicators (SAI Platform 2003).

Water Conservation

Nestlé Corporate Business Principles state the following related to the Nestlé Water Policy (Nestlé 2004):

> Water is a natural resource that is indispensable for life. Nestlé recognizes that the responsible management of world-wide water resources is an absolute necessity. Preserving both the quantity and the quality of water is not only an environmental challenge, but also one that spans agricultural, economic, political, social, cultural and emotional considerations.

As a leading food and beverage company, Nestlé considers water to be a key priority for the manufacturing of its food products, for their preparation by consumers, and for bottled waters. To play its part in assuring a long-term, high-quality, adequate global water supply, Nestlé supports the sustainable use of water, strictly controls its use in the Company's activities, and strives for continuous improvement in the management of water resources.

In its water management report (Nestlé 2007b, Ederer and Goldberg 2007), Nestlé pledges to do the following:

2. The SAI Platform (SAI = Sustainable Agriculture Initiative) www.saiplatform.org is an industry organization with the goal of promoting sustainable agriculture based on the economic, environmental, and social pillars of sustainability.

- Work for ever-lower volumes of water per kilogram in food and beverage production
- Assure that our activities respect local water resources
- Take care that water discharged into the environment is clean
- Engage with agricultural suppliers to promote water conservation among farmers
- Reach out to others to collaborate on water conservation and access, with a particular focus on women and children.

Nestlé has stated that the problem of present and future water scarcity is underestimated by many stakeholders. This is why Nestlé has made water a central theme of its Sustainable Agriculture Initiative (SAI) program. Agriculture consumes 70% of global freshwater withdrawals and is the one link in the food chain that has the biggest impact on water. By transferring upgraded water management techniques to agriculture, Nestlé aims to improve efficiency of water usage in agricultural supply chains (Nestlé 2007b).

In its factories, Nestlé strictly controls water utilization and seeks to improve management of water resources. Globally, Nestlé reduced its own water usage per tonne of product by 59% between 1998 and 2006 (Nestlé 2007b).

For example, the Gorinchem factory in the Netherlands, which produces milk powder, was able to reduce freshwater use by 50% over a five-year period. For certain applications (e.g., precleaning, cooling), freshwater was replaced by the water evaporated during the production of milk powder (Nestlé 2000).

Nestlé's Harrismith factory in South Africa formed a special work team of managers and employees to identify and implement projects to reduce water usage. Each team member was allocated specific tasks, including project coordination, training of employees and contractors, tracking and monitoring costs, water flow meter reading and recordings, supervision of water usage, purchasing of raw materials, and plant inspections. Projects implemented so far include the recovery of water generated by milk evaporation (so-called cows' water), optimization of steam condensate recovery, and the control of municipal water used in the factory. Cows' water, for example, is now used for washing tankers, in cooling towers and ice plants, for cleaning, and for gardening (Nestlé 2003).

In a number of countries, Nestlé supports projects for water education in the framework of its programs to improve conditions in local communities. In Moga, in India's Punjab Province (Nestlé India's main milk district) Nestlé has set up over 100 village school fountains to ensure the supply of clean drinking water for schoolchildren. Nestlé donates 90% of the necessary funds while the local community contributes 10%. This program is accompanied by education sessions organized at schools to inform schoolchildren about risks threatening local water resources (Nestlé India 2007).

Water Pollution Reduction

Agricultural production not only consumes water but also has a potential to contaminate water resources by overuse of pesticides, fertilizers, and inadequate disposal of manure from livestock farming. Because the food industry buys its agricultural raw materials from farmers, it has a unique opportunity through its direct supply chains to influence farmer's activities to improve water use and overall efficiency (Nestlé 2007b).

In its direct relationships with farmers and cooperatives, Nestlé encourages practices to avoid water pollution as well as protecting local water resources from manure contamination or leakage of chemicals. This is again an integral part of sustainable agricultural practices, communicated through farm assurance manuals.

Many local efforts by Nestlé have aimed to reduce water pollution. For example, by focusing on dry-cleaning methods prior to wet washing of equipment, the Elst factory in the Netherlands was able to reduce the biochemical oxygen demand (BOD—a measure of load) in its wastewater by 70% over six years. At the Tongala factory in Australia, Nestlé took steps to reduce the nutrient loading of wastewater as a result of local environmental conditions. This led to a significant reduction in

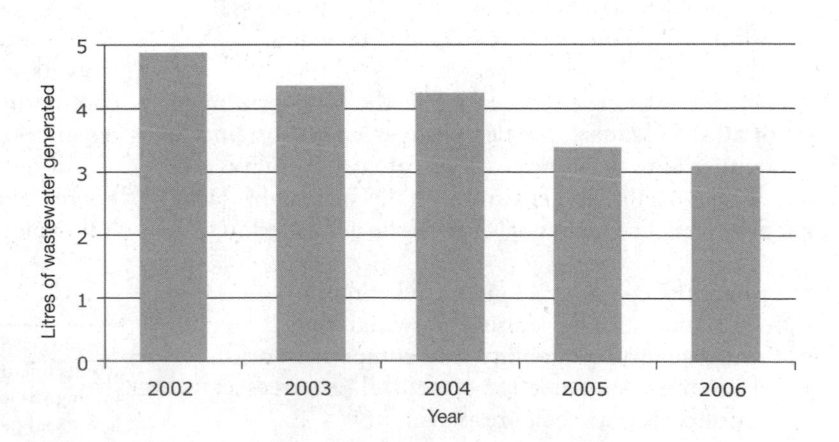

Figure 9.3. Wastewater generated by Nestlé factories to produce 1 kg of product.
Source: The Nestlé Water Management Report, Nestlé S.A. 2006.

both phosphorus and BOD. At the Nanjangud factory in India, all the treated wastewater from the factory is used to irrigate plantations on the factory premises instead of discharging into waterways. Once barren and rocky terrain is now a greenbelt where over four hundred varieties of trees and plants thrive (Nestlé 2000).

Nestlé invests in wastewater treatment plants in developing countries even when there is no legislation. The water treatment plant of Agbara Factory in Nigeria has earned Nestlé the Most Environmentally Proactive Industry Award from the local governor and serves as a model for another treatment plant at Tema Factory in Ghana (Nestlé 2007b).

Soil Conservation

Along with water, soil is the ultimate basis of agricultural production. Soil loss can seriously depress productivity on grassland and fodder crops, as well as food crops. So soil conservation is an essential element of sustainable agriculture, and good practices are disseminated through Nestlé's farm assurance manuals.

Nestlé South Africa engages in a "Work for Water Project" that promotes water-saving techniques in agriculture, such as nighttime irrigation, introduction of computerized irrigation systems, pastures with lower water demand, and the use of oxidation dams to use recycled water on pastures. In addition, farmers receive advice on minimum tillage practices. The project is sponsored by the National Department of Water Affairs and Forestry (Nestlé 2007b).

Reforestation

Trees are an essential element of sustainability. As well as protecting and providing for biodiversity, they can enhance water resources by increasing local precipitation, reducing runoff, increasing soil organic matter that can retain water, and allowing a gradual release of water to springs and streams. They can also reduce soil erosion by slowing the force of rainfall and holding soil in place with their roots.

Deforestation is one of the major reasons for soil erosion and reduction in renewable water resources. Hence it is vitally important to encourage farmers to increase tree cover on their land through agroforestry measures where trees are intermingled with agriculture and livestock operations, or through reforestation. Nestlé is encouraging efforts in this direction, for example, through its farm assurance manuals.

Case Study: Reforestation on Ranches in Tropical Regions of Mexico

The tropical areas of the states of Veracruz, Chiapas, Campeche, and Tabasco supply fresh milk to Nestlé Mexico. The availability of resources for cattle feed and the use of grazing systems are key to the development of milk herds in these regions.

Unfortunately clearing of land for cattle grazing and lack of ecological awareness have resulted in the destruction of woodland areas and tropical rainforests. To combat this trend and help preserve the ecological balance, for a number of years various species of trees and shrubs have been planted to create "living fences" to divide up grazing land and separate estates from one another. Species planted included *Gliricidia sepium*, *Bursera grandiflora*, *Leucaena leucocephala*, *Acacia* spp., *Brosimum* spp., *Casuarina*, and the pigeon pea (*Cajanus cajan*).

Three quarters of Nestlé's milk suppliers in Mexico's tropical regions have already adopted the living fence system. However, they now also need to begin reforesting part of their ranches. Reforestation will sustain local flora and fauna, collect rainwater, and restore springs, and thereby improve groundwater levels to provide a sustainable water supply for their cattle.

The Nestlé factory in Coatepec, Veracruz, is supplied by 2823 farmers, whereas the factory in Chiapa de Corzo has 876 suppliers. Of the farmers supplying milk to Nestlé factories in these tropical regions of Mexico 2774 are already using living fences to divide their pastures, and 903 farmers have begun reforesting part of their estates under the Sustainable Agriculture Initiative Nestlé (SAIN) (Nestlé 2005).

The target is that by the year 2010, 94% of supplying farmers will have adopted the living fence system and more than 45% will have reforested part of their land, thereby helping to improve the aquifer and the biological diversity of the area (Nestlé 2005).

Energy Conservation and Reduction of Greenhouse Gas Emissions

Energy conservation is valuable in itself as a cost reduction measure. However, it also helps to conserve resources, reduce deforestation for fuel wood, and reduce greenhouse gas emissions.

Globally, in its own operations, Nestlé reduced energy use per tonne of product by 45% between 1998 and 2006. Nestlé factories have achieved this through numerous initiatives, for example, switching to more efficient fuels or to renewable energy resources (e.g., coffee grounds generated in Nestlé coffee factories), generating energy from combined production of steam and electricity and methane recovery from wastewater treatment plants (Nestlé 2008).

Energy savings have also been achieved in the fresh milk transport chain through the introduction of computer-based milk collection systems that define the route taken by each vehicle in the fleet. The systems keep delivery times, and kilometers driven, to a minimum. As a result of the implementation of this program in Mexico, Nestlé has achieved a 10% reduction in driving distance. Overall, transport costs have been reduced by 12%, unloading congestion has been avoided by better planning of arrival times, and delivery speeds have been increased, thus ensuring milk quality. The computerized milk-collection systems clearly offer a triple advantage in terms of their environmental, economic, and quality impact (Nestlé 2000).

Nestlé also encourages energy conservation efforts by its milk suppliers, for example, by promoting biogas plants in China.

Case Study: Biogas Digesters in China

Nestlé's Shuangcheng milk district in Heilongjiang Province has been growing with the strong demand for milk in the highly competitive Chinese market. Growth means not only more income for all parties involved but also raises challenges. One issue is manure storage. Most farmers still compost manure outside the farm wall and apply it to the fields in spring and autumn (Nestlé 2005). However, as the amount of manure increases, adequate manure storage is becoming a necessity.

Because modern manure storage systems require high investments with no immediate financial benefits, the incentive for farmers to construct proper storage is limited. Biogas is a sustainable solution to this problem because it can reduce energy costs, bring extra income, and reduce manure storage costs at the same time. Nestlé China and local authorities considered low-cost and adequately sized biogas digesters as a possible solution with measurable financial benefits. This seemed daring considering that in the region's cold winters temperatures can get down to –30°C. After a successful trial of a small biogas digester (10 m^3) in the winter of 2004–05, more than 400 units were constructed on small and medium-sized farms. The gas production is sufficient for cooking three warm meals per day, even in the coldest winter months. This allows a partial replacement of coal, gas, and corn stems that are common sources of fuel (Nestlé 2005).

Equipment such as the gas cooking stoves, gas lamps, special boilers, and even gas-driven generators producing electricity are now available locally. The most recent development is the successful trial of a large biogas digester (100 m^3), which produces sufficient gas for heating, cooking, and even some electrical power for the farm. To date more than three thousand small biogas digesters have been constructed in the Nestlé Shuangcheng milk district, strongly supported by the local authorities.

Nestlé Shuangcheng Ltd is presently collecting milk from more than 27,000 suppliers. In order to demonstrate this technology to all suppliers, the Chinese authorities have agreed that 74 Nestlé demonstration farms will be equipped with such a unit. An additional one to two thousand units were planned for 2006–07 (Nestlé 2005).

Integrated Sustainable Operations

Nestlé farm assurance schemes cover many aspects of environmental sustainability. A good example is the Préférence approach applied in France, a quality assurance partnership with dairy farmers (Nestlé 2005). Together with Nestlé experts, farmers reviewed all aspects of milk production and then set up and implemented best agricultural practices. A vital part of these Préférence best practices covers the protection and careful use of water, and appropriate field irrigation. For example, all the following requirements are imperative:

- Farm effluents must be stored in waterproof installations.
- Silage fluids must be collected, stored, and disposed of according to regulations.
- Wastewater from milking parlors must be recovered.

Préférence has led to a constructive dialogue to improve food safety and quality, and has encouraged farmers to engage in a continuous improvement program to make dairy farming more sustainable. Préférence has proven to be a concrete answer to how to produce definitive results related to quality, traceability, and sustainable agricultural production methods (Nestlé 2003).

Similar programs supporting sustainable agriculture, including the sustainable usage of water, have also been conducted by Nestlé in other countries such as the United Kingdom, the Netherlands, Spain, Australia, Chile, and Brazil. These programs include the following components:

- Fertilizer and nitrates use are limited to the effective needs of plants.
- Uncovered soil should be planted with crops in order to reduce washing out and to trap nitrogen.
- Equipment for pesticide application must be serviced every year.
- After treatment, any small quantities of remaining pesticides and rinsing water must be emptied on the treated field.
- The quality of water used for irrigation must be verified annually. The quantity of water needed must be calculated according to crop needs, then carefully measured and registered.
- Impact of water withdrawal on groundwater level must be controlled.
- Irrigation must take into account fertilization plans and pesticide application programs (Nestlé 2003).

Improving Productivity and Technology

Nestlé has helped farmers to increase the productivity of dairy production systems in its milk districts in many countries. Adapted solutions have been developed and disseminated in a number of regions in developing countries (Nestlé 2005). The following case from Colombia may serve as an example of these initiatives.

Case Study: Developing Dairy Farming Productivity in Colombia

Caqueta is a remote region in the South of Colombia, and a good example of productivity improvement at dairy farms through Nestlé technical assistance. Nestlé arrived in the region 30 years ago and built a milk plant, that capacity of which has been increased several times since then. Milk cooling equipment was deployed at the farm level, and by 1999, with the help of credit institutions, farmers owned 170 cooling tanks. After an average growth of milk production of 7% per year, the region's 2500 dairy farmers are now producing 400 tonnes of fresh milk each day (Nestlé 2006).

Nestlé employs 28 technicians to provide milk farmers with technical assistance regarding cattle genetics, animal nutrition, farm development, and even farm road construction. To ensure that cattle consume sufficient nutrients, particularly protein, Nestlé and the International Center for Tropical Agriculture promoted the cultivation of Brachiaria grass, assisted by the introduction of legumes to increase soil nitrogen. To compensate for the Brachiaria's nutritional deficiencies, Nestlé helped to develop a phosphorus-rich mineral salt, which is now used as supplementary feed.

Nestlé supported the development of a cattle breed adapted to the region's humid tropical climate. By crossing Brahman and Girolando breeds, a new breed with high environmental adaptability and better milk production was created.

Through loans to farmers, Nestlé supported the construction of electric fences, allowing land to be subdivided into smaller plots, so land productivity could be increased through improved pasture management. Nestlé also encouraged the construction of small dams to provide drinking water for cattle throughout the pasture area.

With all these changes, cows now produce on average 9 to 10 liters of milk per day, where 30 years ago the production was only 2 liters per animal.

Conclusions

Gaps

Nestlé has implemented a comprehensive and worldwide strategy and process to improve the environmental performance of its factories and has achieved significant results in reducing the impact of its food processing operations on natural resources.

In its upstream supply chains, Nestlé businesses in many countries have carried out projects to ensure the quality and safety of agricultural raw materials, to combat animal diseases, and to improve the productivity and environmental impact of dairy farming. Although these supply chain projects and their results are impressive, the challenge for the future will be to move beyond the many individual initiatives to an overall strategy and process throughout Nestlé's global supply chain (Nestlé 2006).

In general, farm assurance schemes, while ensuring food safety and quality, still show certain limitations in addressing the environmental and social sustainability of farming:

- Most schemes currently focus more on raw material safety and quality than on the environmental or social impact of farming. Though environmental requirements are already included, they will need to be gradually expanded and further disseminated.
- The multitude of existing farm assurance schemes and sustainability standards in different countries leads to a lack of transparency for farmers and processors and makes it difficult to decide on the most appropriate scheme. Different processors may ask the farmer to implement different schemes. Coordination within the food industry is an answer to this problem.
- Indicators used are sometimes qualitative rather than quantitative (e.g., survey questions ask, "Do you have a nutrient management plan?" more usually than "How much nitrogen fertilizer per area unit do you apply?").
- The concept is relatively static. It implements recognized good farming practices, which in themselves are an innovative approach, but beyond this it does not specifically promote innovative new technologies. Therefore additionally to farm assurance schemes, there is a need for pilot projects testing innovations in reducing the environmental impact of farming.
- Farm assurance schemes for livestock farms may cover the direct environmental impact of the farm itself, but generally not the environmental impacts created through the production of purchased feed concentrates. This can only be ensured through regulatory requirements, or programs of the feed industry.
- There are limits to telling farmers in the mainstream sector what they have to do. Farmers need to be convinced that it makes sense to introduce the recommended practices.
- Sometimes the schemes are not based on joint stakeholder initiatives; consequently, farmers may oppose them and not take real ownership.

An additional concern is that farm assurance schemes are only partly implemented in the food industry. Currently they are in place in a number of developed countries or for supply to developed countries and for specific groups of raw materials. In some countries, there are government or industrywide schemes, but in others, the schemes are mainly implemented by a few big companies that have a higher exposure to or higher concern about food safety risks.

In addition, farm assurance schemes are not feasible everywhere. It is hardly possible for the food industry to introduce and maintain lots of sophisticated requirements for large numbers of small-scale farmers. In such cases, more appropriate solutions need to be sought (e.g., farmer education and working through farmer cooperatives).

Many of the Nestlé initiatives described in this chapter focus on direct procurement—wherein Nestlé buys agricultural raw materials directly from farmers or farmer cooperatives. So far there has been less focus on cases when primary-processed raw materials are purchased from other companies or through the trade. The challenge will be to ensure best farming and sourcing practices not only in Nestlé's direct supply chains from farms but also in the more complex supply chains of commodities and other primary-processed raw materials. This can be achieved through food industry initiatives (e.g., SAI Platform) and through Nestlé's strategic partnerships with suppliers. The Nestlé Supplier Code is addressing this issue and demanding that suppliers improve the sustainability of their operations (Nestlé 2008).

It is practically impossible for a food company to completely audit complex upstream supply chains between its factories and supplying farmers. Nestlé does audit all its own direct raw material suppliers but to a lesser extent the suppliers of its suppliers. When the food industry approaches traders to implement additional demands on upstream supply chains, there is a risk that it may face resistance or demands for significant price premiums.

Ongoing and Future Developments and Responses

As described earlier, Nestlé has already started to respond to the different gaps and challenges by focusing on the following actions and topics:

- Elaboration and implementation of farm assurance schemes in additional countries where Nestlé sources milk from farmers: there are ongoing activities in Spain, Argentina, Brazil, Mexico, and South Africa.
- Building more and more sustainability requirements into the farm assurance manuals and farmer communications.
- Setting standards for semiprocessed raw materials and discussing these standards with suppliers.
- Engaging in new projects to optimize the usage of water in agriculture and to protect water sources such as springs.
- Elaborating plans about how to address the challenge of reducing greenhouse gas emissions caused by livestock farming.
- Founding and promoting industry organizations that address the challenges related to sustainability of livestock farming (e.g., active participation in the SAI Platform Working Group on Dairy).

SAI Platform—Cross-Sector Partnerships for Sustainability

The Sustainable Agriculture Initiative Platform (SAI Platform) is a food industry association to promote sustainable agriculture, addressing the economic, environmental, and social aspects of agricultural production. The aim is to make sustainable agriculture mainstream, working from a continuous improvement approach, building on existing initiatives, and promoting economically viable production.

Following an agreement between the CEOs of Nestlé, Danone, and Unilever, the SAI Platform was created by the three companies in June 2002. Since then, more international food companies have joined, raising the number of members to 23 in early 2008. To ensure close cooperation with the Confederation of the Food and Drink Industries of the European Union (CIAA), the SAI Platform is based at CIAA's premises in Brussels.

SAI Platform has launched working groups for specific raw materials (e.g., milk, cereals, coffee, fruits, and vegetables) to elaborate guidelines and indicators for sustainable agricultural practices. For coffee, pilot projects have been implemented to test the guidelines, while projects for other raw materials have also started. The next step will be to learn from the results of the pilot projects and to develop practical tools and indicators to measure the impact of changing practices. The Common Code for the Coffee Community (4C) is one example of such a scheme already working, whereas livestock-related initiatives have not yet reached such a stage.

Through active communication and workshops, the SAI Platform is engaging with a broad range of external stakeholders. The SAI Platform has redefined its role in promoting sustainable agriculture. Its focus will include not only the elaboration and promotion of commodity-specific sustainable practices but also the coordination of positions and pilot projects of the participating food companies in the context of sustainable agriculture throughout the upstream supply chain. The SAI Platform will also evaluate existing approaches for sustainable agriculture and determine if they are in accordance with the principles and practices elaborated by the commodity-specific SAI Platform Working Groups. The overall aim is to become the central reference center for the food industry on sustainable agriculture.

Importance of Stakeholder Initiatives

Industry initiatives (such as SAI Platform) or dairy sector stakeholder initiatives must play a crucial role in setting standards and applying them throughout the industry, as well as in verifying agricultural systems and farm assurance schemes that meet high requirements on raw material quality and sustainability.

In order to measure progress in achieving sustainable production, indicators and measurable objectives will need to be developed, agreed, and implemented throughout the industry and throughout the farming sector. Education of farmers and awareness creation are also needed to promote sustainable farming methods. Providing incentives for sustainable production and setting of minimum standards may be the next steps.

The food industry alone is not able to address all the various issues related to environmental impacts of livestock farming. Therefore all the different stakeholders need to work together to elaborate and implement sustainable farming practices. For this purpose, the SAI Platform Working Group Dairy is in continuous discussion with the International Dairy Federation and farmers' organizations to reach a common understanding of sustainable dairy farming practices and strategies for implementing them.

References

Ederer P. and R. Goldberg. 2007. *Nestlé Water Management Strategy: A New Competitive Advantage*. Wageningen: Wageningen University.

Goldberg, R. and K. Herman. 2005. *Nestlé's Milk District Model: Economic Development for a Value-Added Food Chain and Improved Nutrition*. Boston: Harvard Business School.

Heer, J. 1966. *World Events 1866–1966: The First Hundred Years of Nestlé*. Lausanne: Imprimeries Réunies.

IDF and FAO. 2004. *Guide to Good Dairy Farming Practice*. Rome: FAO.

Nestec. 2008. The Nestlé Supplier Code. Cited November 2008. Availableathttp://www.Nestlé.com/AllAbout/Suppliers/Introduction.htm.

Nestlé. 2000. Environment Progress Report 2000. Cited 17 March 2008. Available at http://www.Nestlé.com/Resource.axd?Id=F3D57766-6E90-4539-867C-5A507FD4D219.

Nestlé. 2003. Nestlé and Water: Sustainability, Protection, Stewardship. Cited 17 March 2008. Available at http://www.Nestlé.com/Resource.axd?Id=F3D57766-6E90-4539-867C-5A507FD4D219.

Nestlé India. 2007. Nestlé India Creating Shared Value. Cited December 2007. Available at http://www.Nestlé.com/SharedValueCSR/Overview.htm.

Nestlé S.A. 2004. Nestlé Corporate Business Principles, Third Edition. Cited September 2004. Available at http://www.Nestlé.com/Resource.axd?Id=70014B84-A4FC-4F82-BFA0-23939DC52E9D.

Nestlé S.A. 2005. Nestlé's Sustainable Agriculture Initiatives—A Contribution to Sustainable, Profitable Growth. Internal Nestlé papers.

Nestlé S.A. 2006. The Nestlé Concept of Corporate Social Responsibility as Implemented in Latin America. Cited March 2006. Available at http://www.Nestlé.com/Resource.axd?Id=7F757F03-B601-49EA-8395-E24095231754.

Nestlé S.A. 2007a. Driving Sustainable Profitable Growth, Nestlé Management Report 2006. Cited March 2007. Available at http://www.Nestlé.com/Resource.axd?Id=9C91D28D-E55C-427D-839C-1EB9F9E695E5.

Nestlé S.A. 2007b. The Nestlé Water Management Report. Cited March 2007. Available at http://www.Nestlé.com/Resource.axd?Id=F6CDF592-666D-4084-8A3E-D6EE81287686.

Nestlé S.A. 2008. The Nestlé Creating Shared Value Report. Cited March 2008. Available at www.Nestlé.com/csv.

SAI Platform. 2003. Sustainable Agriculture Initiative Information Pack. Cited November 2006. Available at http://www.saiplatform.org/about-us/PDF/SAI-Infopack.pdf.

Steinfeld, H., et al. 2006. *Livestock's Long Shadow—Environmental Issues and Options*. Rome: FAO.

10

Cross-Cutting Observations and Conclusions

Pierre Gerber, Harold A. Mooney, Jeroen Dijkman, Shirley Tarawali, and Cees de Haan

The livestock revolution remains one of the main features of the global livestock sector. All the chapters in this volume show evidence of some, if not all, of the main features of the livestock revolution (i.e., demand-led growth and structural change) (Delgado et al., 1999; Steinfeld et al., 2006). The countries and regions analyzed are at different stages of this revolution. The largest increases in production and per capita consumption were experienced in East and Southeast Asia, especially China, which saw strong growth and rapid structural changes, especially in pork and poultry production and consumption. Similar trends are recorded in Latin America, also for pork and poultry, and in India for poultry, although with more limited structural change. Some countries have now passed the peak of the livestock revolution (e.g., China and, for a longer time, most OECD countries). In general, countries also host different stages of the process at the same time, with some areas or food chains more engaged in the structural change it implies, and others less so.

But the livestock revolution is not universal, and in some of the less developed areas described in this volume it may never take place. The driver of increasing demand is found almost everywhere, but various factors prevent some or all of the features of the revolution from appearing. These include cheap imports competing with local production for urban markets, the remoteness and lack of infrastructure found in specific areas, and also agroecological factors, for example, climate variability, abundance of land, or poor land fertility—both of the latter favoring extensive systems.

A cross-cutting observation about the livestock revolution is that it has the potential for both positive and negative social and environmental consequences, which have, however, been notably negative for significant segments of the population. Environmental concerns are pronounced, as highlighted in Denmark, the United States and India (pollution of land and water), and Brazil (deforestation for grazing and animal feed production). Negative environmental consequences are largely related to the lack of effective institutions and policies to guide and regulate the sector. Authors of the chapters on the Horn of Africa, India, the United States, and West Africa also note that the sector's economic growth has not reached smallholders, and they raise doubts about the distributive effects of the prevailing patterns of livestock growth. It remains, however, relatively unclear if small-scale livestock keepers are being driven out of production against their will, or if they voluntarily leave the sector as the attractiveness of keeping a few livestock diminishes in comparison with nonfarm employment opportunities. On the positive side, for the urban consumers who have driven the process, the livestock revolution has resulted in stable availability of cheap animal products. The chapter on China also observes that the structural change the sector is undergoing allows for greater resource use efficiency and cleaner production.

The chapters in this volume recognize the key role played by the private sector in the livestock revolution. Indeed, the observed structural changes are essentially a result of the private sector's response to demand growth. The chapter on Nestlé illustrates how private companies react to diet changes and related changes in consumers' requirements for the "functional and emotional" content of products. The limited participation of smallholders in sector growth is one of the main issues commonly associated with the livestock revolution. However, the Nestlé chapter illustrates in detail how, in some cases, functional business models can be set up to connect smallholder producers to international markets. Although this type of setup is more viable in the dairy sector than in less labor-intensive animal production activities (e.g., poultry), the Nestlé chapter shows that large companies can provide solutions for smallholders too. On the environmental side the Nestlé chapter also shows that sustainability is of increasing importance on the private sector agenda, pushed by public policies and consumer preferences as well as by the need to secure supply of raw agricultural products.

Competition for land is a growing issue, already intense in almost all parts of the world. Several chapters in the volume draw attention to the consequences for the livestock sector. In Africa, where agricultural intensification is limited, this seems to be most severe. Both chapters on Africa show how increasing pressures on land, combined with land tenure and water access issues, are substantially reducing the mobility of pastoralists, often leading to conflicts and land degradation. In the highlands of the Horn of Africa, subdivisions of land have made productive units very small such that they are hardly able to support livestock. In Europe, milk quotas and the requirement for farmers to comply with manure-spreading regulations are among the drivers pushing land prices up, whereas land shortage in China results in growing feed imports: already a net soy importer since the 1980s, China has now also become a net maize importer. Increased pressure on land also fosters the conversion of natural areas, as seen in parts of Latin America where the land area for livestock production expands into forests, with negative environmental implications, including a major contribution to global warming.

Land shortage is already a serious challenge and poised to increase in the future, as food demand and new land-using sectors such as biofuels continue to increase the demands on land. It can be anticipated that land resources will play a major role in shaping tomorrow's livestock sector. Where livestock production is market oriented and competes with other land users, land availability will become a limiting factor to growth and thus affect countries' relative competitiveness and the location of the industry. On the other hand, where livestock uses marginal lands, access to land and the security of user rights will have significant implications in terms of smallholders' involvement and environmental impacts. In all cases, the sector will most probably continue its evolution toward more land-intensive systems, yielding increasing amounts of animal products per unit of land. Similar trends may actually arise for water as property rights are clarified and the price of this resource increases.

As mentioned in the introduction, the focus of this volume is on responses. Responses are the actions and interventions by which stakeholders (e.g., public sector, NGOs, producers, supermarkets) change the sector so it better responds to societies' needs for safe and affordable animal products but also to other development objectives such as poverty reduction, environmental protection, and so forth. The chapters concur that public policies have often neglected or been ineffective in guiding the livestock sector and particularly in addressing the negative consequences of livestock sector changes. The authors point to lack of awareness among policy makers of the needs of the sector, shortcomings in policy design, and rampant lack of enforcement. The latter has been at times a consequence of the former, for example, in the case of overambitious objectives or lack of phasing. At other times, insufficient enforcement has been caused by lack of resources or political will to implement measures that are unpopular among farmers or vested interests. Unaddressed environmental issues are reported in almost all chapters, and the social impacts identified in the India, US, and Africa chapters are also not being addressed adequately. On a more specific note, the Horn of Africa chapter explains that regional efforts to control or eradicate animal diseases have been of limited success, whereas the quality control schemes developed by several countries have been too centralized, and often too remote from to consumer needs, to be effective.

However the chapters also illustrate some positive results of public policies. In Costa Rica, a change in policy mixes—especially the development of payment for environmental services and the discontinuation of direct subsidies to production—has effectively resulted in a reversal of the deforestation rate. In Denmark, two decades of policy development closely involving the private sector and focusing on restoring land–livestock balances as a key principle have been successful in reducing pollution loads in streams, aquifers, and coastal areas. In the Horn of Africa, the participatory planning of communal land use at the district level (e.g., in Tanzania and Uganda) is proving an effective way to allocate and manage natural resources. Finally, in China, land degradation is being significantly reduced through range management control and support for stall-feeding.

Several policy lessons can be drawn from the responses analyzed throughout this volume. Regarding the policy process, the successful examples show that designing and implementing policies is a continuous trial and error effort. Policies need to be continuously adapted to changing environmental and socioeconomic conditions, farming systems, and societal expectations. The highly complex context in which policies are designed is a further challenge forcing legislators to try "best guesses," monitor results, and adjust the policy mix accordingly. This requires resources, strong analytical skills, continuity in the policy making effort, and ultimately strong institutions. It also requires the support of civil society. Studies suggest that for some types of environmental issues, impacts increase as poor economies begin to develop and then decrease as economies become rich and civil society requires a cleaner environment. This inverted U-shaped relationship between income and environmental damages is frequently referred to as the environmental Kuznets curve (Hartmana and Kwon 2005; Cole 2004).

Awareness and willingness to bear the costs of action, however, develop slowly, and crises are often reached before action is taken, as shown by the chapters on deforestation (Central America) and water pollution (EU).

The cases of Denmark and the Horn of Africa highlight the need to involve the private sector or existing community representatives from the onset of the policy development work, to ground the policy design on a shared understanding of problems and objectives, and to improve the social acceptance of policy measures. The authors of the Danish chapter further identify the allocation of public resources and the collaboration of all ministries as obvious but indispensable elements for success.

Regarding policy instruments, authors in the volume generally recognize the need to combine measures into balanced and enforceable policy mixes, as for example adding in positive incentives to ease enforcement of regulations. In the beginning of the policy process, as the first measures are being enforced and as institutions are still weak, there is also a need to carefully identify the targets of policies and rely more on positive incentives and voluntary approaches. It is only progressively and often for new operations that more stringent regulations may be enforced. Lessons learned regarding policy tools are necessarily specific, given the need to tailor the design of policy measures to local objectives and context. The volume provides a variety of experience in the practical implementation of policy tools; for example, the use of communal land tenure and communal regulation of the access to land in the Horn of Africa, wastewater discharge standards and mandatory range management in China, payment for environmental services in Costa Rica, or watershed management programs in India.

The livestock landscape depicted in this volume is one of complexity, where livestock interact with a variety of natural resources, social issues, and development objectives. It is also one of superimposed patterns, where farming techniques and management systems of different standards coexist, and where local endogenous development processes are increasingly influenced by the intrusion of international trade. It is thus not surprising that the experience emerging from the volume calls for tailored responses, progressive policy development processes relying on multidisciplinary analysis, and the need to carefully balance development objectives when guiding the livestock sector.

References

Hartmana, R. and O. S. Kwon. 2005. Sustainable growth and the environmental Kuznets curve. *Journal of Economic Dynamics and Control* 29: 1701–1736.

Cole, M. A. 2004. Trade, the pollution haven hypothesis and the environmental Kuznets curve: examining the linkages. *Ecological Economics* 48: 71–81.

Delgado, C., M. Rosegrant, H. Steinfeld, S. Ehui, and C. Courbois, 1999. *Livestock to 2020: The Next Food Revolution*. International Food Policy Research Institute, Food and Agriculture Organisation of the United Nations, International Livestock Research Institute. Washington.

Steinfeld, H., P. Gerber, T. Wassenaar, V. Castel, M. Rosales, and C. de Haan. 2006. *Livestock's Long Shadow: Environmental Issues and Options*. FAO. Rome.

Acronyms and Abbreviations

ACU	adult cattle units
ANPP	aboveground net primary production
BSE	bovine spongiform encephalitis (mad cow disease)
CAADP	Comprehensive Africa Agriculture Development Program of NEPAD
CAP	Common Agricultural Policy (of the EU)
CATIE	Centre Agronómico Tropical de Investigación y Enseñenza
CBD	Convention on Biological Diversity
CBPP	contagious bovine pleuropneumonia
CDM	clean development mechanism
CER	certified emissions reduction
CILSS	Comité Permanent Inter-Etats de Lutte Contre la Sécheresse dans le Sahel
CIRAD	Centre de coopération internationale en recherche agrononique pour le développement (French Agricultural Research Centre for International Development)
CSF	classical swine fever
ECOWAS	Economic Community of West African States
EMPRES	Emergency Prevention System for Transboundary Animal and Plant Pests and Diseases
ET	evapotranspiration
EU	European Union
FAO	Food and Agriculture Organization, United Nations
FMD	foot-and-mouth disease
GDP	gross domestic product
GHGs	greenhouse gases
GNI	gross national income
GNP	gross national product
GLEWS	Global Early Warning and Response System for Major Animal Diseases, including Zoonoses
GREP	Global Rinderpest Eradication Programme
HPAI	highly pathogenic avian influenza
IAPS	industrialized (intensive) animal production systems
ILRI	International Livestock Research Institute
INRA	Institut National de la Recherche Agronomique (French National Institute for Agricultural Research)
IPCC	Intergovernmental Panel on Climate Change
IRD	Institut de Recherche pour le Développement (Research Institute for Development [France])
ISRA	Institut Sénégalais de Recherches Agricoles (Senegalese Institute for Agricultural Research)
IUCN	International Union for Conservation of Nature
IVM	integrated vector management
KES	Kenyan Shillings
LEAD	Livestock, Environment and Development Initiative (FAO)
MAP	mean annual precipitation
MDG	Millennium Development Goal
NDVI	Normalized Difference Vegetative Index
NEPAD	New Partnership for Africa's Development
NGO	nongovernmental organization
NPP	net primary production
OECD	Organisation for Economic Cooperation and Development
OIE	World Organisation for Animal Health (from original name in French)
PACE	Pan-African Programme for the Control of Epizootics
PARC	Pan-African Rinderpest Campaign
PAU	Politique Agricole de l'UEMOA
PES	payment for environmental services
PPP	purchasing power parity
PPR	Peste des Petits Ruminants
REDD	reduced emissions from deforestation and degradation
SARS	severe acute respiratory syndrome
SCOPE	Scientific Committee on Problems of the Environment
SHL	Swiss College of Agriculture
SPS	sanitary and phytosanitary
TB	Bovine tuberculosis
TFP	total factor productivity
TLU	tropical livestock unit

TRQ	tariff rate quota
UEMOA	Union Economique et Monétaire Ouest-Africaine
UNCCD	United Nations Convention to Combat Desertification
UNDP	United Nations Development Programme
UNEP	United Nations Environment Programme
UNFCCC	United Nations Framework Convention on Climate Change
URAA	Uruguay Round Agreement on Agriculture
USAID	U.S. Agency for International Development
USDA	U.S. Department of Agriculture
USDA-ARS	U.S Department of Agriculture–Agricultural Research Service
USGS	United States Geological Survey
WFP	World Food Programme

Chemical Symbols, Compounds, and Units of Measurement

CH_4	methane
CO	carbon monoxide
CO_2	carbon dioxide
GtC-eq	gigatons of carbon equivalent
N	nitrogen
N_2O	nitrous oxide
NOx	nitrogen oxides
ppmv	parts per million by volume
SO_2	sulfur dioxide
Teragram	10^{12} grams
M–Mega	SI system of units denoting a factor of 10 to the sixth power, or 1,000,000 (one million)

Glossary

The note number following each term indicates the source reference at the end of the glossary for the term and definition.

Adaptive management[4] The mode of operation in which an intervention (action) is followed by monitoring (learning), with the information then being used in designing and implementing the next intervention (acting again) to steer the system toward a given objective or to modify the objective itself.

Agroecological classification[1] Based on length of available growing period (LGP), which is defined as the period (in days) during the year when rainfed available soil moisture supply is greater than half potential evapotranspiration (PET). It includes the period required for evapotranspiration of up to 100 mm of available soil moisture stored in the soil profile. It excludes any time interval with daily mean temperatures less than 5°C.

- Arid LGP less than 75 days
- Semiarid LGP in the range 75–180 days
- Subhumid LGP in the range 180–270 days
- Humid LGP greater than 270 days

Agropastoralism[2] A production system where all of the family and livestock are sedentary.

Arid zones[1] The areas where the growing period is less than 75 days, too short for reliable rainfed agriculture. The coefficient of variation of the annual rainfalls is high, up to 30%. Abiotic factors, especially rainfall, determine the state of the vegetation. The *nonequilibrium* theory applies in this environment. The main systems found in these zones are the mobile systems on communal lands. Some cases of ranching are present.

Benefits transfer approach Economic valuation approach in which estimates obtained (by whatever method) in one context are used to estimate values in a different context.

Biodiversity[4] The variability among living organisms from all sources including terrestrial, marine, and other aquatic ecosystems and the ecological complexes of which they are part; this includes diversity within and among species and diversity within and among ecosystems.

Biomass[4] The mass of living tissues in ecosystems.

Biosolids[3] Organic solids resulting from wastewater treatment that can be usefully recycled.

Capacity building[4] A process of strengthening or developing human resources, institutions, or organizations.

Capital value (of an ecosystem)[4] The present value of the stream of future benefits that an ecosystem will generate under a particular management regime. Present values are typically obtained by discounting future benefits and costs; the appropriate rates of discount are often a contested issue, particularly in the context of natural resources.

Change in productivity approach[4] Economic valuation techniques that value the impact of changes in ecosystems by tracing their impact on the productivity of economic production processes. For example, the impact of deforestation could be valued (in part) by tracing the impact of the resulting changes in hydrological flows on downstream water uses such as hydroelectricity production, irrigated agriculture, and potable water supply.

Characteristic scale[4] The typical extent or duration over which a process is most significantly or apparently expressed.

Constituents of well-being[4] The experiential aspects of well-being, such as health, happiness, and freedom to be and do, and, more broadly, basic liberties.

Cultural services[4] The nonmaterial benefits people obtain from ecosystems through spiritual enrichment, cognitive development, reflection, recreation, and aesthetic experience, including, for example, knowledge systems, social relations, and aesthetic values.

Decision maker[4] A person whose decisions and actions can influence a condition, process, or issue under consideration.

Decomposition[4] The ecological process carried out primarily by microbes that leads to a transformation of dead organic matter into inorganic mater; the converse of biological production. For example, the transformation of dead plant material, such as leaf litter and dead wood, into carbon dioxide, nitrogen gas, and ammonium and nitrates.

Driver[4] Any natural or human-induced factor that directly or indirectly causes a change in an ecosystem (system).

Driver, direct[4] A driver that unequivocally influences ecosystem (system) processes and can therefore be identified and measured to differing degrees of accuracy.

Driver, endogenous[4] A driver whose magnitude can be influenced by the decision maker. The endogenous or exogenous characteristic of a driver depends on the organizational scale. Some drivers (e.g., prices) are exogenous to a decision maker at one level (a farmer) but endogenous at other levels (the nation-state).

Driver, exogenous[4] A driver that cannot be altered by the decision maker. See also *endogenous driver*.

Driver, indirect[4] A driver that operates by altering the level or rate of change of one or more direct drivers.

Ecological footprint[4] The area of productive land and aquatic ecosystems required to produce the resources used and to assimilate the wastes produced by a defined population at a specified material standard of living, wherever on Earth that land may be located.

Ecosystem[4] A dynamic complex of plant, animal, and microorganism communities and their nonliving environment interacting as a functional unit.

Ecosystem approach[4] A strategy for the integrated management of land, water, and living resources that promotes conservation and sustainable use in an equitable way. An ecosystem approach is based on the application of appropriate scientific methodologies focused on levels of biological organization, which encompass the essential structure, processes, functions, and interactions among organisms and their environment. It recognizes that humans, with their cultural diversity, are an integral component of many ecosystems.

Ecosystem function[4] An intrinsic ecosystem characteristic related to the set of conditions and processes whereby an ecosystem maintains its integrity (such as primary productivity, food chain, biogeochemical cycles). Ecosystem functions include such processes as decomposition, production, nutrient cycling, and fluxes of nutrients and energy.

Ecosystem health[4] A measure of the stability and sustainability of ecosystem functioning or ecosystem goods and services that depends on an ecosystem being active and maintaining its organization, autonomy, and resilience over time. Ecosystem health contributes to human well-being through sustainable ecosystem services and conditions for human health.

Ecosystem interactions[4] Exchanges of materials and energy among ecosystems.

Ecosystem properties[4] The size, biodiversity, stability, degree of organization, internal exchanges of materials and energy among different pools, and other properties that characterize an ecosystem.

Ecosystem services[4] The benefits people obtain from ecosystems. These include provisioning services such as food and water; regulating services such as flood and disease control; cultural services such as spiritual, recreational, and cultural benefits; and supporting services such as nutrient cycling that maintain the conditions for life on Earth. The concept of ecosystem goods and services is synonymous with ecosystem services.

Extensive[5] A livestock production system that uses predominantly noncommercial inputs to the system.

Externality[4] A consequence of an action that affects someone other than the agent undertaking that action and for which the agent is neither compensated nor penalized. Externalities can be positive or negative.

Grazing system[1] The grazing system is predominantly dependent on the natural productivity of grasslands and is therefore defined largely by the agroecological zone. The populations relying on these systems are generally referred to as pastoralist groups, with their main differences defined by their mobility in response to environmental variability. At one extreme the nomadic groups are highly mobile, living in areas with major differences in both seasonal and annual climatic patterns. At the other end agropastoralists and ranchers operate sedentary systems where seasonal and annual climatic variations are minor.

Grazing systems[1] Livestock systems in which more than 90% of dry matter fed to animals comes from rangelands, pastures, annual forages, and purchased feeds and less than 10% of the total value of production comes from nonlivestock farming activities. Annual stocking rates are less than 10 livestock units per hectare of agricultural land. Grazing systems are described for each of the following regions arid, semiarid, subhumid and humid, temperate, and tropical highlands.

Indicator[4] Information based on measured data used to represent a particular attribute, characteristic, or property of a system.

Industrial livestock system[1] Industrial systems are primarily directed at producing high-quality animal protein and other animal products for the urban markets. As a result of this market demand, intensive animal production and processing often take place near urban areas, while primary feed production takes place in distant rural areas. Industrial systems are generally considered modern and efficient, requiring a high level of knowledge and skill. Production techniques are more or less independent of the agroecological zone and of the climate, which explains the worldwide occurrence of the industrial system. These systems have average stocking rates greater than 10 livestock units per hectare of agricultural land, and <10% of the dry matter fed to livestock is produced on the farm. (This is similar to Seré and Steinfeld's classification Landless Livestock Production Systems.) Industrial livestock production systems are associated with a concentration of animals into large units, generally concentrating on a single species. They produce large volumes of waste material, can lead to high animal and human health risks, and pay less attention to animal welfare. Industrial production also occurs in small units operated by specialized smallholders as part of the mixed livestock system.

Intensive[5] A livestock production system that relies on commercial inputs and trade.

Land cover[4] The physical coverage of land, usually expressed in terms of vegetation cover or lack of it. Influenced by but not synonymous with *land use*.

Land use[4] The human utilization of a piece of land for a certain purpose (such as irrigated agriculture or recreation). Influenced by but not synonymous with *land cover*.

Landscape[4] An area of land that contains a mosaic of ecosystems, including human-dominated ecosystems. The term *cultural landscape* is often used when referring to landscapes containing significant human populations.

Livestock mobility[2] Can be divided into macromobility, which refers to large movements (such as transhumance between dry season and wet season pastures) and micromobility, referring to daily or frequent movement between microniches within the same pasture.

Livestock unit (LU)[3] A unit used to compare or aggregate numbers of animals of different species or categories. Equivalences are defined on the feed requirements (or sometimes nutrient excretion). For example for the EU, one 600 kg dairy cow producing 3000 liters of milk per year equals 1 LU, a sow equals 0.45 LU, and a ewe equals 0.18 LU.

Manure supernatant[3] The upper liquid fraction after sedimentation of liquid waste or liquid manure.

Mixed systems[1] Farming systems conducted by households or by enterprises where crop cultivation and livestock rearing are more or less integrated components of one single farming system. The more integrated systems are characterized by interdependency between crop and livestock activities. They are basically resource driven, aiming at an optimal circulation of locally available nutrients (nutrient circulation systems); for example, ecological farming and some, but not all, low external input agriculture (LEIA) systems. The less integrated systems are those where crop and livestock activities can make use of, but not rely on, each other. In general one or both activities are demand driven, supported by external inputs (nutrient throughput systems); for example, high external input agriculture (HEIA) systems. Mixed systems are those systems in which more than 10% of the dry matter fed to livestock comes from crop by-products and/or stubble or more than 10% of the value of production comes from nonlivestock farming activities.

Nomad[2] Production system that is highly mobile but does not necessarily return to a "base" every year and does not include cultivation (e.g., nomads of the Sahara). See also *pastoralism*.

Organic residues[3] Organic material resulting from dead plant material or by-products from processing organic materials of the food industry or other industry.

Organic wastes[3] A general term for any wastes from organic rather than inorganic origin and so containing carbon (e.g., livestock manure, sewage sludge, abattoir wastes).

Pastoralism[2] Predominantly extensive production system that depends on livestock for more than 50% of income; includes nomads, transhumants, and semitranshumants.

Precautionary principle[4] The management concept stating that in cases "where there are threats of serious or irreversible damage, lack of full scientific certainty shall not be used as a reason for postponing cost-effective measures to prevent environmental degradation," as defined in the Rio Declaration.

Prediction (or forecast)[4] The result of an attempt to produce a most likely description or estimate of the actual evolution of a variable or system in the future. See also *projection* and *scenario*.

Primary production[4] Assimilation (gross) or accumulation (net) of energy and nutrients by green plants and by organisms that use inorganic compounds as food.

Private costs and benefits[4] Costs and benefits directly felt by individual economic agents or groups as seen from their perspective. (Externalities imposed on others are ignored.) Costs and benefits are valued at the prices actually paid or received by the group, even if these prices are highly distorted. Sometimes termed "financial" costs and benefits. Compare *social costs and benefits*.

Projection A potential future evolution of a quantity or set of quantities, often computed with the aid of a model. Projections are distinguished from "predictions" in order to emphasize that projections involve assumptions concerning, for example, future socioeconomic and technological developments that may or may not be realized; they are therefore subject to substantial uncertainty.

Provisioning services[4] Products obtained from ecosystems, including, for example, genetic resources, food and fiber, and freshwater.

Resilience[4] The capacity of a system to tolerate impacts of drivers without irreversible change in its outputs or structure.

Responses[4] Human actions, including policies, strategies, and interventions, to address specific issues, needs, opportunities, or problems. In the context of ecosystem management, responses may be of a legal, technical, institutional, economic, and behavioral nature and may operate at a local or micro, regional, national, or international level and at various time scales.

Scale[4] The physical dimensions, in either space or time, of phenomena or observations.

Scenario[4] A plausible and often simplified description of how the future may develop, based on a coherent and internally consistent set of assumptions about key driving forces (e.g., rate of technology change, prices) and relationships. Scenarios are neither predictions nor projections and sometimes may be based on a "narrative storyline." Scenarios may be derived from projections but are often based on additional information from other sources.

Semitranshumant[2] Production system where only part of the family and/or livestock is seasonally mobile and the rest is sedentary in one of the seasonal bases, practicing cultivation (e.g., Dinka tribe of Sudan and Karimojong tribe of Uganda).

Social costs and benefits Costs and benefits as seen from the perspective of society as a whole. These differ from private costs and benefits in being more inclusive (all costs and benefits borne by some member of society are taken into account) and in being valued as social opportunity costs rather than market prices, where these differ. Sometimes termed "economic" costs and benefits.

Solid manure[3] Manure from housed livestock that does not flow under gravity, cannot be pumped but can be stacked in a heap. May include manure from cattle, pigs, poultry, horses, sheep, goats, camelids, and rabbits. There are several types of solid manure arising from different types of livestock housing, manure storage, and treatment.

Stakeholder[4] An actor having a stake or interest in a physical resource, ecosystem service, institution, or social system, or someone who is or may be affected by a public policy.

Stocking rate[3] The number of livestock (or livestock units) per unit area of land.

Sustainability[4] A characteristic or state whereby the needs of the present and local population can be met without compromising the ability of future generations or populations in other locations, in order to meet their needs.

Transhumant[2] Production system that is highly mobile but moves between definite seasonal bases every year (e.g., Samburu of Kenya); it may include a nonsedentary form of cultivation (e.g., Zaghawa of Chad). See also *pastoralism*.

Valuation[4] The process of expressing a value for a particular good or service in a certain context (e.g., of decision making), usually in terms of something that can be counted, often money, but also through methods and measures from other disciplines (sociology, ecology, and so on).

Well-being[4] A context- and situation-dependent state, comprising basic material for a good life, freedom and choice, health, good social relations, and security.

Notes

[1]Excerpts from the LEAD Toolkit (http://www.fao.org/lead/) and Seré and Steinfeld (1996). World livestock production systems: current status, issues and trends. Animal Production and Health Paper No. 127. FAO, Rome.

[2]*Source:* Extract from *Pastoral development in Africa*. Proceedings of the first technical consultation of donor and international development agencies, Paris, December 1993. UNSO/UNDP, 1994.

[3]*Source: RAMIRAN Glossary of terms on manure management 2003.*

[4]*Source:* Millennium Ecosystem Assessment, 2005. *Ecosystems and human well-being: Synthesis*. Island Press, Washington, DC.

[5]Derived from LCL Consultation working group discussion.

List of Editors and Contributors

Editors

Pierre Gerber
Livestock Policy Officer
Livestock Information
Sector Analysis and Policy Branch
FAO

Harold A. Mooney
Professor of Environmental Biology
Stanford University

Jeroen Dijkman
Livestock Development Officer
Pro-Poor Livestock Policy Initiative
Sector Analysis and Policy Branch
FAO

Shirley Tarawali
Director
People, Livestock, and the Environment Theme
ILRI

Cees de Haan
Retired Senior Livestock Development Advisor
Consultant, World Bank

Contributors

Ke Bingsheng
Research Center for Rural Economy
Ministry of Agriculture, China

C. T. Chacko
Livestock Consultant
Piravom, Kerala, India

Abdou Fall
Senior Researcher
ILRI
Bamako, Mali

Gopikrishna
Sociologist
Belgaum, Karnataka, India

Muhammad Ibrahim
Research Professor and Leader of Livestock and Environmental Management Program
CATIE

Torben Moth Iversen
National Environmental Research Institute
University of Aarhus

Brian H. Jacobsen
Institute of Food and Resource Economics
University of Copenhagen

Hans Jöhr
Nestec S.A.

Erastus K. Kang'ethe
Department of Public Health
Pharmacology and Toxicology
University of Nairobi

Søren S. Kjær
Deputy Permanent Secretary
Ministry of Environment, Denmark

Cheikh Ly
Senior Researcher & Lecturer
EISMV

Joseph M. Maitima
ILRI

Neal P. Martin
Director, US Dairy Forage Research Center
United States Department of Agriculture-Agricultural Research Service

Rogerio Martins Mauricio
Research Professor
Fundação Ezequiel Dias e Pontifícia Universidade Católica de Minas Gerais (PUC-Minas)

Søren A. Mikkelsen
University of Aarhus

Manfred Noll
Nestlé India Ltd.

Iheanacho Okike
Scientist
ILRI

Padmakumar
Senior Subject Matter Specialist
CALPI (SDC-IC)

Roberto Porro
Researcher and Coordinator of Latin American Agroforestry Program,
ICRAF

J. Mark Powell
Soil Scientist
United States Department of Agriculture-Agricultural Research Service

Manitra A. Rakotoarisoa
ILRI

Vidya Ramesh
Environmentalist
Centre for Environment Education

Michael P. Russelle
Soil Scientist
United States Department of Agriculture-Agricultural Research Service
Plant Science Research and US Dairy Forage Research Center

Sheilendra Tiwari
Director
Natural Resource Development Unit

Index